Advances in Sustainable Valorization of Natural Waste and Biomass

Advances in Sustainable Valorization of Natural Waste and Biomass

Editor

Abu Naser Md Ahsanul Haque

MDPI

Basel • Beijing • Wuhan • Barcelona • Belgrade • Novi Sad • Cluj • Manchester

Editor
Abu Naser Md Ahsanul Haque
Institute for Frontier Materials
Deakin University
Geelong
Australia

Editorial Office
MDPI
St. Alban-Anlage 66
4052 Basel, Switzerland

This is a reprint of articles from the Special Issue published online in the open access journal *Sustainability* (ISSN 2071-1050) (available at: www.mdpi.com/journal/sustainability/special_issues/natural_waste_biomass).

For citation purposes, cite each article independently as indicated on the article page online and as indicated below:

Lastname, A.A.; Lastname, B.B. Article Title. *Journal Name* **Year**, *Volume Number*, Page Range.

ISBN 978-3-7258-0306-4 (Hbk)
ISBN 978-3-7258-0305-7 (PDF)
doi.org/10.3390/books978-3-7258-0305-7

Contents

About the Editor . . . vii

Preface . . . ix

Abu Naser Md Ahsanul Haque and Maryam Naebe
Waste Wool Powder for Promoting Plant Growth by Moisture Retention
Reprinted from: *Sustainability* **2022**, *14*, 12267, doi:10.3390/su141912267 . . . 1

Olatunji V. Oni, Michael A. Lawrence, Mark E. Zappi and William M. Chirdon
A Review of Strategies to Enhance the Water Resistance of Green Wood Adhesives Produced from Sustainable Protein Sources
Reprinted from: *Sustainability* **2023**, *15*, 14779, doi:10.3390/su152014779 . . . 9

Lelia Lawson, Lauren M. Degenstein, Bronwyn Bates, Wade Chute, Dan King and Patricia I. Dolez
Cellulose Textiles from Hemp Biomass: Opportunities and Challenges
Reprinted from: *Sustainability* **2022**, *14*, 15337, doi:10.3390/su142215337 . . . 30

Shubham Sharma, P. Sudhakara, Jujhar Singh, Sanjay M. R. and S. Siengchin
Fabrication of Novel Polymer Composites from Leather Waste Fibers and Recycled Poly(Ethylene-Vinyl-Acetate) for Value-Added Products
Reprinted from: *Sustainability* **2023**, *15*, 4333, doi:10.3390/su15054333 . . . 46

Walid Abotbina, S. M. Sapuan, R. A. Ilyas, M. T. H. Sultan and M. F. M. Alkbir
Preparation and Characterization of Black Seed/Cassava Bagasse Fiber-Reinforced Cornstarch-Based Hybrid Composites
Reprinted from: *Sustainability* **2022**, *14*, 12042, doi:10.3390/su141912042 . . . 96

Dina Aboelela, Habibatallah Saleh, Attia M. Attia, Yasser Elhenawy, Thokozani Majozi and Mohamed Bassyouni
Recent Advances in Biomass Pyrolysis Processes for Bioenergy Production: Optimization of Operating Conditions
Reprinted from: *Sustainability* **2023**, *15*, 11238, doi:10.3390/su151411238 . . . 109

Abu Naser Md Ahsanul Haque, Nigar Sultana, Abu Sadat Muhammad Sayem and Shamima Akter Smriti
Sustainable Adsorbents from Plant-Derived Agricultural Wastes for Anionic Dye Removal: A Review
Reprinted from: *Sustainability* **2022**, *14*, 11098, doi:10.3390/su141711098 . . . 139

Fatemeh Shahmoradi Ghaheh, Aminoddin Haji and Elaheh Daneshvar
Sustainable Dyeing Process for Nylon 6 Fabrics by Rhubarb Flower Using Different Bio-Mordants
Reprinted from: *Sustainability* **2023**, *15*, 9232, doi:10.3390/su15129232 . . . 164

Meral Özomay
Sustainable and Environmental Dyeing with MAUT Method Comparative Selection of the Dyeing Recipe
Reprinted from: *Sustainability* **2023**, *15*, 2738, doi:10.3390/su15032738 . . . 179

Nusrat Jahan and Sharfun Nahar Arju
A Sustainable Approach to Study on Antimicrobial and Mosquito Repellency Properties of Silk Fabric Dyed with Neem (*Azadirachta indica*) Leaves Extractions
Reprinted from: *Sustainability* **2022**, *14*, 15071, doi:10.3390/su142215071 . . . 193

Shahid Adeel, Fozia Anjum, Muhammad Zuber, Muhammad Hussaan, Nimra Amin and Meral Ozomay
Sustainable Extraction of Colourant from Harmal Seeds (*Peganum harmala*) for Dyeing of Bio-Mordanted Wool Fabric
Reprinted from: *Sustainability* **2022**, *14*, 12226, doi:10.3390/su141912226 . . . . **214**

Shahid Adeel, Maryam Habiba, Shumaila Kiran, Sarosh Iqbal, Shazia Abrar and Ch Moazzam Hassan
Utilization of Colored Extracts for the Formulation of Ecological Friendly Plant-Based Green Products
Reprinted from: *Sustainability* **2022**, *14*, 11758, doi:10.3390/su141811758 . . . . **225**

About the Editor

Abu Naser Md Ahsanul Haque

Dr. Abu Naser Md. Ahsanul Haque currently serves as an Associate Research Fellow at the Institute for Frontier Materials at the Geelong Waurn Ponds Campus of Deakin University. He received his Doctor of Philosophy degree (Institute for Frontier Materials) from Deakin University in 2021 with the prestigious Alfred Deakin Medal. He obtained his Master of Science in Textile Engineering from Mawlana Bhashani Science and Technology University in 2014 and his Bachelor of Science in Textile Technology from the University of Dhaka in 2008.

Preface

With the increasing global awareness of the need for a sustainable world, the careful processing of different natural wastes and biomass has become an important area of research. The valuation of these resources is not only economically viable but also ensures the minimization of waste in the environment, making our world cleaner. This reprint brings together some recent advances in the sustainable re-design of natural waste and biomass in different material productions for both textile and non-textile applications.

These include a wide array of topics, such as the following: the use of textile waste fibers, i.e., wool as a fertilizer to promote plant growth; the potential of cellulose textiles from hemp biomass; the use of green wood adhesives from sustainable proteins; optimization of the pyrolysis process of biomass; the use of natural wastes in composites preparation, including cellulosic waste as well as leather waste; dye adsorbents prepared from cellulosic wastes; some new methods of environmentally friendly dyeing; and imparting functionality to textiles by using natural ingredients.

Abu Naser Md Ahsanul Haque
Editor

sustainability

MDPI

ommunication

Vaste Wool Powder for Promoting Plant Growth by Moisture Retention

bu Naser Md Ahsanul Haque and Maryam Naebe *

Institute for Frontier Materials, Deakin University, Waurn Ponds, VIC 3216, Australia
* Correspondence: maryam.naebe@deakin.edu.au; Tel.: +61-3-522-72783

Abstract: Natural wastes are widely used as composts for plant growth. However, wool waste has received little attention in this regard, despite its nitrogen-rich chemical structure owing to amide groups. A few studies have been conducted for soil amendment using wool, mostly in raw or pellet form. However, despite the possible consistent mixing and more uniform effect of powders inside soil, wool has never been implemented in powder form in soil for improving moisture. This study demonstrates the effectiveness of using wool as a powder, facilitating better mixing and spreading in soil. Results show that wool powders are more effective in retaining soil moisture compared to wool pellets and are comparable to commercial fertiliser. The findings further indicate that a balanced amount of wool is required to maintain a proper moisture level (not too wet or dry) to promote actual plant growth.

Keywords: wool powder; wool pellet; soil amendment; keratin waste; plant growth; sustainability

check for updates

itation: Haque, A.N.M.A.; Naebe, . Waste Wool Powder for ·omoting Plant Growth by Moisture etention. *Sustainability* **2022**, *14*, :267. https://doi.org/10.3390/ 141912267

cademic Editor: Ramchandra Pode

eceived: 20 August 2022
ccepted: 26 September 2022
ublished: 27 September 2022

ublisher's Note: MDPI stays neutral ith regard to jurisdictional claims in ıblished maps and institutional affil- tions.

1. Introduction

Wool is a natural keratin fibre sourced from sheep in numerous countries, and more than 1.16 million tonnes are produced each year worldwide [1]. The countries that produce wool are widespread and from different continents, including Australia, New Zealand, China, the United States, the United Kingdom, Iran, Turkey, Argentina, and India. The estimated amount of wool production is mainly the total of the clean or traded wool, while the amount of waste wool is rarely counted. According to a report from 2019, it is predicted that around 10–15% of wool is wasted during its sorting and cleaning, and an added 12–15% is lost during spinning and weaving [2]. The annual combined generation of wool waste from the nine key countries listed above is estimated at around 317 thousand tonnes per year, which is mostly the pile of low-grade wool that is not possible to spin into textiles. These wastes are frequently discarded without any care, thus often causing environmental pollution. For example, the disposal of wool waste through incineration and landfilling causes air and soil pollution when the amount of wool is not wisely controlled [3,4].

To combat this pollution and utilise this massive waste as a resource, waste wool has been proposed over the years for numerous applications, such as insulation [5], building materials [6], adsorbents [7], and composites [8]. However, wool is rich in nitrogen due to its amide groups [1], and it possesses a moisture retention ability 3.5 times its weight [9]; both are beneficial for plant growth. There are some other resources that have been proposed for soil moisture retention, such as the use of biochar from coconut fibre [10], walnut shells [11], water hyacinths [12], mixed wood [13], superabsorbent nanocomposites from silica [14], *Sphagnum* moss [15], compost [16–19], and minerals including vermiculite [20], bentonite, and zeolite [21]. However, the advantage of wool in comparison with other resources is its amide structure, which can simultaneously benefit the soil with moisture retention, such as by nitrogen mineralisation. Wool is also biodegradable within six months in ideal conditions [22].

Though commercial fertilisers are widely used for plant cultivation, often, they include heavy metal impurities, such as cadmium, mercury, and lead [23], that can accumulate in the soil or even be transferred into the ultimate plant products [24]. Further, there is no previous evidence on the release of nitrogen oxides or nitric acid from wool to soil while nitrogen is mainly transformed as non-hazardous nitrogen–ammonia (NO_3–N) or ammonium nitrogen (NH_4–N). Therefore, the use of wool for plant growth is sustainable and can benefit the soil if sensibly used, which can reduce the fertiliser amount to a minimum, if not totally replacing it. Each year, around 317 thousand tonnes of wool waste are discarded in major wool-growing countries around the world. The majority of these wool wastes (including consumed wool garments) currently enter landfill [25], and their proper use in plant cultivation has received little attention.

A few studies have previously shown the prospects of wool in soil amendment. An increased yield of basil, thorn apple, peppermint, and garden sage when using raw wool fibre was reported, where wool was found to raise the NH_4–N and NO_3–N content in soil mainly related to the nitrogen mineralisation from wool [26]. Around 40 g wool waste was observed to be sufficient for 6 kg of soil, supporting 2–3 harvests. Further, nitrogen mineralisation in plant tissue was also evident, e.g., nitrogen in thorn apple and mint tissue increased from 19.5 to 44.3 g/kg and 13.2 to 30.3 g/kg, respectively, when using wool waste, and soil microbial biomass also increased [26]. Wool has also been reported as an organic fertiliser for tomato and broccoli in the form of pellets [9,27]. However, using wool in powder form is more advantageous than using its raw form (fibre) or pellets. The use of pellets can only provide a localised benefit in soil, while the fibres can hinder proper spreading in soil due to entanglement. The use of wool powders could deliver more convenient and consistent mixing in soil and evenly enhance the soil properties.

In the past, authors have demonstrated several methods of preparing powders from different keratin fibres, such as wool [1] and alpaca [28]. Powders have been prepared from these resources from a coarse level (particle size around 70 μm) to an ultra-fine size of particles, such as 2.6 μm [28]. To develop these powders, both mechanical [28] and chemical methods [1] were used. However, the preparation of ultra-fine powders from wool requires either prolonged mechanical processing, such as ball milling and air jet milling [28], or intensive chemical treatment [1]. Nevertheless, coarse powder preparation from wool is rather quick, sustainable, and less energy-intensive, while the prepared powders are still suitable for handling and uniform mixing with soil.

The aim of this preliminary work was to evaluate the effectiveness of coarse wool powder for soil moisture retention compared to wool pellets. The morphology of the powder and pellets was analysed and the moisture retention ability was observed over a 5-day period and compared with the results obtained with commercial fertiliser and non-fertilised soil. The further impact of these soil variants on the plant growth was also assessed by a 40-day experiment on actual plants.

2. Materials and Methods

Waste wool fibre sourced from a local farm was collected from the Commonwealth Scientific and Industrial Research Organisation (CSIRO), Geelong Waurn Ponds Campus, Australia. Garden soil mix (Australian Standard AS4454), all-purpose NPK fertiliser and Stock Marionette (*Matthiola incana*) seedlings were purchased from the Bunnings Warehouse, Waurn Ponds, Australia.

Waste wool fibres were ground to a powder by a cutting mill (Pulverisette 19, Fritsch, Idar-Oberstein, Germany) using a 200 μm sieve. Part of the wool powder (WP) was converted into pellets using a hydraulic pellet press (Atlas Auto touch, Specac, Orpington, UK) using 8T pressure. The weight of each pellet was 165 ± 1 mg. The morphology of wool powder and pellets was observed using a DP71 optical microscope (Olympus, Tokyo, Japan).

Six separate 10 × 10 cm garden pots were each filled with around 150 ± 2 g soil. The soils differed in terms of mixing wool powder, wool pellets, and commercial fertiliser, and were named as per Table 1. One of the soil samples was kept untreated as the control. The amount of the fertiliser was chosen based on the recommended levels provided on the commercial fertiliser package (10 g per 40 cm pot).

Table 1. Combinations of soil prepared with wool powder and pellets and commercial fertiliser, and control sample with no fertiliser.

Sample	Name	Fertiliser (mg)	Wool Powder (mg)	Wool Pellets (mg)
Commercial fertiliser	CF	2500	0	0
No fertiliser	NF	0	0	0
Wool powder	W-3PW	0	165	0
	W-7PW	0	385	0
Wool pellets	W-3PL	0	0	165 (3 pellets)
	W-7PL	0	0	385 (7 pellets)

The soils were watered until they all reached a moisture rating of 7 (measured by a soil moisture meter, Brunnings, Australia) [29]. The soil was then kept unwatered for the next 4 days and changes in the rating were measured. The overall process was conducted in three batches, and average and standard deviation values reported. The statistical significance of the difference in datasets was calculated using a two-tailed t test, where $p \leq 0.05$ indicated a significant difference and $p > 0.05$ indicated no significant difference.

To investigate the impact of wool powder and pellets on plant growth, a further six soil samples were prepared using the same combinations as in Table 1 to identify the impact of the wool amount and form of wool in relation to no fertiliser and commercial fertiliser. Six Marionette seedlings of the same height (~7 cm) were planted and their growth up to 40 days was monitored. The plants were watered every 5 days, by spraying with water (20 ± 1 mL). The garden pots were kept indoors under a transparent roof, receiving ~2000 lumen during daytime. The average temperature was recorded at a maximum of 21–24 °C and a minimum of 11–13 °C. The average relative humidity and daylight hours were 68% and 12.3 h, respectively.

3. Results

3.1. Surface Morphology of Wool Powder and Pellets

Figure 1 shows the surface morphology of the wool powder and the prepared pellet. The wool powder (Figure 1a) retained the fibrous structure of the wool, although there was a combination of random lengths. The diameter of the wool mostly remained similar (mostly around 20–23 µm), though some fibres at a lower limit were also seen (such as 11 µm). The length of the fibre widely varied due to the cutting operation, starting from around 82 µm to 690 µm. On the other hand, the morphology of the wool pellet (Figure 1b) showed a dense, fibrous structure, where ribbon-shaped fibres were attached to each other. The aggregation of the particles in the pellet was presumably much higher, compared to the randomly distributed fibres in the wool powder.

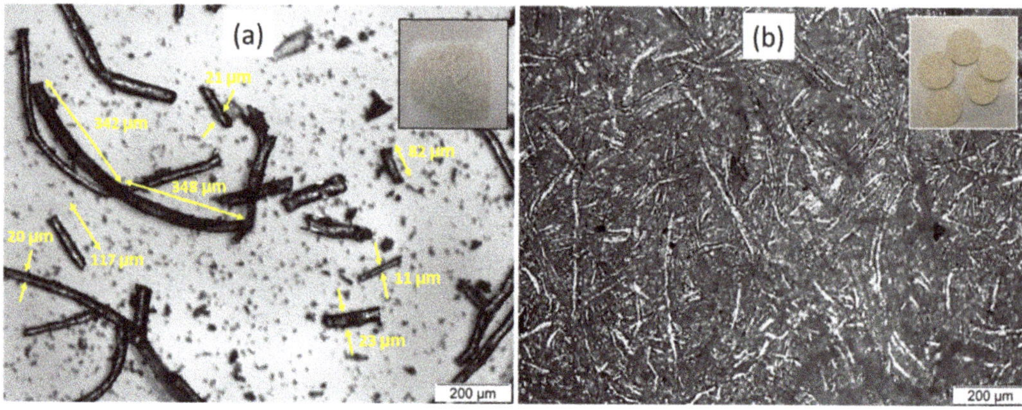

Figure 1. Morphology of the prepared (**a**) wool powder and (**b**) wool pellet under optical microscope.

3.2. Moisture Retention Ability of Wool Powder and Pellets

Figure 2 shows the moisture rating of the soil prepared with the samples over a 5-day period. The moisture meter provides data on three regions: wet, moist, and dry. The wet region (moisture rating 8–10) is suitable for aquatic plants, the dry region (moisture rating 0–3) is suitable for low-water plants such as Cactus and Sanseviera, and the moist region (moisture rating 4–7) is ideal for most common plants [29]. The results showed that the commercial fertiliser (CF) was more effective in holding the moisture and remained in the ideal region even after 5 days. This was probably because of some chemical elements present in the composition, such as ammonium (7.8%) [24], responsible for the high hygroscopicity [30]. Both the wool powders (W-3PW and W-7PW) also retained the moisture in the ideal region for all 5 days, although the amount was lower than CF. The moisture gained in soil due to the presence of wool powders was probably related to the lower particle size and higher surface area of the powders. Powders prepared from keratinous resources commonly show a higher moisture absorption ability compared to raw fibre (an 18% increase was reported), due to the enhanced surface area, which provides more space for the moisture to accumulate [28]. A smaller size leads to a higher surface area and thus can accumulate more moisture. In the current experiment, the soil with no fertiliser or wool (NF) radically dried up over this period and mostly stayed in the dry region in the last 3 days, identical to the result obtained from three wool pellets (W-3PL). Comparing the wool pellets and wool powder, powders were more effective in moisture retention, while the pellets allowed the soil to reach to dry region mostly after 2–3 days. This was probably because the wool powder was spread throughout the soil and completely mixed with it, while the pellets were mainly concentrated in specific places. The decrease in the soil moisture was found insignificant ($p > 0.05$) among days 3, 4, and 5 for W-3PW and W-7PW samples, showing a consistent water-holding ability. However, the samples with wool pellets, as well as the NF sample, showed a drastic soil moisture reduction over the 5-day period ($p \leq 0.05$). A higher amount of wool in either powder or pellet form was more effective in moisture retention compared with a lower amount of wool. For example, W-7PW maintained the soil moisture at an average rating of 5.8 after 5 days, while W-3PW was at the average rating of 4.8 after the same period.

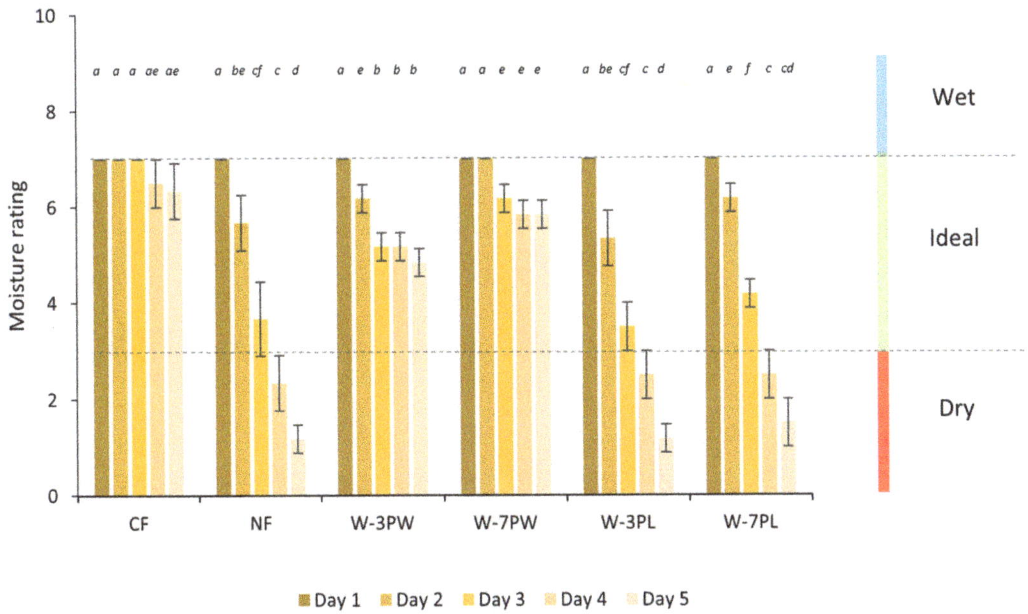

Figure 2. Moisture rating over 5-day period for soils prepared from wool powder (PW) and pellets (PL) of different amounts, 165 mg (3) and 385 mg (7), compared to commercial fertiliser (CF) and no fertiliser (NF). Error bars show standard deviation; different superscripts indicate that data are significantly different ($p \leq 0.05$).

3.3. Effect on Actual Plant Growth by Wool Powder and Pellets

Figure 3 shows the growth of plants over 40 days using the six types of soil. Common plants such as stock flowers need a balanced, moist soil for appropriate growth. In line with the findings in Figure 2, although the moisture retention from W-3PW was lower than that of W-7PW and CF, it showed a more balanced effect (not excessive low/high moisture retention). This could be the reason that W-3PW showed the greatest influence on the plant growth (around 43% increase in 10 days and around 160% increase over 40 days), compared to other samples. The lowest value was obtained for NF, which showed around a 14% increase in 10 days and around a 114% increase in 40 days. The samples with pelletised wool, CF and W-7PW, showed moderate growth. It is probable that the pelletised wool was not as effective as the wool powders in terms of moisture retention and allowed the sample to dry up sooner. However, the lower value for CF and W-7PW, compared to W-3PW, was probably related to the higher moisture in the soil, caused by greater moisture retention. Although this plant growth process was not replicated in simultaneous batches, the positive impact of wool was steady for all the wool-included samples, and it was consistent with the soil moisture retention findings and with the literature [26]. The difference achieved in the plant height data for the W-3PW sample after 40 days was more than 5%, except for CF and NF (error bars in Figure 3a). The higher impact observed with a lower amount of wool could be explained by the similar result noted for the wool pellets as well. Based on these initial findings, future trials using multiple batches of plants and different species should be performed to further validate this impact.

Figure 3. Plant height (**a**) from first day to up to 40 days, affected by commercial fertiliser (CF), n fertiliser (NF), and wool powders (PW) and pellets (PL) of different amounts, and (**b**) photograph o plant growth after 40 days, when comparing NF with the samples prepared with wool powders o two different amounts, 3PW (165 mg) and 7PW (385 mg). Error bars show 5% deviation of actual data

4. Summary

Due to the nitrogen and protein content of wool waste and its moisture retention abilit its use can lead to sustainable cultivation, partly or fully replacing common fertilisers tha possess the risk of transferring heavy metal impurities to both soil and plants. Wool wast can achieve reasonable use in soil as a low-cost, feasible, and environmentally friendl option. This is likely to be an industrially applicable solution, given the amount of woo waste generated worldwide. This can also potentially eliminate the current additiona costs related to waste wool disposal. This study presents the usefulness of using woo powder for soil moisture retention compared to wool pellets and commercial fertilisers. Th morphology of wool powders showed the fibrous structure of wool, with a combination o random lengths and widths, while the pellets showed a dense, fibrous surface. In 150 g soi the moisture retention performance of both 165 and 385 mg wool powders was comparabl to that of the commercial fertiliser. However, the plant height after 40 days was observe to be higher when using a lower amount of wool (165 mg), due to its balanced moistur retention ability. Therefore, the results show that the use of an optimised amount of woo powder can be effective for both moisture retention ability in soil and the promotion o plant growth. This initial study indicates the potential of wool powders for soil amendmen which should encourage future research, such as changes in the characteristics of particle and pellets by hydration and the presumable nutritional benefits for the soil and plants.

Author Contributions: Conceptualization, M.N.; methodology, A.N.M.A.H.; formal analysis A.N.M.A.H.; investigation, A.N.M.A.H.; writing—original draft preparation, A.N.M.A.H.; writing—review and editing, M.N.; visualization, A.N.M.A.H.; supervision, M.N. All authors have read and agreed to the published version of the manuscript.

Funding: This research received no external funding.

Institutional Review Board Statement: Not applicable.

Informed Consent Statement: Not applicable.

Data Availability Statement: Data is contained within the article.

Acknowledgments: The authors would like to acknowledge the contributions of the late Isuru Jeewantha

Conflicts of Interest: The authors declare no conflict of interest.

References

Haque, A.N.M.A.; Naebe, M.; Mielewski, D.; Kiziltas, A. Thermally stable micro-sized silica-modified wool powder from one-step alkaline treatment. *Powder Technol.* **2022**, *404*, 117517. [CrossRef]

Sharma, S.; Sahoo, A.; Chand, R. Potential use of waste wool in agriculture: An overview. *Indian J. Small Rumin.* **2019**, *25*, 1–12. [CrossRef]

Kumawat, T.K.; Sharma, A.; Sharma, V.; Chandra, S. Keratin waste: The biodegradable polymers. In *Keratin*; IntechOpen: London, UK, 2018.

Marchelli, F.; Rovero, G.; Curti, M.; Arato, E.; Bosio, B.; Moliner, C. An integrated approach to convert lignocellulosic and wool residues into balanced fertilisers. *Energies* **2021**, *14*, 497. [CrossRef]

Cai, Z.; Al Faruque, M.A.; Kiziltas, A.; Mielewski, D.; Naebe, M. Sustainable lightweight insulation materials from textile-based waste for the automobile industry. *Materials* **2021**, *14*, 1241. [CrossRef]

Buratti, C.; Belloni, E.; Merli, F. Water vapour permeability of innovative building materials from different waste. *Mater. Lett.* **2020**, *265*, 127459. [CrossRef]

El-Geundi, M.S. Adsorbents for industrial pollution control. *Adsorpt. Sci. Technol.* **1997**, *15*, 777–787. [CrossRef]

Remadevi, R.; Al Faruque, M.A.; Zhang, J.; Naebe, M. Electrically conductive honeycomb structured graphene composites from natural protein fibre waste. *Mater. Lett.* **2020**, *264*, 127311. [CrossRef]

Ordiales, E.; Gutiérrez, J.I.; Zajara, L.; Gil, J.; Lanzke, M. Assessment of utilization of sheep wool pellets as organic fertilizer and soil amendment in processing tomato and broccoli. *Mod. Agric. Sci. Technol.* **2016**, *2*, 20–35.

Guarnieri, S.F.; Nascimento, E.C.d.; Costa Junior, R.F.; Faria, J.L.B.d.; Lobo, F.d.A. Coconut fiber biochar alters physical and chemical properties in sandy soils. *Acta Sci. Agron.* **2021**, *43*, e51801. [CrossRef]

Wang, D.; Li, C.; Parikh, S.J.; Scow, K.M. Impact of biochar on water retention of two agricultural soils—A multi-scale analysis. *Geoderma* **2019**, *340*, 185–191. [CrossRef]

Xing, X.; Liu, Y.; Garg, A.; Ma, X.; Yang, T.; Zhao, L. An improved genetic algorithm for determining modified water-retention model for biochar-amended soil. *Catena* **2021**, *200*, 105143. [CrossRef]

Ulyett, J.; Sakrabani, R.; Kibblewhite, M.; Hann, M. Impact of biochar addition on water retention, nitrification and carbon dioxide evolution from two sandy loam soils. *Eur. J. Soil Sci.* **2014**, *65*, 96–104. [CrossRef]

Olad, A.; Zebhi, H.; Salari, D.; Mirmohseni, A.; Tabar, A.R. Slow-release NPK fertilizer encapsulated by carboxymethyl cellulose-based nanocomposite with the function of water retention in soil. *Mater. Sci. Eng. C* **2018**, *90*, 333–340. [CrossRef]

Waddington, J.; Lucchese, M.; Duval, T. Sphagnum moss moisture retention following the re-vegetation of degraded peatlands. *Ecohydrology* **2011**, *4*, 359–366. [CrossRef]

Zemánek, P. Evaluation of compost influence on soil water retention. *Acta Univ. Agric. Silvic. Mendel. Brun.* **2011**, *59*, 227–232. [CrossRef]

Neugebauer, M.; Jakubowski, T.; Sołowiej, P.; Wesołowski, M. A fuzzy model of the composting process with simultaneous heat recovery and aeration rate control. In *Renewable Energy Sources: Engineering, Technology, Innovation*; Springer: Berlin/Heidelberg, Germany, 2018; pp. 151–160.

Mbah, B.N.; Odili, P.N. Changes in moisture retention properties of five waste materials during short-term mesophilic composting. *Compost Sci. Util.* **1998**, *6*, 67–73. [CrossRef]

Mylavarapu, R.; Zinati, G. Improvement of soil properties using compost for optimum parsley production in sandy soils. *Sci. Hortic.* **2009**, *120*, 426–430. [CrossRef]

Lai, W.; Ahmad, W.W.; Jasni, J.; Ab Kadir, M. A review on the usage of Zeolite, Perlite and Vermiculite as natural enhancement materials for grounding system installations. In Proceedings of the IEEE 15th Student Conference on Research and Development (SCOReD), Wilayah Persekutuan Putrajaya, Malaysia, 13–14 December 2017; pp. 338–343.

Amani, E.; Ghasemi, A.R.; Nouri, M.R.; Motaghian, H. Effect of Vermiculite, Bentonite and Zeolite on Evaporation and Soil Characteristic Moisture Curve. *J. Water Soil Conserv.* **2021**, *28*, 83–101.

Swan, P. Wool is biodegradable. *International Wool Textile Organisation*. Available online: https://iwto.org/wp-content/uploads/2020/04/IWTO_Wool-is-Biodegradable.pdf (accessed on 19 August 2022).

Oyedele, D.; Asonugho, C.; Awotoye, O. Heavy metals in soil and accumulation by edible vegetables after phosphate fertilizer application. *Electron. J. Environ. Agric. Food Chem.* **2006**, *5*, 1446–1453.

Brunnings, All Purpose NPK Fertiliser: Analysis. Brunnings Garden Products, Australia. 2022. Available online: https://brunnings.com.au/product/all-purpose-npk-fertiliser-500g/ (accessed on 1 March 2022).

Wiedemann, S.; Biggs, L.; Nebel, B.; Bauch, K.; Laitala, K.; Klepp, I.; Swan, P.; Watson, K. Environmental impacts associated with the production, use, and end-of-life of a woollen garment. *Int. J. Life Cycle Assess.* **2020**, *25*, 1486–1499. [CrossRef]

Zheljazkov, V.D. Assessment of wool waste and hair waste as soil amendment and nutrient source. *J. Environ. Qual.* **2005**, *34*, 2310–2317. [CrossRef] [PubMed]

Böhme, M.H. Use of bio-waste as fertiliser for the protected vegetable cultivation. *J. Vietnam. Environ.* **2018**, *10*, 27–32. [CrossRef]

Al Faruque, M.A.; Remadevi, R.; Wang, X.; Naebe, M. Preparation and characterisation of mechanically milled particles from waste alpaca fibres. *Powder Technol.* **2019**, *342*, 848–855. [CrossRef]

29. Brunnings, 3 in 1 Soil Meter: Direction of Use. Brunnings Garden Products, Australia. 2022. Available online: https://brunning com.au/product/3-in-1-soil-meter/ (accessed on 1 March 2022).
30. Elzaki, B.I.; Zhang, Y.J. Anti-hygroscopic surface modification of ammonium nitrate (NH_4NO_3) coated by surfactants. *Arab. Chem.* **2020**, *13*, 3460–3473. [CrossRef]

sustainability

eview

A Review of Strategies to Enhance the Water Resistance of Green Wood Adhesives Produced from Sustainable Protein Sources

latunji V. Oni [1], Michael A. Lawrence [2], Mark E. Zappi [1,3] and William M. Chirdon [1,3,*]

1 Department of Chemical Engineering, University of Louisiana at Lafayette, Lafayette, LA 70504, USA
2 Department of Chemistry, University of Louisiana at Lafayette, Lafayette, LA 70504, USA
3 Energy Institute of Louisiana, University of Louisiana at Lafayette, Lafayette, LA 70504, USA
* Correspondence: william.chirdon@louisiana.edu; Tel.: +1-(337)-482-6564

Abstract: The health risks associated with formaldehyde have propelled relevant stakeholders to push for the production of non-toxic wood adhesives. Several countries including the USA, Japan, and Germany have implemented policies mandating manufacturers to reduce the emissions of formaldehyde to lower levels. Protein adhesives stand out due to their sustainability, renewable sources, and biodegradability. However, they are limited by poor wet strength and water resistance, which affect their wide acceptability in the marketplace. Researchers have developed multiple strategies to mitigate these issues to advance protein adhesives so they may compete more favorably with their petroleum-based counterparts. This review paper explores these strategies including cross-linking, modified fillers, and the removal of hydrophilic content while providing insights into the methodological approaches utilized in recent literature with a comparison of the resultant protein adhesives.

Keywords: formaldehyde; wood adhesives; protein adhesives; soybean; water resistance

itation: Oni, O.V.; Lawrence, M.A.; appi, M.E.; Chirdon, W.M. A eview of Strategies to Enhance the ater Resistance of Green Wood dhesives Produced from ıstainable Protein Sources. *ıstainability* **2023**, *15*, 14779. tps://doi.org/10.3390/ 152014779

cademic Editor: Abu Naser Md hsanul Haque

eceived: 29 August 2023
evised: 30 September 2023
ccepted: 8 October 2023
ıblished: 12 October 2023

1. Introduction

The wood adhesive market is estimated to be worth over $6 billion by 2025 [1,2]. Despite the huge market, synthetic adhesives constitute a greater proportion of adhesives due to their superior adhesive performance and low cost [3]. Yet, historically, glues, resins, and composite binders were made using plant- and animal-based proteins until the mid-1900s when petroleum-based products, mainly formaldehyde, urea, and phenolics, became the mainstay feedstocks for industrial adhesives [4].

Urea–formaldehyde resins are the most utilized synthetic adhesives due to their desirable properties, including good bonding strength, clear color, low cost, fast curing, and resistance to moisture [5]. However, the major challenge associated with synthetic adhesives is the emissions of toxic substances including formaldehyde during manufacture and use [6]. Recent years have witnessed an increase in global public sensitivity and awareness about health and environmental concerns [7]. Over time, stricter laws setting new toxic emissions limits for wood panels have been implemented [8,9]. These are incorporated into standards in the United States, Japan, and Europe [7].

This increasingly tightening of regulations and concerns over pollutant volatilization and the manufacturing ecological footprint (sustainability) of these formaldehyde and phenolics-based adhesives creates an opportunity for the development of renewable and "environmentally-friendly" protein adhesives in the market [10]. However, protein adhesives only constitute a small market portion due to a number of challenges preventing their wider application [2,11]. The primary reason protein adhesives cannot be utilized in more applications is their lack of water resistance [4] and wet strength [12,13], yet there is still a growing interest in protein adhesives as evidenced by the increasing rate of publicatio which is shown in Figure 1.

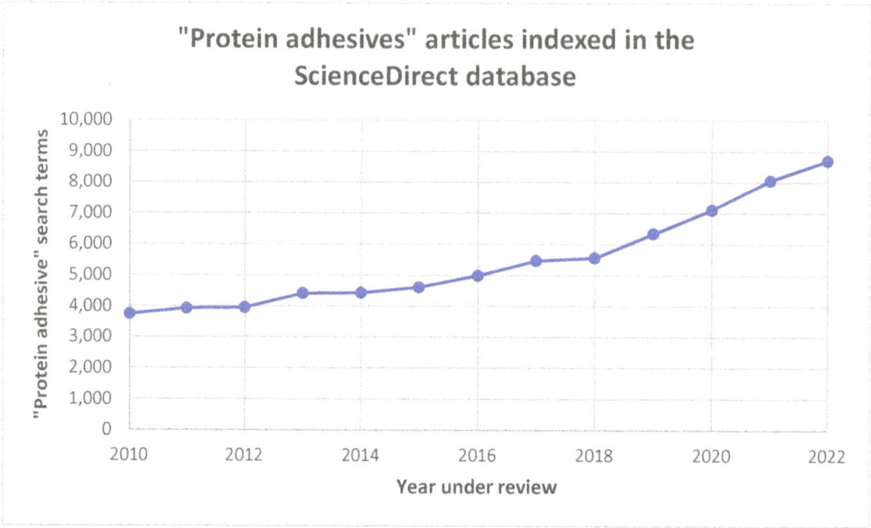

Figure 1. Publication of articles indexed by the ScienceDirect database found under the search tern "Protein adhesives" since 2010.

Proteins are the most abundant class of bio-based adhesive feedstocks [4]; they ar primarily linear polyamides composed of polypeptides linkages of amino acids [14]. Th fundamental structure is determined by the polypeptide structure of the protein [15]. Sid chains of amino acids can form bonds with several substances, including cellulose anc lignocellulose [16]. Yet, certain potentially reactive side chains are not on the surface of th protein. Moreover, the loss of the adhesive bond under moist conditions occurs within th adhesive itself (cohesive failure) and not between the wood and the adhesive [16]. Protein have to be denatured to expose the functional groups, facilitate stronger bonding to wooc and enable the aggregation of protein molecules [16].

The overall mechanism to produce protein-based adhesives is a denaturing proces that involves an aqueous reaction matrix [17]. The goal of the denaturation process for th formation of adhesives is to dissolve the quaternary, tertiary, and secondary structures o the protein [17,18]. The addition of a strong base increases the pH of the proteins above thei isoelectric point [19]. Under this condition, the ions in the solution interfere with the hydro gen bonds and the electrostatic dipole–dipole interactions that allow the proteins to retai their shape and higher-order structures [20]. Many of the natural, covalent cross-linkin groups (including cysteine and disulfide bonds) that maintain the tertiary structure in thes proteins are also vulnerable to disassociation under alkaline conditions [21]. By splitting th cross-linking groups and interfering with the secondary bonding, this denaturation proces exposes the hydrophilic portions of proteins, which allows these adhesive groups to adsorl strongly onto substrates (as a glue) and filler materials (as a binder/resin) [22]. The condi tions commonly used to denature proteins, especially temperature and pH modificatior also catalyze the hydrolysis of the proteins [23]. This results in lower-molecular-weigh protein fragments that suffer a significant loss of shear strength, especially when the chain become too short to entangle with each other [24]. Reaction time, temperature, and pH car be optimized and controlled to achieve denaturation without significant hydrolysis resul ing in high-quality adhesives and a chemical foundation for other polymeric substance (such as resins and coatings) [25].

The quantity of amino acid residue and its position along the polypeptide chain are the primary determinants of the physical and chemical properties of proteins [15]. Figure 2 shows how the primary structure of proteins is naturally folded into secondary structures such as α-helices or β-sheets as triggered by the specific amino acid sequences. This process leads to the presentation of some polar groups that may associate with other polar groups to create a stable tertiary structure. Thus, several varieties of sequences can be formed based on the amino acid's number, chaining, and percentage [4]. The functional groups present in the side chains of the polypeptide structure determine the hydrophilicity or hydrophobicity of the amino acid. It offers interactions with carboxyl or hydroxyl groups in wood coupled with its cross-linking sites [26].

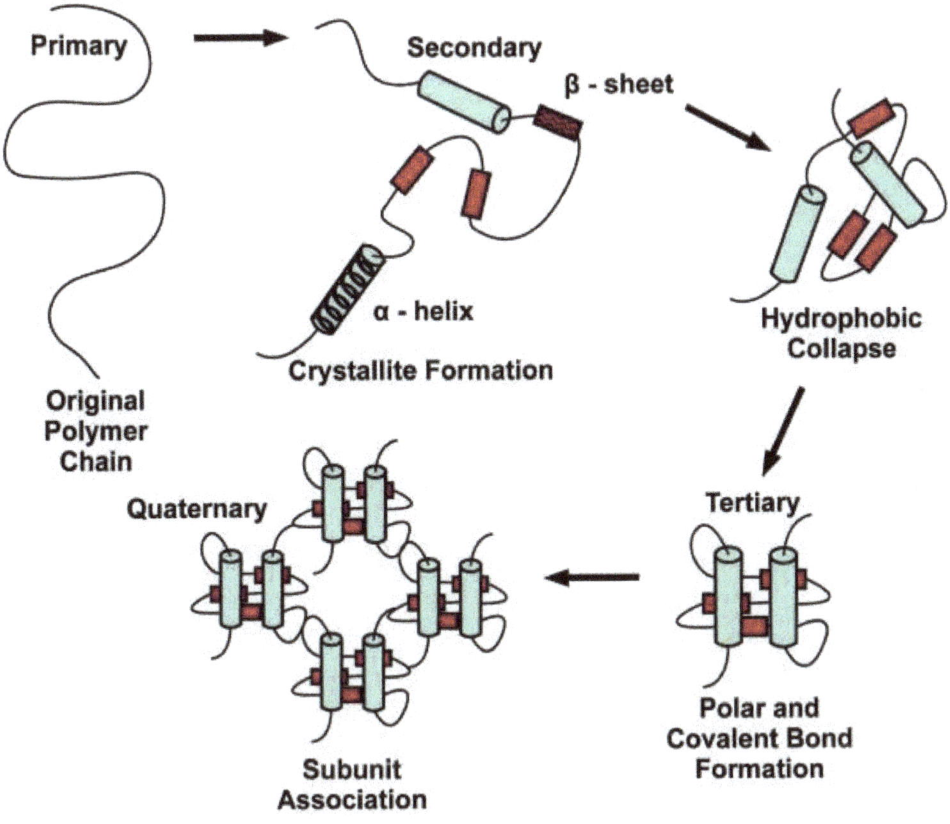

Figure 2. Illustration of levels of protein structure [18] (Reprinted/adapted with permission from Ref. [18]. 2014, American Chemical Society).

Water can degrade protein adhesives by altering the adhesive's physical and chemical characteristics. Protein adhesives are primarily composed of water-sensitive animal or plant proteins that undergo structural changes when exposed to moisture, even after conversion into adhesives, thus the adhesive also suffers from water-exposure-based structural failures [27].

When water interacts with protein adhesives, it can break the hydrogen bonds and other intermolecular interactions that hold the protein structure together [27]. This can result in the adhesive losing its strength, becoming brittle, and losing its ability to adhere to surfaces [27].

The poor water resistance is linked to the multiple polar groups of most of the protei which make the adhesives absorb water and weaken the bond between the two surfaces [2 Another author ascribed the reason to the large constituent of carbohydrates in soy flo which contains abundant hydroxyl groups [18]. Fourier Transform Infrared (FTIR) analys has shown that water quickly forms hydrogen bonds with hydrophilic groups, theret resulting in poor water resistance [29]. The reduction in the carbohydrate content decrease the hydrophilicity [29]. Protein adhesives especially soy-based adhesives are expected witness significant growth with a compound annual growth rate of around 7% from 201 to 2025 [11]. The expansion is due to the environmental benefits offered by the produc which is the major factor for the increasing demand [30].

Protein adhesives have the potential to overtake an increased share of the woo adhesives market in the future if the problem of poor water resistance can be mitigate There have also been some promising results in the use of bio-based adhesives based c tannins [31], lignins [32], and starch [33]. The use of these bio-based materials will also b discussed in this paper, especially as used as additives or modifiers for protein adhesive

2. Strategies to Improve Water Resistance in Protein Adhesives

Generally, the applications of protein adhesives are limited by their low water resi tance. Several strategies are utilized to mitigate or overcome this challenge, includin thermochemical treatment, cross-linking networks, and using modified fillers. This pape will review contemporary strategies including bio-based and synthetic treatments.

2.1. Cross-Linking Networks

Cross-linking is a common method for creating or enhancing a network structur hence improving the protein structural stability and adhesive capabilities of soy protei based adhesives [34]. These materials are added to protein adhesives during their prep ration or before their application [31]. Cross-linking modifications using latex, syntheti resin, polyamide-epichlorohydrin, and isocyanate have been shown to be efficient method for increasing the water resistance of an adhesive [32,33,35,36].

2.1.1. Synthetic Cross-Linking Agent Sources

There are numerous advantages to synthetic sources including their ease of use, wid range of applications, and low cost [37]. Most of these synthetic sources are derive from petroleum and are known for releasing volatile organic compounds that are likely t contaminate the environment and offer possible health problems [38,39]. These sources als do not offer highly sustainable, green production options which have become a priority i many manufacturing sectors, including the adhesives markets.

The effects of the storage stability of polyamidoamine-epichlorohydrin (PAE) resin o the properties of defatted soybean-flour-based adhesives were explored by Gao et al. [36 Polyamidoamine-epichlorohydrin is a water-soluble polymeric resin widely utilized in p per sheets as a wet strength enhancer [40,41]. The team explored different modifications c PAE like a polyamidoamine (PA) without incorporating epichlorohydrin and two sample of epichlorohydrin-modified polyamidoamine with different solid contents of 25% (PAE-2 and 12% (PAE-12), respectively [36].

The blending of polyamidoamine-epichlorohydrin and defatted soybean flour (DSI revealed the cross-linking reactions of both constituents, thereby indicating that the azet dinium units of polyamidoamine-epichlorohydrin can effectively cross-link with the amine group and the DSF's carboxyl groups to form three-dimensional cross-linking structures t enhance the water resistance of the DSF-based adhesives [36]. The wet shear strength c plywood panels bonded with DSF adhesive and fresh PAE-25 and PAE-12 is 1.22 MPa an 1.30 MPa, respectively, and reduces by around 30% after 56 days of storage to 0.79 MP whereas that of plywood bound with adhesive DSF-PAE-12 decreases by around 17% t 1.07 MPa, thereby demonstrating the effective stability of PAE-12 compared to PAE-25 [36

Kan et al. explored the possibility of using both polyamidoamine-epichlorohydrin and melamine–urea–glyoxal resins to formulate double-network structures through their cross-linking to soybean-meal-based adhesive with stable and required water resistance for structural use and several analyses, including Nuclear Magnetic Resonance (NMR), Thermogravimetric Analysis (TGA), X-ray Photoelectron Spectroscopy (XPS), and FTIR, which revealed the double-network structure [42,43]. Co-cross-linking between polyamidoamine-epichlorohydrin and melamine–urea–glyoxal resins by aldehyde-amino and azetidinium-amino reactions expanded the three-dimensional networks even further [42]. The advantage is that more than 35% of the expensive polyamidoamine-epichlorohydrin could be replaced by low-cost melamine–urea–glyoxal resin, thereby resulting in an adhesive at a 16% reduced cost [42]. The co-cross-linker and soybean meal adhesive had superior storage stability than melamine–urea–glyoxal and polyamidoamine-epichlorohydrin resins with a wet strength of around 0.95 MPa after 60 days [42].

The impacts of cross-linked phenol–formaldehyde resin and the reaction between glutamic acid and hydroxymethyl phenol (HPF) on the performance of soy-based adhesives were studied by Wu et al. [33]. The addition of HPF increased the dry shear strength from 0.98 MPa up to 1.81 MPa and the wet strength from zero to 1.18 MPa [33]. HPF-containing formulations prepared by the researchers had a higher content of hydroxymethyl groups, which are the main reactive groups that affect the adhesive properties, compared to a normal PF resin [38]. The results from several experiments based on resultant products derived from model compounds revealed that the soy-protein-based adhesive modified with phenol-to formaldehyde-based resin exhibited improved water resistance, indicating a reaction between soy protein and phenol–formaldehyde resin [33]. Although phenol–formaldehyde resin is hazardous, this research gave light on how aldehyde and its derivatives modify soy-based adhesives.

The introduction of wheat gluten protein hydrolysate for the preparation of non-toxic protein-based adhesives was evaluated by Xi et al. [44]. Glutaraldehyde was modified to prepare the adhesive [44]. Also, they explored polyethyleneimine as the cross-linking agent. Thermomechanical analysis showed that the modulus of the adhesive increased due to the addition of PEI, which also increases the wet shear strength from 0.91 MPa to 2.02 MPa. Glutaraldehyde modification also increased the dry strength from 0.76 MPa to 1.14 MPa. The authors attributed the motivation to use these materials to the water solubility properties of polyethyleneimine because it is very reactive and has been discovered to improve wet strength in paper preparation [45,46]. Aside from improving the bonding performance of the adhesives, polyethyleneimine improves its storage modulus and toughness.

Lei et al. studied the effect of several cross-linkers (melamine–formaldehyde, epoxy resin, and their mixture) on the water resistance of soy protein adhesives. Aside from preparing an adhesive product with excellent water resistance, they also intended to improve and optimize the composition of components of multiple cross-linkers [43]. The results showed that all three cross-linkers enhanced the performance of the protein adhesive [43]. The mixture of both epoxy resin and melamine–formaldehyde cross-linker stood out from the individual product and was confirmed by different mechanical tests as having the best water resistance of the formulation systems tested [43]. The improvement of water resistance modified by these cross-linkers might be due to the reaction between melamine–formaldehyde and –NH, as well as the reaction between epoxy and –OH [43]. Thus, this research was able to prepare an adhesive with excellent water resistance as well as effectively optimize the proportion of its components.

Zeng et al. introduced a tailor-made cross-linker containing flexible long chains and multiple epoxy groups [47]. The multiple epoxy groups joined together with the hydrophilic groups to design dense cross-linking networks, thereby improving the water resistance and wet bonding strength of the protein adhesive [47]. Another cross-linker bis(hexamethylene) triamine was synthesized by adding triethylenetetramine and 1, 6 hexadiol glycidyl ether to a flask followed by continuous stirring under controlled temperature [47]. Additionally, the flexible long chains contributed to the enhancement of the toughness of the adhesive, which increased the dry bonding strength and wet bonding strength from 1.11 MPa and 0.22 MPa to 2.79 MPa and 1.12 MPa, respectively [47]. Thus, this research provided a novel strategy for developing environmentally friendly material and bio-based adhesives with excellent water resistance and bonding strength.

Previous research has shown that developing a hyper-branched, cross-linked structure considerably raises the adhesive's cross-link density in the cured adhesive, thereby blocking moisture intrusion [48]. Zhang et al. improved on this research and subsequently built a flexible long-chain biopolymer with a hyper-branched structure and used it to create a unique, high-performance soymeal adhesive with a highly cross-linked structure and wet strength of 1.00 MPa, which is around 300% higher compared to soymeal adhesive with a wet strength of 0.22 MPa [49]. The team synthesized a long-chain dialdehyde starch grafted with hyper-branched polyamide to produce hyper-branched amino starch [49]. Thus, the resulting adhesive was found to be economical with great water resistance and outstanding mechanical qualities, and it has been successfully put to industrial pilot-scale manufacturing with the resulting plywood meeting the standards for interior usage [49]. Wheat flour adhesives can be made more water-resistant through the simultaneous modification of its two primary components, wheat starch and wheat protein, into a water-resistant three-dimensional, cross-linked network structure [50]. For the most part, raising the temperature causes carbohydrates and proteins to undergo a Maillard reaction, resulting in self-cross-linked network structures [50].

It has been confirmed that thermal treatment of wheat protein is an effective method for unfolding the glutenin protein structure, which enhances the aggregation of the structure and inter-molecular disulfide/sulfhydryl exchange reactions [51]. An anionic surfactant sodium dodecyl sulfate could be utilized to alter the wheat flour's protein and starch content [51]. Because sodium dodecyl sulfate can react with the hydrophobic groups of the wheat protein to obliterate the hydrophobic reactions and efficiently open up the protein structure, as well as interact with the wheat starch component to form starch–sodium dodecyl sulfate complexes, sodium dodecyl sulfate modification may lead to improved mobility and storage stability of starch-based adhesives [52–54].

Thus, Bai et al. worked on investigating the possibility of preparing water-resistant wheat-flour-based adhesives for wood applications by subjecting wheat flour to thermal–sodium dodecyl sulfate (SDS) or thermal–acid treatment, followed by cross-linking with a reactive polymeric diphenylmethane disocyanate (p MDI) resin coupled with the addition of a polyvinyl alcohol solution as an aqueous polymer [55].

Improved cross-linking density from self-aggregation and inter-change reaction between wheat proteins, the Maillard reaction between wheat starch and wheat protein and the cross-linking reaction between thermal–SDS-treated wheat flour and cross-linker p-MDI all contributed to the adhesive's increased water resistance, which increased the wet strength from 0.29 MPa to 1.22 MPa [55]. Nevertheless, the thermal–chemical treatment should be performed in the presence of non-toxic chemical substances. It is noteworthy that many crude/raw sources of proteins also contain appreciable levels of carbohydrates, which offers the Maillard reaction enhancement pathway for the resulting adhesives produced from these protein sources.

The strategies for improving the water resistance of protein adhesives, particularly soy-based product, was investigated by Wang et al. [56]. The team modified defatted soy flour by applying waterborne polyurethane resin and investigated the impact of the waterborne polyurethane on the water resistance of the adhesive through several experimental tests including tensile tests, thermogravimetric analyzer analysis, scanning electron microscopy and Fourier transform infrared spectroscopy [56]. Waterborne polyurethane was selected for this research due to its non-toxicity and excellent chemical and mechanical properties; also, waterborne polyurethane is widely used in adhesive, textile, building materials, paint, and leather processing industries [57]. The results showed that the initial denature temperature of the adhesive with waterborne polyurethane modification was lesser than the initial denature temperature without waterborne polyurethane modification [56]. Thus, this implied that the initial denature temperature with waterborne polyurethane modification showed better water resistance and mechanical performance, which increased the wet shear strength from 0.65 MPa to 1.1 MPa [56]. Aside from the fact that waterborne polyurethane reacts with the reducing sugar and active groups in soy protein structures, it also forms a reaction with the active groups in the wood structure, thereby facilitating the significant improvement of the water resistance [56]. The waterborne polyurethane-modified protein adhesive has shown a potential to replace the conventional formaldehyde-based resin, thereby reducing the dependency on petrol-based products.

Xu et al. explored cottonseed meal as the raw material for protein adhesives [58]. They studied the effects of modifier reaction conditions and additions on the bonding properties of the adhesive as well as the modification mechanism [58]. Cottonseed protein has not been fully utilized by other markets (access product resulting in low cost); thus, it can help reduce the cost of adhesive, as it is cheaper than soy protein. At the same time, this excess supply of cottonseed protein can be used to overcome the demand and supply limits of soybean protein, thereby providing an additional high-value usage of cottonseed meal [59]. Figure 3 shows the cross-linking mechanism between soy protein isolate (SPI), polyacrylamide (PAM), and 1,2,3-propanetriol-diglycidyl-ether (PTGE) through which the components form a larger network structure [60].

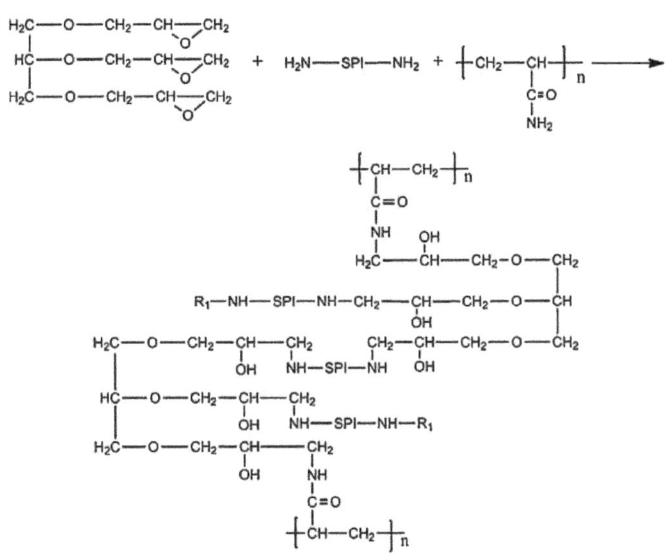

Figure 3. The cross-linking reaction among soy protein isolate (SPI), polyacrylamide (PAM), and 1,2,3-propanetriol-diglycidyl-ether (PTGE) [60]. Reprinted/adapted with permission from Ref. [60]. 2023, Industrial Crops and Products.

The team blended different proportions of isocyanate with cottonseed protein adhesive [58]. The results showed that the solid content and viscosity of the adhesive increased with the addition of isocyanate proportion, and the researchers confirmed that the solid content of adhesive has impacts on the viscosity of the adhesive, thereby affecting the bonding strength, wettability, sizing effect, and amount of glue [58]. The wet bonding strength increased from 0.65 MPa to 1.68 MPa. The increase in the amount of isocyanate translated to a reduction in the application period of the adhesive [58]. Thus, this implies that the preparation process of the constituents (cottonseed protein adhesive and isocyanate) needs to be further explored [58]. Nevertheless, this research provides practical and theoretical guidance and a basis for the preparation of novel low-cost bio-based adhesives and explores a new way for utilizing the waste cottonseed meal.

A summary of the wet strengths found from various proteins combined with cros linking agents is given in Table 1.

Table 1. Summary of wet bond strengths of proteins with their cross-linking agents.

Protein and Cross-Linking Agent	Wet Bonding Strength (MPa)	Reference
Wheat protein GPPEI	0.44 2.02	[44]
Soy protein EPR + MF/SP	0 0.85	[43]
Soy protein EG/SP	0.22 1.12	[47]
Soymeal SM/TGIC/HD	0.22 1.0	[49]
Wheat flour T-SDS-WF	0.29 1.22	[55]
Soy protein WPU/SP	0.65 1.1	[56]
Cottonmeal CM/Isocyanate	0.65 1.68	[58]

Abbreviations: GPPEI—glutaraldehyde-polyethyleneimine-modified wheat protein; EPR + MF—epoxy resi and melamine–formaldehyde modified soy protein; EG/SP—epoxy groups modified soy protein; SM/TGIC HD—hyperbranched triglycidyl isocyanurate modified soybean meal; T-SDS-WF—thermally treated sodiu dodecyl sulfate modified wheat flour; WPU/SP—waterborne epoxy emulsion modified soy protei CM/Isocyanate—isocyanate-modified cottonmeal.

2.1.2. Bio-Based Cross-linking Agent Sources

Bio-based sources are synthesized from biomass and renewable materials. Thus, potentially has the advantages of lower environmental toxicity, renewability, and sustai ability.

Pang et al. successfully prepared waterborne epoxy emulsion through graft polyme ization to generate an oil-in-water emulsion [61]. As determined by different characte ization tests including Fourier Transform Infrared Spectroscopy (FTIR), the synthesize waterborne epoxy emulsion (WEU) had good dispersibility and stability and was deeme an efficient cross-linker for generating multiple stable cross-linking networks with so protein molecules [61]. The results showed a significant water resistance improvement wit a wet shear strength of 0.71 MPa compared to unmodified soy-protein-based adhesive wit a wet shear strength of 0.39 MPa and meet the requirements of interior-usage plywoo which is either greater than or equal to 0.7 MPa [61].

A group of researchers worked on exploring an enzyme complex (*Aspergillus niger*) b using soybean hulls as a substrate for the development of the protein adhesive. An ine pensive *Aspergillus niger* fermentation broth contains an enzyme complex that performe exceptionally well at hydrolyzing the polysaccharides in defatted soy flour [62,63]. Th decreased quantity of water-insoluble materials and the weakened rheological properties c the slurry were evidence that this enzyme complex effectively hydrolyzed polysaccharide in defatted soy flour [64]. In this study, different soy polysaccharides were hydrolyzed int reducing sugars using an enzyme broth made up of *A. niger* which is particularly effectiv for doing so. The exposed functional groups of the modified soy protein reacted with th reducing sugars, significantly enhancing the soy adhesive's water resistance. The reactio between reducing sugars and soy protein increased the strength and water resistance c soy adhesives, which together helped to increase wet bonding strength [64]. Althoug the synthesis process is quite complicated, there is a lot of potential for the self-cros linked soy adhesive created by this practical and affordable technology to replace syntheti wood adhesives.

Due to their catechol composition, marine mussels possess excellent adhesive characteristics [61]. This catechol structure is capable of forming strong hydrogen bonds with the substrate and reacting with amino groups after oxidation to quinones [65,66]. Thus, catechol-structured biomass raw materials appear to offer an excellent option for cross-linking soybean protein adhesives.

Zheng et al. built on previous research by developing a lignin with a catechol structure using a single-step demethylation [67]. Copper ions, lignin, and soybean protein isolate were combined to produce a bio-based adhesive with a triple network structure without the use of an epoxy cross-linking agent derived from petroleum, thereby expanding the potential value of bio-based adhesives in industrial applications and production while increasing the overall sustainability of the potential process [67]. Catechols were randomly oxidized into quinones, and lignin's self-bonding properties made it possible to form both covalent- and hydrogen-bond cross-links with just the addition of lignin [67]. Due to the abundance of metal coordination and hydrogen bonds in the system, which serves as a mechanism for energy dissipation, the adhesive demonstrated the ideal viscosity, outstanding coating characteristics, and tensile strength [67]. The increased number of glue nails and lower viscosity helped create a strong adhesive bond in the wood while the adhesive's water resistance can be attributed to its dense cross-linked structure [67].

Tung oil is extracted from the tung tree's seeds or kernels [68,69]. Most commonly, it is found in inks, coatings, and resins, but it is also used as a natural varnish for wood [68]. About 80% of its fatty acid content is a conjugated, trienoic fatty acid, which gives it its distinctive drying properties [68]. Tung oil's high iodine value, rapid drying time, increased water resistance, and increased hardness as a result of its high level of unsaturation make it a desirable protective coating material [70,71]. He et al. worked on the application of tung oil on adhesives produced from cottonseed protein isolate and water-washed cottonseed meal [72]. The water resistance test was performed by immersing the glued wood specimens in a water bath followed by drying at room temperature for more than 18 h [72]. Water-washed cottonseed meal and cotton-seed protein isolate with tung oil showed improved adhesive strength, compared to tung oil-free adhesive controls [72]. The addition of tung oil to water-washed cottonseed meal and cotton-seed protein isolate also increased their resistance to water by over 40%. Optical microscopic photographs of bond lines of these bonded wood pairs demonstrate that the improvement was mostly attributable to the tung oil's ability to prevent the adhesives from diffusing away from the bond joints [72]. This research has shown that promoting cottonseed protein products as wood adhesives over wheat and soy protein products will alleviate long-term concerns about global food security.

Another bio-based material for cross-linking is tannic acid processed from vegetable tannins [73]. Tannic acid has been widely used as a cross-linking material for different biopolymers like chitin whiskers, chitosan, collagen, casein, albumin, and gelatin [73–79]. Tannic acid is categorized as a tannin which is processed through hydrolysis under mild alkaline or acidic conditions to obtain phenolic acids and carbohydrates [73].

For their research, Ghahri et al. worked on the application of tannic acid for modifying soy flour adhesives, thereby increasing their moisture resistance for wood applications [80]. The tannic acid solution was mixed with the soy slurry and reacted with soy flour to form a cross-linked network, thereby increasing the water resistance and mechanical properties of manufactured wood panels. Figure 4 shows the cross-linking mechanism between amino acids and tannic acids, where the protein–tannin complexes self-associate through hydrogen bonding [80]. The modulus of elasticity was increased from 3213 MPa to 3590 MPa, the modulus of rupture increased from 17 MPa to 18 MPa, and the shear strength increased from 1.43 MPa to 1.96 MPa [31]. The delamination test was successfully completed by plywood manufactured using adhesives produced from tannic-acid-modified soy flour [80]. Adding tannic acid to the soy glue recipe enhanced the plywood's shear strength [80]. Fiberboard tests revealed that the modified soy-based glue improved the resulting board's water resistance and thickness swelling in addition to enhancing its mechanical properties [80]. This has shown to be a good way to bring new eco-friendl and water-resistant soy-based adhesives to the wood composites industry.

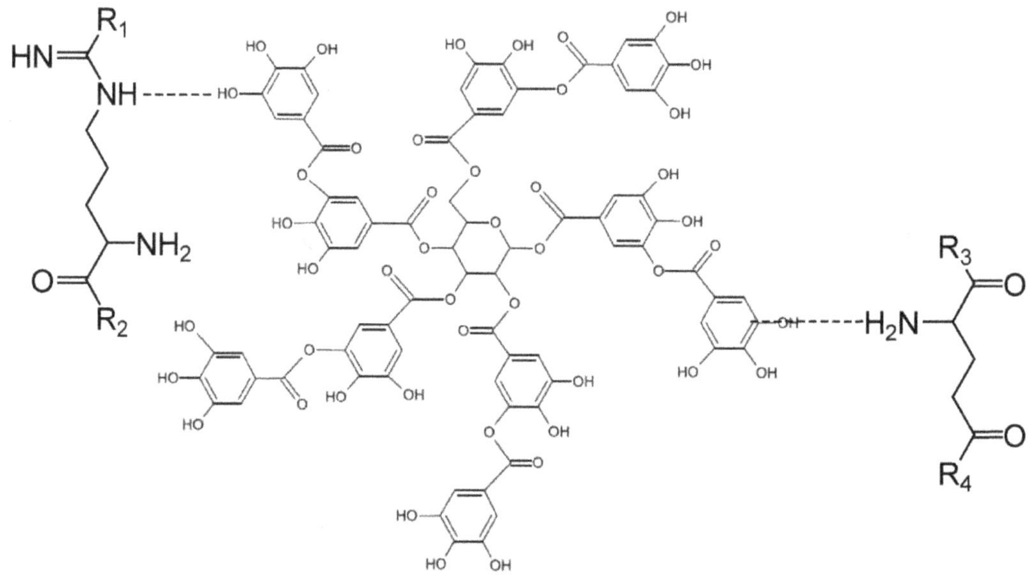

Figure 4. Cross-linking reaction between tannic acids and amino acids [80]. Reprinted/adapted wit permission from Ref. [80]. 2023, Journal of Polymers and the Environment.

Lignin-based resin reacted with itself and protein molecules to form an interpenetra ing and cross-linking network [81]. Lignin exists in plant materials/products like sorghum corncob, and wood and has a cross-linked and aromatic structure comprising of severa building units [81]. Due to the availability of aromatic compounds derived from plant which are renewable products, the de-polymerization of lignin can be applied to mak bio-based polymers [82,83]. Another property of lignin is that some of the compositiona structures have similar features to phenol, thus providing the potential to replace phenol i the phenol–formaldehyde resin preparation [81].

The use of lignin-based resin increased the wet shear strength and the water resistanc of the resulting soymeal-based adhesive [84]. The researchers attributed the successe to the following reasons: the formation of a cross-linking network due to the reactio between the soy protein molecules and the lignin-based resin; the formation of a smoothe fracture surface, thereby preventing the intrusion of moisture; the formation of an interpen etrated network by self-cross-linking lignin-based resin molecules and cross-linked protei molecules; an improvement in the thermal stability of the adhesive; and the resultan appropriate viscosity, thereby benefiting the adhesive distribution and forming a stronge interlock [84].

Pradyawong et al. worked on studying the effects of Kraft lignin on lignin and protei interactions through thermal and rheological properties and the adhesion performance o the protein product [85]. Previous research has shown that Kraft lignin is hydrophobic i nature and has a high content of aliphatic hydroxyl groups allowing for reactive modifica tion and serving as a superb substrate for lignin-derived products [86,87]. Lignin and so protein were mixed at different weight ratios, and the results showed an improvement i water resistance due to the unfolding of the hydrophobic group to the protein structure [85 Another advantage is that the lignin increased the thermal stability and dispersal of th soy protein adhesives [85]. The team explored more research on Kraft lignin by modify ing laccase/TEMPO lignin to study its adhesion properties and performance [88]. Th synthesis process involves depolymerizing kraft lignin by laccase enzyme and TEMPO to enhance the oxidation reaction of both non-phenolic and phenolic compounds [88]. The unique advantage of this process is that it enhances the lignin–protein interaction [88]. The simplified process of mixing the soy protein, kraft lignin, and laccase enzymes increased the wet shear strength by 106% and strengthened the protein structure due to the interaction between the $-NH_2$ and $-COO^-$ group [88]. However, these interactions led to a decrease in the spreadability and flowability [88]. Thus, we can conclude from this research that it revealed a simplified synthesis process for adhesives' preparation and shows that the combination of these materials has great potential to fulfil the massive demand for green products [88].

Another team of researchers led by Zhu also explored depolymerized lignin to improve the water resistance of camelina protein [89]. In this case, the depolymerization of Kraft lignin was completed by H_2O_2-induced oxidation via ultrasound irradiation to reduce the thermal stability and particle size of the lignin and increase the hydroxyl content [89]. The modified lignin contained more hydroxyl groups and smaller particles [89]. Ultrasound-induced oxidation was more effective in inducing the disintegration of the lignin structure and enhancing its reactivity [89]. Thus, the team developed a bio-based wood adhesive from camelina proteins and copolymerized the product with the modified lignin [89]. Also, they compared the product with the most commonly used adhesives like polymeric amine epichlorohydrine, urea-formaldehyde-based adhesives, and $NaHSO_3$-modified soy protein adhesives [89]. The results showed that the oxidation-pretreated lignin-based adhesive produced by this team had the highest water resistance and strongest intermolecular interaction evaluated when compared to the other adhesives [89]. Industrial adhesive uses of the non-edible camelina protein copolymerized with oxidized kraft lignin would be beneficial to both the camelina oil-based energy business and the paper industry [89]. However, the synthesis process should be simplified for large-scale production.

Xiao et al. also worked on improving the water resistance of soy protein adhesives by adding lignin. In this case, the researchers added extruded sorghum lignin and sorghum lignin and investigated the adhesion properties [83]. They added extruded sorghum lignin and sorghum lignin to soy protein adhesives in four different reaction solutions, namely, modified soy protein, soy protein isolate, alkaline solution, and neutral water [83]. The glued wood samples were soaked in tap water at room temperature for 48 h before being air-dried in a fume hood [83]. When the combined amount of sorghum lignin and extruded sorghum lignin grew from 30% to 50%, the water resistance improved [83]. This finding suggests that when sorghum lignin or extruded sorghum lignin is blended into soy protein adhesives at higher concentrations, the advantage of entrapping interaction between lignin molecules and modified soy protein predominates, as opposed to being the driving force behind the blended sorghum lignin and extruded sorghum lignin at lower percentages [83]. Soy protein adhesives combined with extruded sorghum lignin from modified soy protein fared better in water resistance tests than those blended with sorghum lignin, whereas the inverse was true for the range of blending percentages between 30% and 50% [83]. By blending with sorghum lignin, soy protein adhesives' shear strength and water resistance were considerably enhanced, but the improvement with extruded sorghum lignin was more pronounced when the mixed soy protein adhesives were suspended in water [83]. Further tests should be carried out by using other lignin. However, this research further certified the potential of lignin for protein adhesives.

Lignin processed from the waste product of fermenting cellulosic-based sugars into bioethanol was explored for protein adhesive by Xin et al. [90]. The researchers used Fenton oxidation at room temperature in a basic aqueous solution to change the hydroxyl groups in lignin into carbonyl groups (like ketones or aldehydes) [91,92]. Compared to previous research projects, this particular lignin was insoluble in water and most organic solvents due to its original highly branched structure [93]. In organic chemistry, reductive amination is a highly efficient process for obtaining amino substituted products [93]. The team prepared both unmodified lignin and partially depolymerized lignin and used both products for the preparation of lignin amines [90]. The results showed that the amine prepared via th unmodified lignin has a greater effect on the water resistance and adhesion strength of sc protein adhesives [90]. This research demonstrated a novel way of preparing adhesive through the Fenton oxidation and reductive animation method [90]. This process provide competitive advantages of non-toxic, high conversion, environmentally friendly reaction and solvents at room temperatures [90]. The lignin amine might have a potential for usag in other bio-based materials, such as polyurethanes and epoxy resins [90].

There are several issues associated with soy adhesives, such as high viscosity, th requirement of cross-linking modification, and denaturation agent due to its high molecul weight [94]. Due to the use of petrochemical agents, modified soy protein glue is not completely sustainable product [94]. The unstable performance of the soy protein glu restricts its widespread use in the wood panel manufacturing business [94]. Therefore, soy protein adhesive with high bonding strength and water resistance, low cross-link addition, moderate viscosity, and consistent bonding performance must be developed [9

This interest in enhancing the sustainability of soy-protein-based adhesives was th basis for the research performed by Xu et al. to improve the qualities of soy protei glues by using the recombination strategy [95]. Molecular recombination is simply th deterioration of a substance followed by a recombination process to achieve excellent pro erties [96]. Through a process involving both biological and chemical alteration, molecul recombination theory served as a guiding principle for the modification of the soy protei adhesive [96]. Soy protein adhesives of varying molecular weights were manufactured b breaking down soy protein molecules into tiny peptide chains using bromelain of varyin concentrations [96]. Using enzymatic hydrolysis, the protein endonuclease-bromelai disrupted the peptide bonds in the soy protein molecules, resulting in the degradation c the protein into short molecular polypeptide chains and a lower molecular weight [9 Triglycidylamine, a bio-derived cross-linking agent, was used at low concentrations t recombine these polypeptide chains [95]. Using a two-step procedure, the adhesive's pe formance and bond stability were enhanced by creating a consistent and stable cross-linke network structure [95]. Meanwhile, the adhesive's viscosity was lowered by employin bromelain [95]. By replacing the weak hydrogen bond with a chemical bond during th cross-linking step, water resistance in adhesives is enhanced [95]. After enzymatic hy drolysis, the molecular weight of the soy protein was measured [95]. The resulting so protein adhesives were analyzed for their viscosity, adhesive layer homogeneity, therma stability, functional groups, and micro-structure of the fracture surface [95]. Soy protei adhesive plywood samples were fabricated, and their wet shear strength was evaluated [9 These findings provided evidence for the beneficial effects and underlying mechanism c molecular recombination involving soy proteins in the adhesives [95].

Mussel-inspired polydopamine coating utilizing dopamine chemistry has been demon strated to be an efficient method for enhancing the surface reactivity of diverse material including hydrophilic and hydrophobic surfaces as well as inorganic and organic sub strates [97,98]. Also, solutions containing catechols had the highest adhesion energ indicating that adhesives containing catechols should also exhibit exceptional adhesio energy [99]. The catechol groups of dopamine can oxidize to generate a reactive quinon capable of undergoing Schiff base reactions or Michael-type additions with radical couplin and nucleophiles with other amines or catechol in alkaline conditions without the use c complicated equipment [100,101].

Pang et al. worked on the surface functional modification of mussel proteins t enhance the weak surface reactivity of silkworm silk fiber [102]. The next step was th application of the as-fabricated silkworm silk fiber to improve soy-based adhesives, thereb generating a compact structure [102]. Several experimental analyses confirmed that tanni acid, dopamine, and alkali lignin were successfully coated onto silkworm silk fiber [102 The soy protein composite had improvements in properties like toughness and wet shea strength. The results showed that the adhesive sample attained a high wet shear strengt that is far greater than the interior use standards.

Undecylenic acid (UA) is a green chemical generated from castor oil by pressure-cracking and is extensively utilized in the perfume, cosmetic, and pharmaceutical industries [103]. Undecylenic acid is non-soluble in water and has long hydrophobic chains, thereby preventing water from gaining access into the interfacial surface of the wood and adhesive coupled with a significant improvement of the wet shear adhesion strength in the range of 35–62% [104]. Liu et al. focused on the development and characterization of UA-modified soy proteins to enhance their water resistance for the production of adhesives [104]. The team successfully incorporated undecylenic acid into soy protein via the reaction between NH_2 and COOH, which was confirmed by FTIR [104]. Also, it confirmed the reduction in amino group concentration via the use of the ninhydrin reagent [104]. Thus, this formulation created an effective pathway to develop biodegradable and cost-effective adhesives.

2.1.3. Synthetic and Bio-Based Cross-linking Sources

Zhou et al. produced flame retardant and high-strength adhesive by combining the effect of cross-linking strategy and borate chemistry [105]. Here, oxidized soybean polysaccharide was complexed with borax to form an in situ cross-linking skeleton with periodate oxidation–adipic acid dihydrazide via a Schiff base reaction; chitosan was introduced for the purpose of improving the active sites as well as the water resistance; and the process was inspired by borate chemistry in plants and the periodate oxidation–adipic acid dihydrazide cross-linking strategy [105]. Hydrogen bonding and the covalent bonds of acylhydrazone and borate ester are just two of the cross-linking networks that were built into the soy-protein-based glue [105]. The researchers also aimed to develop an environmentally friendly and effective cross-linking technique of polysaccharides, which is crucial for their use in wood adhesives. The strength of borate-chelated polysaccharide adhesive must be enhanced [105]. In recent years, the periodate oxidation–adipic acid dihydrazide cross-linking technique has been applied successfully to the development of robust polysaccharide-based composites, such as cellulose, starch, and pectin [106]. Also, periodate oxidation–adipic acid dihydrazide is environmentally friendly, and the cross-linked polysaccharide material exhibits outstanding biocompatibility and mechanical strength [107]. Thus, this cross-linking strategy has the potential to develop soy protein adhesive with high cohesiveness, which was greatly enhanced by the double-network structure resulting from multiple cross-linking [105]. The results showed that the wet and dry shear strength increased significantly which was accompanied by a reduction in the moisture absorption rate and an increase in the residual rate, thereby signifying excellent water resistance and bonding strength [105]. The novel soy-protein-based adhesive significantly reduced health threat and environmental emissions [105]. Although the synthesis process is quite complicated, the process of producing excellent polysaccharide-based materials does support its use [105].

Phenol-amine chemistry was applied toward the preparation of novel organic–inorganic protein-based adhesives [108]. Li et al. developed a versatile and green strategy for the development of a robust soy protein adhesive through bio-mineralization reinforcement and phenol-amine synergy [108]. Thus, the researchers co-assembled hydroxyapatite and gallic acid via calcium ion phenolic coordination bonds to synthesize gallic-acid-functionalized hydroxyapatite [108]. In addition, a Schiff base reaction was used to graft ε-polylysine with numerous amino groups to the phenolic gallic-acid-functionalized hydroxyapatite nanoparticles [108]. Several researchers have worked on synthesizing hydroxyapatite to prepare strong adhesives. For instance, Yu et al. incorporated hydroxyapatite and polyvinyl alcohol to develop a composite microfiber with significant improvement in flexibility and toughness [109]. In a similar vein, Jiang et al. developed a superior gelatin-based adhesive via the incorporation of polydopamine-designed hydroxyapatite nanoparticles [110]. The disadvantage is that these methods require tedious and complex processes, the use of harmful chemicals and expensive raw materials. Li et al. have been able to mitigate these challenges by developing a sustainable and simple strategy for preparing organic–inorganic biopolymer materials [108]. The results showed that the synergistic effect of phenolic gallic-acid-functionalized hydroxyapatite with amine-rich ε-polylysine and soy protein chains simultaneously increased the hybrid adhesive's cohesion and adhesion strength [108]. Also, the phenol-amine synergy and solid bio-mineralized structure conferred water resistance and thermal stability to the resulting glue [108]. Thus, the biomimetic approach offers a flexible method for producing biopolymer materials for adhesives, films, coatings, and hydrogels [108].

2.2. *Water Resistance from Modified Fillers*

The addition of fillers can significantly impact the thermal and mechanical properties of protein adhesive, and the mechanism involves the interaction between the protein matrix and the filler material [111,112]. The incorporation of sodium montmorillonite (Na MMT) into soy protein adhesives at different concentrations was explored by Qi et al. [113]. Sodium montmorillonite is the most extensively utilized kind of silicate clay in polymer nanocomposites with properties, such as thermal and chemical stability, natural abundance and non-toxicity [114]. The material is widely employed as a reinforcing and nano-filler material to produce nanocomposites owing to its high aspect ratio and unusual layered and nanoscale structure [115]. Thus, Qi et al. worked on developing a novel soy protein and clay system with excellent flowability and strong adhesion at high protein content [113]. Sodium bisulfite ($NaHSO_3$) was used in these formulations, and hydroxyethyl cellulose was used as a suspension agent [113]. $NaHSO_3$ can be utilized as a reducing agent to break the disulfide bonds in protein molecules, thereby resulting in an increase in surface hydrophobicity, solubility, and flexibility [116]. Another researcher, Zhang and Sun, corroborated this claim by using $NaHSO_3$ to break the disulfide bonds of soy glycinin to increase the surface hydrophobicity [117].

The results showed that the addition of Na MMT significantly improved the adhesion strength of the soy protein adhesives due to the adsorption of the soy protein molecules on the surface of the interlayer of Na MMT via electrostatic interaction and hydrogen bonding [113]. Thus, the water resistance of the soy protein/Na MMT increased to 4.3 MPa compared to the 2.9 MPa of control SP at 8% Na MMT and the dry shear strength from 5.7 MPa to 6.38 MPa [113]. This research shows an innovative way to develop an adhesive with excellent adhesion with the incorporation of silicate clay materials.

Ciannamea et al. prepared soy protein concentrate (SPC)-based adhesives and rice husks (RCs) to produce particleboards with the main goal of upgrading the final water resistance and mechanical properties of RH-SPC particleboards via the alkali treatment of soy protein concentrate and rice husks, coupled with bleaching of rice husks with hydrogen peroxide [118]. This facilitated chemical interactions via hydrogen bonds between the more exposed hydroxyl groups of cellulose from rice husks and the polar groups of the unfolded proteins of soy protein concentrate treated with alkali [118]. The particleboards met the mechanical properties requirements for commercial use consideration but failed to achieve the minimum requirements for water resistance as recommended by US Standard ANSI/A208.1 [118,119]. The limitation was linked to the increase in the amorphous content of the cellulose after dispersing rice husk in NaOH for a short period [118]. This drawback is counterbalanced by the adhesive's lack of formaldehyde and the use of complete rice husks in particleboard production, which eliminates milling and screening procedures, resulting in cost savings [118].

2.3. *Removal of Hydrophilic Content*

Gui et al. centered their research on the preparation of water-resistant soy flour adhesives through the reduction in the water-soluble constituents [120]. Previous research has shown that poor water resistance is mainly caused by feedstock water-soluble components [29]. The design approach involved suspending the defatted soy flour in water followed by the adjustment of the dispersion pH at different temperatures and time points to 4.5 by adding NaOH and HCl solution, respectively [120]. Then, the sample with less water-soluble constituents was separated through a centrifugation [120]. The application of the modified soy flour adhesive on poplar plywood shows that it had a wet strength of 1.02 MPa [120]. The remaining multi-level structures of soy protein contributed positively to soy adhesives' water resistance [120]. Although this is a simple and novel way of developing water-resistant soy adhesives, the product is still limited by low solid content and fluidity compared to formaldehyde adhesives [120]. Thus, further research is required to enhance these properties.

Zhang et al. subjected defatted soybean flour (DSF) adhesive to thermal treatment at different test temperatures to improve its water resistance and investigate the effects of the thermal pretreatment on increasing the water-insoluble content, crystalline degree, and chemical structure [121]. The team also tested the thermal stabilities and bonding qualities of soy adhesives made from thermal treatment DSF (T-DSF) and cross-linker epichlorohydrin-modified polyamide (EMPA) [121]. The test result showed that the thermal treatment facilitated the increase in the acetaldehyde value and water-soluble content of T-DSF [121]. Thus, thermal treatment can enhance protein–carbohydrate Maillard reactions, protein–protein self-cross-linking, and protein–EMPA cross-linking by unfolding the globular form of soy protein and releasing hidden functional groups [121]. Uncertain are the quantitative contributions of protein–protein self-cross-linking, protein–carbohydrate Maillard processes, and protein–EMPA cross-linking, as well as their impact on the water resistance of T-DSF-based adhesives [121].

Qi et al. investigated the effect of liquid 2-octen-1-ylsuccinic anhydride (OSA) on soy protein adhesives [122]. The OSA possesses a long alkyl chain and oily nature coupled with a succinylation reaction that can help enhance protein adhesion strength [122]. Thus, the team studied the adhesive properties of soy protein adhesives modified by OSA at different concentrations as well as characterized its physicochemical properties like morphological, rheological, thermal, and turbidity properties [122]. The OSA modification increased the wet shear strength in plywood samples up to 3.2 MPa with up to 60% wood cohesive failure in comparison to 1.8 MPa of wet strength for the control. They found that the modification of the soy protein adhesives with OSA facilitated the introduction of hydrophobic materials to the protein structures [122]. Due to OSA's hydrophobic properties, hollow cavities could not form since water could not penetrate the interface between the wood surface and the adhesive [122]. The researchers stated that this could be the primary factor for the soy protein adhesives' significantly improved water resistance [122].

Zhang et al. worked on the modification of soy protein adhesives using epoxidized oleic acid and prepared the chemically modified adhesive soy protein and rice straw formulation to produce a fiberboard that might serve as a viable substitute for wood-based fiberboard [123]. The utilized epoxidized oleic acid contains many free epoxy groups that could have a reaction with the functional amino groups in the SPI molecule. Epoxidized oleic acid was utilized to react with the amino groups in SPI molecules to increase the mechanical and water resistance qualities of the adhesive [123]. The team investigated modified-SPI adhesive addition, the effects of NaOH concentration, and fiberboard density on the water resistance and mechanical properties of rice straw fiberboards [123]. The results showed that the fiberboards have optimal water resistance and mechanical performance, which is due to the reaction between the soy protein and the epoxidized oleic acid, and the removal of the wax layer using NaOH with a water resistance value of around 64% [123]. The advantage is that the raw materials are low-cost and easily biodegradable; thus, these fiberboards are an excellent substitute for petroleum-based panels and can be utilized in indoor furniture and decoration.

3. Conclusions

This review paper has outlined recent strategies for improving the water resistance of protein adhesives, which are categorized into cross-linking networks, separation techniques, and modifications with different reagents. In each case, the researchers explored the use of synthetic, bio-based, or a combination of both.

Cross-linking improves the water resistance and adhesive strength of protein adh sives [124]. However, this strategy can increase the complexity of the manufacturing proce while some of the agents are harmful to the environment [33]. The use of naturally occu ring modified fillers has shown to be effective in improving the water resistance of woc adhesives; however, the challenge is that they may require additional processing steps [11 The most promising strategy is the removal of hydrophilic content because it does n require a complex synthesis process and is suitable for large-scale applications [120].

The trends have shown that many researchers have started to explore the possibi ties of bio-based sources for improving water resistance in protein adhesives due to the sustainability and renewability [125]. However, the researchers should try to simpli the synthesis process for large-scale applications. The combination of both tried to op mize its advantages but came with a more complicated synthesis process and expensiv raw materials.

An epoxy cross-linking agent has been shown to be efficient for interior plywoo use ($\geq$0.7 MPa) like flooring, ceiling, etc., but its exterior usage has not been fully esta lished [124]. The complete replacement of petrol-based wood adhesives with protein is n feasible unless the limitations of protein adhesives are eliminated. Thus, it is essential fc the industry stakeholders to continue to develop viable ways to overcome these challeng to move from a petroleum-based economy to a more sustainable economy and to minimiz the use of formaldehyde-based adhesives.

While continuing to improve water resistance and wet strength, future research shoul also include reducing the complexity and reaction times of these water-resistance amen ments because wood composite industries need to make products using fewer steps wit short press times in order to be cost-competitive. The combination of proteins with starche lignin, tannins, and other biomass sources is likely to be a continued area of interest wit great potential. Further research should also be conducted to assess the environment impacts of these novel modifications. One would expect that if the protein adhesives ar made more water-resistant, the biodegradability will be affected. The likely environment fate of the modified proteins, cross-linking agents, and other additives should also b determined before commercializing on a large scale.

Author Contributions: Conceptualization, W.M.C., O.V.O. and M.E.Z.; writing—original draft prep ration, O.V.O., M.A.L., W.M.C. and M.E.Z.; writing—review and editing, W.M.C. and M.E.Z.; superv sion, W.M.C. and M.E.Z.; funding acquisition, M.E.Z. and W.M.C. All authors have read and agree to the published version of the manuscript.

Funding: This review was funded by the Energy Institute of Louisiana, the University of Louisian and the Industrial Ties Research Subprogram (contract number: LEQSF (2022-25)-RD-B-06) of th Louisiana Board of Regents. The APC was funded by the Energy Institute of Louisiana.

Acknowledgments: The authors thank Isabel Dacdac for her assistance with the literature search.

Conflicts of Interest: The authors declare no conflict of interest.

References

1. Grand View Research. Wood Adhesives Market Growth & Trends. Grand View Research. Available online: https://ww grandviewresearch.com/press-release/global-wood-adhesives-market (accessed on 21 March 2023).
2. Pradyawong, S.; Li, J.; He, Z.; Sun, X.S.; Wang, D.; Cheng, H.N.; Klasson, K.T. Blending Cottonseed Meal Products with Differer Protein Contents for Cost-Effective Wood Adhesive Performances. *Ind. Crop. Prod.* **2018**, *126*, 31–37. [CrossRef]
3. Antonio Pizzi, K.M. *Wood Adhesives*, 1st ed.; CRC Press: London, UK, 2011; p. 462. [CrossRef]
4. Raydan, N.D.V.; Leroyer, L.; Charrier, B.; Robles, E. Recent Advances on the Development of Protein-Based Adhesives for Woo Composite Materials—A Review. *Molecules* **2021**, *26*, 7617. [CrossRef]
5. Wibowo, E.S.; Lubis, M.A.R.; Park, B.D.; Kim, J.S.; Causin, V. Converting Crystalline Thermosetting Urea–Formaldehyde Resir to Amorphous Polymer Using Modified Nanoclay. *J. Ind. Eng. Chem.* **2020**, *87*, 78–89. [CrossRef]
6. He, Z.; Zhang, Y.; Wei, W. Formaldehyde and VOC Emissions at Different Manufacturing Stages of Wood-Based Panels. *Buil Environ.* **2012**, *47*, 197–204. [CrossRef]
7. Kumar, R.N.; Pizzi, A. Environmental Aspects of Adhesives—Emission of Formaldehyde. *Adhes. Wood Lignocellul. Mater.* **2019**, 293–315. [CrossRef]

Congress. Formaldehyde Standards for Composite Wood Products. Public Law, 2010, Title VI (II). Available online: https://www.congress.gov/111/plaws/publ199/PLAW-111publ199.pdf (accessed on 21 March 2023).

Kristak, L.; Antov, P.; Bekhta, P.; Lubis, M.A.R.; Iswanto, A.H.; Reh, R.; Sedliacik, J.; Savov, V.; Taghiyari, H.R.; Papadopoulos, A.N.; et al. Recent Progress in Ultra-Low Formaldehyde Emitting Adhesive Systems and Formaldehyde Scavengers in Wood-Based Panels: A Review. *Wood Mater. Sci. Eng.* **2023**, *18*, 763–782. [CrossRef]

Podlena, M.; Böhm, M.; Saloni, D.; Velarde, G.; Salas, C. Tuning the Adhesive Properties of Soy Protein Wood Adhesives with Different Coadjutant Polymers, Nanocellulose and Lignin. *Polymers* **2021**, *13*, 1972. [CrossRef] [PubMed]

Grand View Research. Wood Adhesives Market Size. Grand View Research. Available online: https://www.grandviewresearch.com/industry-analysis/wood-adhesives-market (accessed on 21 March 2023).

Cholewinski, A.; Yang, F.; Zhao, B. Algae-Mussel-Inspired Hydrogel Composite Glue for Underwater Bonding. *Mater. Horiz.* **2019**, *6*, 285–293. [CrossRef]

North, M.A.; Del Grosso, C.A.; Wilker, J.J. High Strength Underwater Bonding with Polymer Mimics of Mussel Adhesive Proteins. *ACS Appl. Mater. Interfaces* **2017**, *9*, 7866–7872. [CrossRef] [PubMed]

Tsujimoto, Y. Molecular Biology of Cell Death. *Jpn. J. Cancer Chemother.* **1994**, *21*, 591–595.

Diani, J.; Gall, K. Finite Strain 3D Thermoviscoelastic Constitutive Model. *Polym. Eng. Sci.* **2006**, *46*, 486–492. [CrossRef]

Vnučec, D.; Kutnar, A.; Goršek, A. Soy-Based Adhesives for Wood-Bonding—A Review. *J. Adhes. Sci. Technol.* **2017**, *31*, 910–931. [CrossRef]

Chirdon, W.M. Utilization of Biorefi Nery Waste Proteins as Feed, Glues, Composites, and Other Co-Products. In *Algal Biorefineries Volume 2: Products and Refinery Design*; Springer: Berlin/Heidelberg, Germany, 2015; pp. 367–392. [CrossRef]

Frihart, C.R.; Birkeland, M.J. Soy Properties and Soy Wood Adhesives. *ACS Symp. Ser.* **2014**, *1178*, 167–192. [CrossRef]

Shi, Q.; Zhou, Y.; Sun, Y. Influence of PH and Ionic Strength on the Steric Mass-Action Model Parameters around the Isoelectric Point of Protein. *Biotechnol. Prog.* **2005**, *21*, 516–523. [CrossRef] [PubMed]

Zhang, D.W.; Zhao, X.; Hou, J.L.; Li, Z.T. Aromatic Amide Foldamers: Structures, Properties, and Functions. *Chem. Rev.* **2012**, *112*, 5271–5316. [CrossRef] [PubMed]

Abedi, E.; Pourmohammadi, K. Physical Modifications of Wheat Gluten Protein: An Extensive Review. *J. Food Process Eng.* **2021**, *44*, e13619. [CrossRef]

Dunky, M. Wood Adhesives Based on Natural Resources: A Critical Review: Part I. Protein-Based Adhesives. *Rev. Adhes. Adhes.* **2020**, *8*, 199–332. [CrossRef]

Tavano, O.L. Protein Hydrolysis Using Proteases: An Important Tool for Food Biotechnology. *J. Mol. Catal. B Enzym.* **2013**, *90*, 1–11. [CrossRef]

Bekard, I.B.; Asimakis, P.; Bertolini, J.; Dunstan, D.E. The Effects of Shear Flow on Protein Structure and Function. *Biopolymers* **2011**, *95*, 733–745. [CrossRef] [PubMed]

Hemmilä, V.; Adamopoulos, S.; Karlsson, O.; Kumar, A. Development of Sustainable Bio-Adhesives for Engineered Wood Panels—A Review. *RSC Adv.* **2017**, *7*, 38604–38630. [CrossRef]

Braun, R.; Seebeck, T. RNA Metabolism. In *Cell Biology Physarum Didymium*; Academic Press: Cambridge, MA, USA, 1982; pp. 393–435. [CrossRef]

Hettiarachchy, N.S.; Kalapathy, U.; Myers, D.J. Alkali-Modified Soy Protein with Improved Adhesive and Hydrophobic Properties. *J. Am. Oil Chem. Soc.* **1995**, *72*, 1461–1464. [CrossRef]

Wei, Y.; Yao, J.; Shao, Z.; Chen, X. Water-Resistant Zein-Based Adhesives. *ACS Sustain. Chem. Eng.* **2020**, *8*, 7668–7679. [CrossRef]

Chen, N.; Lin, Q.; Rao, J.; Zeng, Q. Water Resistances and Bonding Strengths of Soy-Based Adhesives Containing Different Carbohydrates. *Ind. Crop. Prod.* **2013**, *50*, 44–49. [CrossRef]

Gui, C.; Zhu, J.; Zhang, Z.; Liu, X. Research Progress on Formaldehyde-Free Wood Adhesive Derived from Soy Flour. In *Adhesives—Applications and Properties*; Intechopen: London, UK, 2016. [CrossRef]

Lamaming, S.Z.; Lamaming, J.; Rawi, N.F.M.; Hashim, R.; Kassim, M.H.M.; Hussin, M.H.; Bustami, Y.; Sulaiman, O.; Amini, M.H.M.; Hiziroglu, S. Improvements and Limitation of Soy Protein-Based Adhesive: A Review. *Polym. Eng. Sci.* **2021**, *61*, 2393–2405. [CrossRef]

Li, H.; Kang, H.; Zhang, W.; Zhang, S.; Li, J. Physicochemical Properties of Modified Soybean-Flour Adhesives Enhanced by Carboxylated Styrene-Butadiene Rubber Latex. *Int. J. Adhes. Adhes.* **2016**, *66*, 59–64. [CrossRef]

Wu, Z.; Xi, X.; Lei, H.; Liang, J.; Liao, J.; Du, G. Study on Soy-Based Adhesives Enhanced by Phenol Formaldehyde Cross-Linker. *Polymers* **2019**, *11*, 365. [CrossRef] [PubMed]

Xu, Y.; Han, Y.; Chen, M.; Luo, J.; Shi, S.Q.; Li, J.; Gao, Q. Constructing a Triple Network Structure to Prepare Strong, Tough, and Mildew Resistant Soy Protein Adhesive. *Compos. Part B Eng.* **2021**, *211*, 108677. [CrossRef]

Hand, W.G.; Robert Ashurst, W.; Via, B.; Banerjee, S. Curing Behavior of Soy Flour with Phenol-Formaldehyde and Isocyanate Resins. *Int. J. Adhes. Adhes.* **2018**, *87*, 105–108. [CrossRef]

Gao, D.; Fan, B.; Zhang, B.; Mi, Y.; Zhang, Y.; Gao, Z. Storage Stability of Polyamidoamine-Epichlorohydrin Resin and Its Effect on the Properties of Defatted Soybean Flour-Based Adhesives. *Int. J. Adhes. Adhes.* **2019**, *91*, 92–101. [CrossRef]

Zhao, L.F.; Liu, Y.; Xu, Z.D.; Zhang, Y.Z.; Zhao, F.; Zhang, S.B. State of Research and Trends in Development of Wood Adhesives. *For. Stud. China* **2011**, *13*, 321–326. [CrossRef]

38. Souza, A.M.; Nascimento, M.F.; Almeida, D.H.; Lopes Silva, D.A.; Almeida, T.H.; Christoforo, A.L.; Lahr, F.A.R. Wood-Base Composite Made of Wood Waste and Epoxy Based Ink-Waste as Adhesive: A Cleaner Production Alternative. *J. Clean. Prod.* **201** *193*, 549–562. [CrossRef]
39. Jin, S.; Li, K.; Gao, Q.; Zhang, W.; Chen, H.; Li, J.; Shi, S.Q. Multiple Crosslinking Strategy to Achieve High Bonding Strength an Antibacterial Properties of Double-Network Soy Adhesive. *J. Clean. Prod.* **2020**, *254*, 120143. [CrossRef]
40. Bai, Y.Y.; Lei, Y.H.; Shen, X.J.; Luo, J.; Yao, C.L.; Sun, R.C. A Facile Sodium Alginate-Based Approach to Improve the Mechanic Properties of Recycled Fibers. *Carbohydr. Polym.* **2017**, *174*, 610–616. [CrossRef] [PubMed]
41. Li, K.; Peshkova, S.; Geng, X. Investigation of Soy Protein-Kymene® Adhesive Systems for Wood Composites. *J. Am. Oil Chen Soc.* **2004**, *81*, 487–491. [CrossRef]
42. Kan, Y.; Sun, B.; Bai, Y.; Gao, Z. Double-Network Strategy for a Cost-Effective Soybean Meal-Based Adhesive with Required an Stable Water Resistance for Structural Use. *Compos. Part B Eng.* **2022**, *235*, 109744. [CrossRef]
43. Lei, H.; Du, G.; Wu, Z.; Xi, X.; Dong, Z. Cross-Linked Soy-Based Wood Adhesives for Plywood. *Int. J. Adhes. Adhes.* **2014**, *5* 199–203. [CrossRef]
44. Xi, X.; Pizzi, A.; Gerardin, C.; Liao, J.; Amirou, S.; Abdalla, S. Glutaraldehyde-Wheat Gluten Protein Adhesives for Wood Bondin *J. Adhes.* **2021**, *97*, 88–100. [CrossRef]
45. Li, K.; Geng, X. Formaldehyde-Free Wood Adhesives from Decayed Wood. *Macromol. Rapid Commun.* **2005**, *26*, 529–53 [CrossRef]
46. Liu, Y.; Li, K. Development and Characterization of Adhesives from Soy Protein for Bonding Wood. *Int. J. Adhes. Adhes.* **2007**, *2* 59–67. [CrossRef]
47. Zeng, Y.; Xu, P.; Yang, W.; Chu, H.; Wang, W.; Dong, W.; Chen, M.; Bai, H.; Ma, P. Soy Protein-Based Adhesive with Superic Bonding Strength and Water Resistance by Designing Densely Crosslinking Networks. *Eur. Polym. J.* **2021**, *142*, 110128. [CrossRe
48. Zhang, Y.; Zhang, M.; Chen, M.; Luo, J.; Li, X.; Gao, Q.; Li, J. Preparation and Characterization of a Soy Protein-Based Hig Performance Adhesive with a Hyperbranched Cross-Linked Structure. *Chem. Eng. J.* **2018**, *354*, 1032–1041. [CrossRef]
49. Zhang, Y.; Zhang, J.; Chen, M.; Luo, J.; Shi, S.Q.; Gao, Q.; Li, J. A Tough, Water-Resistant, High Bond Strength Adhesive Derive from Soybean Meal and Flexible Hyper-Branched Aminated Starch. *Polymers* **2019**, *11*, 1352. [CrossRef] [PubMed]
50. Zhang, B.; Li, J.; Kan, Y.; Gao, J.; Zhang, Y.; Gao, Z. The Effect of Thermo-Chemical Treatment on the Water Resistance of Defatte Soybean Flour-Based Wood Adhesive. *Polymers* **2018**, *10*, 955. [CrossRef]
51. Li, J.; Zhang, B.; Li, X.; Yi, Y.; Shi, F.; Guo, J.; Gao, Z. Effects of Typical Soybean Meal Type on the Properties of Soybean-Base Adhesive. *Int. J. Adhes. Adhes.* **2019**, *90*, 15–21. [CrossRef]
52. Remya, R.; Jyothi, A.N.; Sreekumar, J. Effect of Chemical Modification with Citric Acid on the Physicochemical Properties an Resistant Starch Formation in Different Starches. *Carbohydr. Polym.* **2018**, *202*, 29–38. [CrossRef] [PubMed]
53. Zheng, X.; Cheng, L.; Gu, Z.; Hong, Y.; Li, Z.; Li, C. Effects of Heat Pretreatment of Starch on Graft Copolymerization Reactio and Performance of Resulting Starch-Based Wood Adhesive. *Int. J. Biol. Macromol.* **2017**, *96*, 11–18. [CrossRef] [PubMed]
54. Chen, L.; Xiong, Z.; Din, Z.-U.; Nawaz, A.; Xiong, H.; Cai, J. Interfacial Modification of Starch at High Concentration by Sodiur Dodecylsulfate as Revealed by Experiments and Molecular Simulation. *J. Mol. Liq.* **2020**, *310*, 113190. [CrossRef]
55. Bai, Y.; Zhao, F.; Shen, J.; Zhang, Y. Improvement of Water Resistance of Wheat Flour-Based Adhesives by Thermal–Chemica Treatment and Chemical Crosslinking. *J. Appl. Polym. Sci.* **2021**, *138*, 50458. [CrossRef]
56. Wang, Y.; Fan, Y.; Deng, L.; Li, Z.; Chen, Z. Properties of Soy-Based Wood Adhesives Enhanced by Waterborne Polyurethan Modification. *J. Biobased Mater. Bioenergy* **2017**, *11*, 330–335. [CrossRef]
57. Zia, F.; Zia, K.M.; Zuber, M.; Ahmad, H.B.; Muneer, M.I. Glucomannan Based Polyurethanes: A Critical Short Review of Recen Advances and Future Perspectives. *Int. J. Biol. Macromol.* **2016**, *87*, 229–236. [CrossRef] [PubMed]
58. Xu, J.; Luo, Z.; Luo, L.; Yin, X.; Li, Q.; Mao, A. Study on Protein Based Adhesives Blending-Modified by Isocyanate. *Am. J. Enviror Sci. Eng.* **2022**, *6*, 62. [CrossRef]
59. Liu, C.; Zhang, Y.; Li, X.; Luo, J.; Gao, Q.; Li, J. A High-Performance Bio-Adhesive Derived from Soy Protein Isolate an Condensed Tannins. *RSC Adv.* **2017**, *7*, 21226–21233. [CrossRef]
60. Xu, F.; Dong, Y.; Zhang, W.; Zhang, S.; Li, L.; Li, J. Preparation of Cross-Linked Soy Protein Isolate-Based Environmentally-Friendl Films Enhanced by PTGE and PAM. *Ind. Crop. Prod.* **2015**, *67*, 373–380. [CrossRef]
61. Lee, H.; Lee, B.P.; Messersmith, P.B. A Reversible Wet/Dry Adhesive Inspired by Mussels and Geckos. *Nature* **2007**, *448*, 338–34 [CrossRef]
62. Al Loman, A.; Ju, L.K. Towards Complete Hydrolysis of Soy Flour Carbohydrates by Enzyme Mixtures for Protein Enrichment: Modeling Approach. *Enzym. Microb. Technol.* **2016**, *86*, 25–33. [CrossRef]
63. Islam, S.M.M.; Loman, A.A.; Ju, L.K. High Monomeric Sugar Yields from Enzymatic Hydrolysis of Soybean Meal and Effects o Mild Heat Pretreatments with Chelators. *Bioresour. Technol.* **2018**, *256*, 438–445. [CrossRef] [PubMed]
64. Zheng, P.; Chen, N.; Mahfuzul Islam, S.M.; Ju, L.K.; Liu, J.; Zhou, J.; Chen, L.; Zeng, H.; Lin, Q. Development of Self-Cross-Linke Soy Adhesive by Enzyme Complex from Aspergillus Niger for Production of All-Biomass Composite Materials. *ACS Sustair Chem. Eng.* **2019**, *7*, 3909–3916. [CrossRef]
65. Lo Presti, M.; Rizzo, G.; Farinola, G.M.; Omenetto, F.G. Bioinspired Biomaterial Composite for All-Water-Based High-Performanc Adhesives. *Adv. Sci.* **2021**, *8*, 2004786. [CrossRef]

- Qian, Y.; Zhou, Y.; Lu, M.; Guo, X.; Yang, D.; Lou, H.; Qiu, X.; Guo, C.F. Direct Construction of Catechol Lignin for Engineering Long-Acting Conductive, Adhesive, and UV-Blocking Hydrogel Bioelectronics. *Small Methods* **2021**, *5*, 2001311. [CrossRef] [PubMed]
- Liu, Z.; Liu, T.; Li, Y.; Zhang, X.; Xu, Y.; Li, J.; Gao, Q. Performance of Soybean Protein Adhesive Cross-Linked by Lignin and Cuprum. *J. Clean. Prod.* **2022**, *366*, 132906. [CrossRef]
- Dyer, J.M.; Chapital, D.C.; Kuan, J.C.W.; Shepherd, H.S.; Tang, F.; Pepperman, A.B. Production of Linolenic Acid in Yeast Cells Expressing an Omega-3 Desaturase from Tung (Aleurites Fordii). *JAOCS J. Am. Oil Chem. Soc.* **2004**, *81*, 647–651. [CrossRef]
- Zappi, M.E.; Zappi, A.; Revellame, E.; Sharp, W.; Fortela, D.L.; Hernandez, R.; Chambers, T.; Ritter, K.; Gang, D. An Assessment of the Potential to Produce Commercially Valuable Lipids on Highway Right-of-Way Land Areas Located within the Southeastern United States. *Sustainability* **2020**, *12*, 5225. [CrossRef]
- Huang, Y.; Pang, L.; Wang, H.; Zhong, R.; Zeng, Z.; Yang, J. Synthesis and Properties of UV-Curable Tung Oil Based Resins via Modification of Diels-Alder Reaction, Nonisocyanate Polyurethane and Acrylates. *Prog. Org. Coat.* **2013**, *76*, 654–661. [CrossRef]
- Thanamongkollit, N.; Miller, K.R.; Soucek, M.D. Synthesis of UV-Curable Tung Oil and UV-Curable Tung Oil Based Alkyd. *Prog. Org. Coat.* **2012**, *73*, 425–434. [CrossRef]
- He, Z.; Chapital, D.C.; Cheng, H.N.; Thomas Klasson, K.; Olanya, O.M.; Uknalis, J. Application of Tung Oil to Improve Adhesion Strength and Water Resistance of Cottonseed Meal and Protein Adhesives on Maple Veneer. *Ind. Crop. Prod.* **2014**, *61*, 398–402. [CrossRef]
- Božič, M.; Gorgieva, S.; Kokol, V. Homogeneous and Heterogeneous Methods for Laccase-Mediated Functionalization of Chitosan by Tannic Acid and Quercetin. *Carbohydr. Polym.* **2012**, *89*, 854–864. [CrossRef]
- Koupantsis, T.; Pavlidou, E.; Paraskevopoulou, A. Glycerol and Tannic Acid as Applied in the Preparation of Milk Proteins—CMC Complex Coavervates for Flavour Encapsulation. *Food Hydrocoll.* **2016**, *57*, 62–71. [CrossRef]
- Xu, F.; Weng, B.; Gilkerson, R.; Materon, L.A.; Lozano, K. Development of Tannic Acid/Chitosan/Pullulan Composite Nanofibers from Aqueous Solution for Potential Applications as Wound Dressing. *Carbohydr. Polym.* **2015**, *115*, 16–24. [CrossRef]
- Rubentheren, V.; Ward, T.A.; Chee, C.Y.; Tang, C.K. Processing and Analysis of Chitosan Nanocomposites Reinforced with Chitin Whiskers and Tannic Acid as a Crosslinker. *Carbohydr. Polym.* **2015**, *115*, 379–387. [CrossRef] [PubMed]
- Sionkowska, A.; Kaczmarek, B.; Lewandowska, K. Modification of Collagen and Chitosan Mixtures by the Addition of Tannic Acid. *J. Mol. Liq.* **2014**, *199*, 318–323. [CrossRef]
- Oh, H.; Hoff, J.E.; Armstrong, G.S.; Haff, L.A. Hydrophobic Interaction in Tannin-Protein Complexes. *J. Agric. Food Chem.* **1980**, *28*, 394–398. [CrossRef]
- Van Buren, J.P.; Robinson, W.B. Formation of Complexes between Protein and Tannic Acid. *J. Agric. Food Chem.* **1969**, *17*, 772–777. [CrossRef]
- Ghahri, S.; Mohebby, B.; Pizzi, A.; Mirshokraie, A.; Mansouri, H.R. Improving Water Resistance of Soy-Based Adhesive by Vegetable Tannin. *J. Polym. Environ.* **2018**, *26*, 1881–1890. [CrossRef]
- Yang, S.; Yuan, T.Q.; Li, M.F.; Sun, R.C. Hydrothermal Degradation of Lignin: Products Analysis for Phenol Formaldehyde Adhesive Synthesis. *Int. J. Biol. Macromol.* **2015**, *72*, 54–62. [CrossRef]
- Zhang, W.; Ma, Y.; Wang, C.; Li, S.; Zhang, M.; Chu, F. Preparation and Properties of Lignin-Phenol-Formaldehyde Resins Based on Different Biorefinery Residues of Agricultural Biomass. *Ind. Crop. Prod.* **2013**, *43*, 326–333. [CrossRef]
- Xiao, Z.; Li, Y.; Wu, X.; Qi, G.; Li, N.; Zhang, K.; Wang, D.; Sun, X.S. Utilization of Sorghum Lignin to Improve Adhesion Strength of Soy Protein Adhesives on Wood Veneer. *Ind. Crop. Prod.* **2013**, *50*, 501–509. [CrossRef]
- Luo, J.; Luo, J.; Yuan, C.; Zhang, W.; Li, J.; Gao, Q.; Chen, H. An Eco-Friendly Wood Adhesive from Soy Protein and Lignin: Performance Properties. *RSC Adv.* **2015**, *5*, 100849–100855. [CrossRef]
- Pradyawong, S.; Qi, G.; Li, N.; Sun, X.S.; Wang, D. Adhesion Properties of Soy Protein Adhesives Enhanced by Biomass Lignin. *Int. J. Adhes. Adhes.* **2017**, *75*, 66–73. [CrossRef]
- Doherty, W.O.S.; Mousavioun, P.; Fellows, C.M. Value-Adding to Cellulosic Ethanol: Lignin Polymers. *Ind. Crop. Prod.* **2011**, *33*, 259–276. [CrossRef]
- El Mansouri, N.E.; Salvadó, J. Structural Characterization of Technical Lignins for the Production of Adhesives: Application to Lignosulfonate, Kraft, Soda-Anthraquinone, Organosolv and Ethanol Process Lignins. *Ind. Crop. Prod.* **2006**, *24*, 8–16. [CrossRef]
- Pradyawong, S.; Qi, G.; Sun, X.S.; Wang, D. Laccase/TEMPO-Modified Lignin Improved Soy-Protein-Based Adhesives: Adhesion Performance and Properties. *Int. J. Adhes. Adhes.* **2019**, *91*, 116–122. [CrossRef]
- Zhu, X.; Wang, D.; Li, N.; Sun, X.S. Bio-Based Wood Adhesive from Camelina Protein (a Biodiesel Residue) and Depolymerized Lignin with Improved Water Resistance. *ACS Omega* **2017**, *2*, 7996–8004. [CrossRef] [PubMed]
- Xin, J.; Zhang, P.; Wolcott, M.P.; Zhang, J.; Hiscox, W.C.; Zhang, X. A Novel and Formaldehyde-Free Preparation Method for Lignin Amine and Its Enhancement for Soy Protein Adhesive. *J. Polym. Environ.* **2017**, *25*, 599–605. [CrossRef]
- Ma, R.; Guo, M.; Zhang, X. Selective Conversion of Biorefinery Lignin into Dicarboxylic Acids. *ChemSusChem* **2014**, *7*, 412–415. [CrossRef] [PubMed]
- Passauer, L.; Fischer, K.; Liebner, F. Activation of Pine Kraft Lignin by Fenton-Type Oxidation for Cross-Linking with Oligo(Oxyethylene) Diglycidyl Ether. *Holzforschung* **2011**, *65*, 319–326. [CrossRef]
- Shokrolahi, A.; Zali, A.; Keshavarz, M.H. Reductive Amination of Aldehydes and Ketones by NaBH 4 Using Carbon-Based Solid Acid (CBSA) as Catalyst. *Green Chem. Lett. Rev.* **2011**, *4*, 195–203. [CrossRef]

94. Song, F.; Zhang, L.M. Gelation Modification of Soy Protein Isolate by a Naturally Occurring Cross-Linking Agent and Its Potenti Biomedical Application. *Ind. Eng. Chem. Res.* **2009**, *48*, 7077–7083. [CrossRef]
95. Xu, Y.; Han, Y.; Shi, S.Q.; Gao, Q.; Li, J. Preparation of a Moderate Viscosity, High Performance and Adequately-Stabilized So Protein-Based Adhesive via Recombination of Protein Molecules. *J. Clean. Prod.* **2020**, *255*, 120303. [CrossRef]
96. Wu, F.; Yang, Z.; Su, X.; Gong, Y.; Kuang, T. Molecular Reorganization Induced by Ca2+ of Plant Photosystem I Reconstituted in Phosphatidylglycerol Liposomes. *Chem. Phys. Lipids* **2005**, *136*, 73–82. [CrossRef] [PubMed]
97. Lyu, Q.; Hsueh, N.; Chai, C.L.L. Direct Evidence for the Critical Role of 5,6-Dihydroxyindole in Polydopamine Deposition an Aggregation. *Langmuir* **2019**, *35*, 5191–5201. [CrossRef] [PubMed]
98. Hu, W.; Lu, S.; Zhang, Z.; Zhu, L.; Wen, Y.; Zhang, T.; Ji, Z. Mussel-Inspired Copolymer-Coated Polypropylene Mesh wi Anti-Adhesion Efficiency for Abdominal Wall Defect Repair. *Biomater. Sci.* **2019**, *7*, 1323–1334. [CrossRef]
99. Chirdon, W.M.; O'Brien, W.J.; Robertson, R.E. Adsorption of Catechol and Comparative Solutes on Hydroxyapatite. *J. Biome Mater. Res.-Part B Appl. Biomater.* **2003**, *66*, 532–538. [CrossRef] [PubMed]
100. Subair, R.; Tripathi, B.P.; Formanek, P.; Simon, F.; Uhlmann, P.; Stamm, M. Polydopamine Modified Membranes with in Sit Synthesized Gold Nanoparticles for Catalytic and Environmental Applications. *Chem. Eng. J.* **2016**, *295*, 358–369. [CrossRef]
101. Zhang, C.; Xiang, L.; Zhang, J.; Gong, L.; Han, L.; Xu, Z.K.; Zeng, H. Tough and Alkaline-Resistant Mussel-Inspired Wet Adhesic with Surface Salt Displacement via Polydopamine/Amine Synergy. *Langmuir* **2019**, *35*, 5257–5263. [CrossRef] [PubMed]
102. Pang, H.; Zhao, S.; Mo, L.; Wang, Z.; Zhang, W.; Huang, A.; Zhang, S.; Li, J. Mussel-Inspired Bio-Based Water-Resistant Sc Adhesives with Low-Cost Dopamine Analogue-Modified Silkworm Silk Fiber. *J. Appl. Polym. Sci.* **2020**, *137*, 48785. [CrossRef]
103. Van der Steen, M.; Stevens, C.V. Undecylenic Acid: A Valuable and Physiologically Active Renewable Building Block from Cast Oil. *ChemSusChem* **2009**, *2*, 692–713. [CrossRef] [PubMed]
104. Liu, H.; Li, C.; Sun, X.S. Improved Water Resistance in Undecylenic Acid (UA)-Modified Soy Protein Isolate (SPI)-Based Adhesive *Ind. Crop. Prod.* **2015**, *74*, 577–584. [CrossRef]
105. Zhou, Y.; Zeng, G.; Zhang, F.; Luo, J.; Li, K.; Li, X.; Li, J.; Fang, Z. High Strength and Flame Retardant Soybean Polysaccharid Based Wood Adhesive Produced by Borate Chemistry and Crosslinking Strategy. *Eur. Polym. J.* **2022**, *164*, 110973. [CrossRef]
106. Xue, F.; Zhang, H.; Hu, J.; Liu, Y. Hyaluronic Acid Nanofibers Crosslinked with a Nontoxic Reagent. *Carbohydr. Polym.* **2021**, *25* 117757. [CrossRef] [PubMed]
107. Yang, B.; Song, J.; Jiang, Y.; Li, M.; Wei, J.; Qin, J.; Peng, W.; Lasaosa, F.L.; He, Y.; Mao, H.; et al. Injectable Adhesive Self-Healin Multicross-Linked Double-Network Hydrogel Facilitates Full-Thickness Skin Wound Healing. *ACS Appl. Mater. Interfaces* **202** *12*, 57782–57797. [CrossRef] [PubMed]
108. Li, K.; Jin, S.; Zeng, G.; Zhou, Y.; Zhang, F.; Li, J.; Shi, S.Q.; Li, J. Biomimetic Development of a Strong, Mildew-Resistant Sc Protein Adhesive via Mineral–Organic System and Phenol-Amine Synergy. *Ind. Crop. Prod.* **2022**, *187*, 115412. [CrossRef]
109. Yu, Y.; He, Y.; Mu, Z.; Zhao, Y.; Kong, K.; Liu, Z.; Tang, R. Biomimetic Mineralized Organic–Inorganic Hybrid Macrofiber wi Spider Silk-Like Supertoughness. *Adv. Funct. Mater.* **2020**, *30*, 1908556. [CrossRef]
110. Jiang, Y.; Pan, X.; Yao, M.; Han, L.; Zhang, X.; Jia, Z.; Weng, J.; Chen, W.; Fang, L.; Wang, X.; et al. Bioinspired Adhesive and Tumc Microenvironment Responsive NanoMOFs Assembled 3D-Printed Scaffold for Anti-Tumor Therapy and Bone Regeneratio *Nano Today* **2021**, *39*, 101182. [CrossRef]
111. Liu, J.; Li, Y.; Mo, H.; Xie, E.; Fang, J.; Gan, W. Current Utilization of Waste Biomass as Filler for Wood Adhesives: A Revie *J. Ind. Eng. Chem.* **2022**, *115*, 48–61. [CrossRef]
112. Hasan Faris, A. Fillers in Wood Adhesives. In *Fillers*; Intechopen: London, UK, 2021. [CrossRef]
113. Qi, G.; Li, N.; Wang, D.; Sun, X.S. Development of High-Strength Soy Protein Adhesives Modified with Sodium Montmorilloni Clay. *JAOCS J. Am. Oil Chem. Soc.* **2016**, *93*, 1509–1517. [CrossRef]
114. Yang, Z.; Peng, H.; Wang, W.; Liu, T. Crystallization Behavior of Poly(ε-Caprolactone)/Layered Double Hydroxide Nanocompo ites. *J. Appl. Polym. Sci.* **2010**, *116*, 2658–2667. [CrossRef]
115. Pojanavaraphan, T.; Magaraphan, R.; Chiou, B.S.; Schiraldi, D.A. Development of Biodegradable Foamlike Materials Based o Casein and Sodium Montmorillonite Clay. *Biomacromolecules* **2010**, *11*, 2640–2646. [CrossRef]
116. Qi, G. Modified Soy Protein Based Adhesives and Their Physicochemical Properties. Ph.D. Thesis, Kansas State Universit Manhattan, KS, USA, 2011.
117. Zhang, L.; Sun, X.S. Effect of Sodium Bisulfite on Properties of Soybean Glycinin. *J. Agric. Food Chem.* **2008**, *56*, 11192–1119 [CrossRef]
118. Ciannamea, E.M.; Stefani, P.M.; Ruseckaite, R.A. Medium-Density Particleboards from Modified Rice Husks and Soybean Prote Concentrate-Based Adhesives. *Bioresour. Technol.* **2010**, *101*, 818–825. [CrossRef]
119. *ANSI A208.-1*; Particleboard. American National Standards Institute: Washington, DC, USA, 2016.
120. Gui, C.; Zhu, J.; Liu, X.; Zhang, Z. Preparation of Water-Resistant Adhesives from Soy Flour with Less Water-Soluble Componen for Wood Bonding. *Pigment. Resin Technol.* **2017**, *46*, 253–258. [CrossRef]
121. Zhang, B.; Fan, B.; Li, M.; Zhang, Y. Effects of Thermal Treatment on the Properties of Defatted Soya Bean Flour and Its Adhesic to Plywood. *R. Soc. Open Sci.* **2018**, *5*, 180015. [CrossRef] [PubMed]
122. Qi, G.; Li, N.; Wang, D.; Sun, X.S. Physicochemical Properties of Soy Protein Adhesives Modified by 2-Octen-1-Ylsuccin Anhydride. *Ind. Crop. Prod.* **2013**, *46*, 165–172. [CrossRef]

23. Zhang, W.; Sun, H.; Zhu, C.; Wan, K.; Zhang, Y.; Fang, Z.; Ai, Z. Mechanical and Water-Resistant Properties of Rice Straw Fiberboard Bonded with Chemically-Modified Soy Protein Adhesive. *RSC Adv.* **2018**, *8*, 15188–15195. [CrossRef] [PubMed]
24. Pang, H.; Zhao, S.; Wang, Z.; Zhang, W.; Zhang, S.; Li, J. Development of Soy Protein-Based Adhesive with High Water Resistance and Bonding Strength by Waterborne Epoxy Crosslinking Strategy. *Int. J. Adhes. Adhes.* **2020**, *100*, 102600. [CrossRef]
25. Arias, A.; González-Rodríguez, S.; Vetroni Barros, M.; Salvador, R.; de Francisco, A.C.; Moro Piekarski, C.; Moreira, M.T. Recent Developments in Bio-Based Adhesives from Renewable Natural Resources. *J. Clean. Prod.* **2021**, *314*, 127892. [CrossRef]

sustainability

MDPI

Perspective

Cellulose Textiles from Hemp Biomass: Opportunities and Challenges

Lelia Lawson [1,*], Lauren M. Degenstein [1], Bronwyn Bates [2], Wade Chute [3], Dan King [1] and Patricia I. Dolez [2]

1 Davey Textile Solutions, Edmonton, AB T5P 4Y7, Canada
2 Department of Human Ecology, University of Alberta, Edmonton, AB T6G 2N1, Canada
3 Techfibre Industries, Ardrossan, AB T8G 2C4, Canada
* Correspondence: llawson@daveytextiles.com

Abstract: Worldwide demand for man-made cellulosic fibres (MMCF) are increasing as availabilit of cotton fibre declines due to climate change. Feedstock for MMCF include virgin wood, agricultur residues (e.g., straw), and pre- and post-consumer cellulosic materials high in alpha-cellulose conter Lyocell MMCF (L-MMCF) offer large advantages over other MMCF processes in terms of bot environmental and social impacts: the solvent for cellulosic dissolution, n-methyl-morpholine- oxide, can be recycled, and the process utilizes non-toxic chemicals and low amounts of water. Hem can be a preferential cellulosic feedstock for L-MMCF as hemp cultivation results in carbon dioxid sequestration, and it requires less water, fertilizers, pesticides, and herbicides than other L-MMC feedstock crops. These factors contribute to hemp being an environmentally conscious crop. Th increased legalization of industrial hemp cultivation, as well as recent lifts on cannabis restriction worldwide, allows accessibility to local sources of cellulose for the L-MMCF process. In additio hemp biomass can offer a much larger feedstock for L-MMCF production per annum than othe cellulosic sources, such as eucalyptus trees and bamboo. This paper offers perspectives on th agricultural, manufacturing, and economic opportunities and challenges of utilizing hemp biomas for the manufacturing of L-MMCF.

Keywords: man-made cellulosic fibres; lyocell process; hemp biomass; environmental impact; regenerate cellulose fibres

check for updates

Citation: Lawson, L.; Degenstein, L.M.; Bates, B.; Chute, W.; King, D.; Dolez, P.I. Cellulose Textiles from Hemp Biomass: Opportunities and Challenges. *Sustainability* **2022**, *14*, 15337. https://doi.org/10.3390/su142215337

Academic Editor: Abu Naser Md Ahsanul Haque

Received: 2 November 2022
Accepted: 16 November 2022
Published: 18 November 2022

1. Introduction

Worldwide fibre production has nearly quadrupled since 1970, leading to numerou environmental impacts including air and water pollution, textile waste and microplasti creation [1]. Of the over 100 millions tonnes of fibres produced annually, polyester—a energy intensive, petrochemical-based fibre—and cotton—a cultivated fibre requiring larg volumes of water and chemical resources to grow—account for the greatest share of fibr volume at 54% and 22%, respectively [1–3]. Lyocell, a man-made cellulosic fibre (MMCF is created by directly dissolving cellulose in n-methyl-morpholine-n-oxide (NMMO), non-derivative, non-toxic solvent of which 99.5% can be recovered in the Lyocell MMC (L-MMCF) process [4,5]. Feedstock for the L-MMCF process can come from any hig alpha-cellulose-based material including wood, plants, and even pre- and post-consume textiles (e.g., cotton garments). However, the use of virgin resources, such as eucalyptu trees or bamboo, as L-MMCF feedstock has been linked to deforestation [6–8]; farming methods can use large amounts of water, fertilizers, and herbicides, as well as lead t soil erosion [9]; and the use of pre- and post-consumer cellulosics can be challenging du to textile sorting, removal of dyes and finishing additives, having consistent feedstock and reduction in fibre strength after regeneration [10]. Therefore, while the L-MMC process is itself more sustainable, improvements can be made in terms of utilizing a mor environmentally conscious source of feedstock to produce the L-MMCF dissolving pulp.

One such feedstock is hemp, which has been cultivated for textile applications for millennia. Hemp has a high alpha-cellulose content and offers higher cellulose yield than wood sources [11], which is ideal for the L-MMCF process. It also offers an alternative and potentially preferable source of cellulose for L-MMCF as it requires less water, fertilizers, pesticides, and herbicides to grow than other crops such as cotton. Hemp cultivation also aids in carbon dioxide (CO2) sequestration [12] and is considered carbon neutral [13], leading to regenerative agricultural practices [14]. Unfortunately, hemp cultivation and subsequent fibre usage is sparse in the current textile landscape. Prohibition, restrictions on growing cannabis, and a general stigma towards hemp have halted the sowing of this crop. However, the legalization of industrial hemp and cannabinoid cultivation on a global scale has allowed society to once again capitalize on this multi-purpose plant [3,15,16].

Currently, there are no North American (NA) sources of filament or staple L-MMCF despite growing demand for the fibre. It is estimated that global demand for L-MMCF will grow to a $2B industry by 2027 (745,900 tonne/year) [17] due to the vast consumer and commercial market applications of this high tenacity, absorbent, and biodegradable fibre [12]. On the other hand, NA hemp is a readily accessible commodity that is being farmed primarily for oil and seed applications. L-MMCF from residual hemp oilseed straw would assist in whole plant utilization, thus allowing the cultivation of industrial hemp for multiple purposes to be more economically sustainable. In addition, blending the properties of hemp bast and hurd at various ratios may alter the degree of polymerization and thus the properties of the extruded fibre based on desired end-use.

This article gives a critical look at the opportunities and challenges of MMCF, L-MMCF, and L-MMCF from hemp biomass as an alternative to cotton and other natural fibers. It considers different perspectives: environmental, agricultural, manufacturing, and economic. Additional research is also suggested to address the challenges with manufacturing L-MMCF from hemp biomass, as well as how manufacturers can take advantage of the opportunities it provides.

2. Environmental Challenges and MMCF

As of 2021, global textile consumption has risen to an average of 15 kg per capita, with petroleum-based synthetic fibres accounting for the largest segment of textile purchasing at 64% [18]. Consumption is estimated to further increase by 3% every year until 2030, with 70% utilized towards consumer textiles (e.g., clothing, household) due to their versatility and low market price [19,20]. Petroleum-based synthetic fibre usage leads to high emissions of greenhouse gas (GHG), bioaccumulation of microplastics, and contamination of water, amongst other detrimental environmental and social impacts [20]. At end-of-life stage, it is estimated that over 275 million metric tonnes of petroleum-based textiles were landfilled in 2010, with 4.8 to 12.7 million metric tonnes entering oceans [21]. These synthetic textiles are non-biodegradable and require hundreds to thousands of years before they begin to break down [22]; unfortunately, petroleum-based synthetic textile breakdown leads to further bioaccumulation of microplastics within aquatic and terrestrial habitats.

Due to the rising global population and enhanced awareness of "sustainable" natural cellulosic fibres, fibres such as cotton are increasing in demand [23]. However, cotton cannot be grown in northern climates, such as in Canada or the northern United States (US), and its cultivation is highly reliant on pesticides and large quantities of water [24], resulting in loss of arable land throughout the world. Haemmerle estimates that by 2030, the maximum cotton production, based on available arable land, will be approximately 26 million tonnes (Figure 1) [23]. This maximum available production will eventually decline as global arable landmass decreases by an estimated 0.8% to 4.4% due to climate change (e.g., increase in global temperatures, reduced water availability, frequent weather events, and increasing population) [25]. The per capita consumption of cellulosic fibres is estimated to increase from 3.7 kg to 5.4 kg in 2030 [23]. As the world population is expected to reach 8.5 billion by 2030 [26], this will result in a staggering cellulose fibre demand of 45.9 million tonnes. With cellulose fibre demands rising, it is estimated that this would lead to a 19 million tonne cotton fibre deficit due to unavailability of arable land (i.e., considering that all availab land for cotton cultivation is currently being used). Moreover, cotton cultivation consum "11% of the world's pesticides while it is grown on only 2.4% of the world's arable land" [

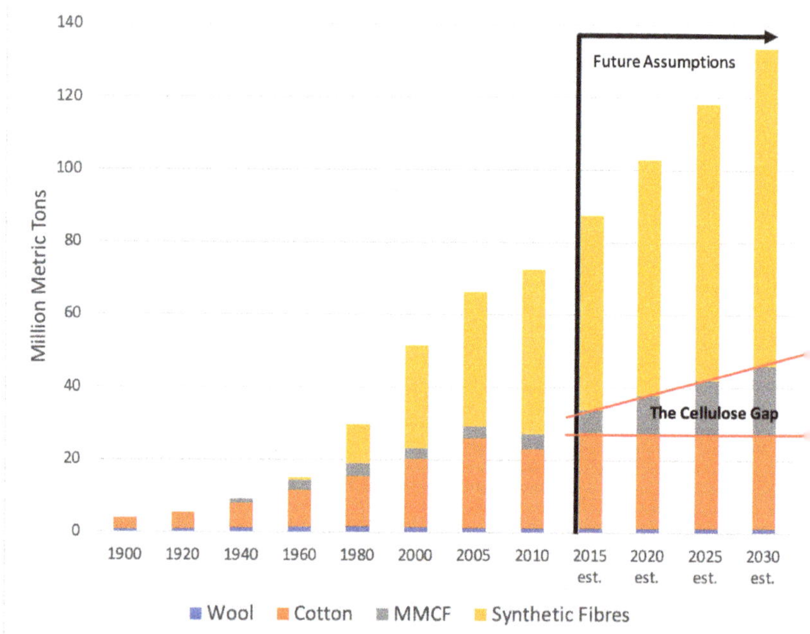

Figure 1. The Cellulose Gap: Textile Fibre Consumption 1900–2030 (data from Haemmerle [23]).

The World Wildlife Foundation has stated that cotton cultivation results in soil erosio and degradation (e.g., loss of topsoil from conventional tillage), pollution (e.g., use c herbicides and pesticides), and water contamination (e.g., nitrate-contaminated leacha from fertilizers) [27]. Furthermore, with every centimeter of top soil erosion, cotton fib yield reduces by approximately 4% [28]. While there are initiatives to reduce these e vironmental concerns, including cover crops, conservation tillage, and crop rotations reduce soil erosion, as well as implementation of water conservation measures, cotton st requires extensive resources to cultivate and process into textiles [28,29]. 70% of the world consumption of water is used in the agricultural industry [30], with cotton requiring large volume of water to grow, as well as process into end-use textiles. For instance, takes nearly 2700 litres of water to produce one cotton t-shirt [31]. Considering that t-shirt weighs approximately 150 g, this equates to 18,000 litres of water to produce 1 kg c cotton textile. In comparison, the water footprint of industrial hemp textiles is 2820 m^3/to (3108 L/kg) (Table 1) [32]. While this number is considerably lower than cotton, it is high than the water footprint for MMCF since these fibres do not have to go through addition water processes (e.g., chemical retting, cottonization).

Table 1. Water footprint of various fibres (data from [32,33]).

Staple Fibre	Water Footprint (L/kg)
Polyester	84–143
Cotton	4342–6902
Hemp (bast)	3108
MMCF	351–520
L-MMCF	290
Hemp L-MMCF	<290 (estimate)

In a life cycle assessment study on MMCF, Shen and Patel [33] determined that the water footprint for staple fibre production was between 84–143 L/kg for petroleum-based synthetics, between 351–520 L/kg for MMCF, 290 L/kg for L-MMCF, and between 4342 and 6902 L/kg for cotton (Table 1). These numbers indicate the opportunity for MMCF in the consumer-based cellulosics market as they have a water footprint that is less than 10% than that of cotton. Shen and Patel [33] noted that they did not include data for finished textile products as finishing processes vary considerably depending on the end product. As such, the water footprint would be higher when including additional processes such as bleaching, dyeing, and finishing. Finally, Shen and Patel [33] based their study on typical MMCF feedstock (i.e., wood and eucalyptus trees), which may require irrigation. The water footprint for L-MMCF from hemp would likely be lower than for L-MMCF from other sources as irrigation is not required for hemp cultivation (Table 1).

Increased natural fibre demand, in addition to an international shortage of cellulosic fibres for consumer and commercial demand (i.e., the Cellulose Gap [19,23], Figure 1), is encouraging manufacturers to seek alternatives to naturally cultivated cellulosic fibres. MMCF provide an alternative to natural cellulosic fibres, such as cotton and flax. Cellulosic feedstock for MMCF production is abundantly available, through agricultural crops, agricultural residues and wastes, pre- and post-consumer textiles, wood wastes, etc.

Unfortunately, manufacturing processes for some MMCF lead to additional environmental and social implications. A report published by Changing Markets in 2017 titled Dirty Fashion: How pollution in the global textiles supply chain is making viscose toxic, outlines the environmental and social impacts linked to viscose fibre production [34]. The report details how chemicals used in the production of viscose (Figure 2), such as carbon disulphide and hydrogen sulphide, have been linked to function impairment, neurological changes, and even death in factory workers. In addition, the viscose process uses large amounts of sodium hydroxide and sulphuric acid, which are considered highly toxic. The report further states that "20–30 g of carbon disulphide and 4–6 g of hydrogen sulphide are emitted" for every kilogram of viscose produced (p. 14). Furthermore, viscose fibre manufacturing leads to water and air pollution, and has been linked to acute aquatic toxicity, as well as increased cancer rates in individuals living in proximity to viscose manufacturing facilities.

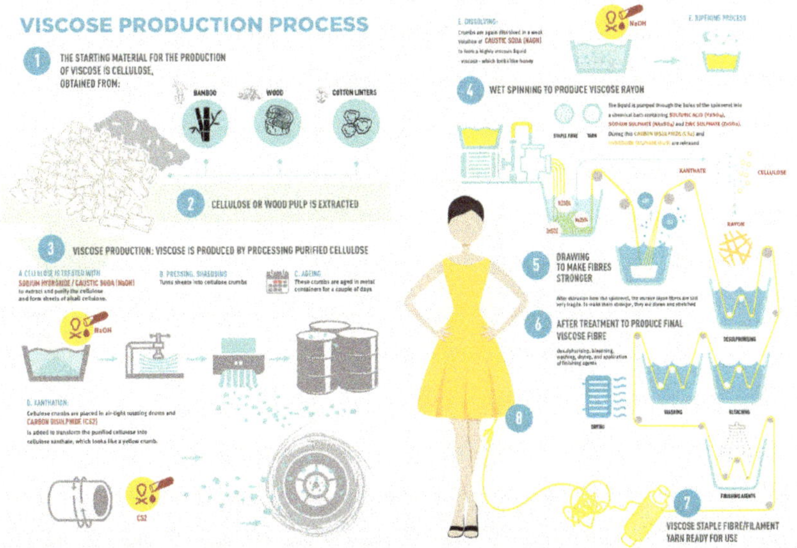

Figure 2. Viscose production process (Reproduced from [34] with permission from Changing Markets).

Viscose MMCF is produced via a derivatizing process, where xanthate is formed (i.e xanthation) by the reaction of cellulose with carbon disulphide [5,8,35,36]. The origina cellulosic molecule is modified for the regeneration process, where the derivative chemical are dissolved following regeneration [5,36]. These derivative waste chemicals end up ente ing waterways, air, and soil, leading to the environmental and social impacts mentione above [37]. The viscose process is used to manufacture viscose rayon, cellulose acetat modal, and triacetate MMCF. Cellulose sources for the viscose process include bambo wood, and cotton linters [35].

Some sources of cellulose such as bamboo are considered more environmentally sustai able than others [38]. Bamboo is a rhizome that regenerates growth after harvest and doe not require irrigation, pesticides, or herbicides for cultivation [38,39]. In addition, a bambo rhizome stand can replenish every three to four years and can survive for many decade While there are varieties of bamboo that grow in temperate climates, the majority of larg scale bamboo cultivation is in tropical and warmer temperature climactic zones [40,41], an whole plant utilization is not common. Moreover, while the use of feedstock such as bambo in the viscose process creates sustainability from an agricultural perspective, all the othe environmental and social implications from the viscose process still arise.

Direct dissolution is another process for producing MMCF, whereby cellulose is di solved in an organic solvent, producing a dope solution without the formation of intermed ate compounds [5,8]. Cuprammonium MMCF [42] and lithium chloride/dimethylacetamic (LiCl/DMAc) [43] are two examples of this process; unfortunately, both manufacturin processes do not provide any advantages over the viscose process regarding environmenta or social impacts.

A third example of direct dissolution is the production of L-MMCF via dissolutio in NMMO [44], which was first developed in 1969 by Eastman Kodak and later con mercialized by both Courtauld and Lenzing in the 1990's. This process is sustainabl environmentally friendly, and socially responsible. L-MMCF is made by dissolution i the NMMO solvent, which is greater than 99% recoverable, making it a closed-loop sys tem [39,45]. While the dissolution of cellulose in the L-MMCF process does take longer an is more expensive than the viscose process, exposure to toxic chemicals and developmen of toxic effluents are avoided [4]. The viscose MMCF process and L-MMCF process ar illustrated side by side in Figure 3.

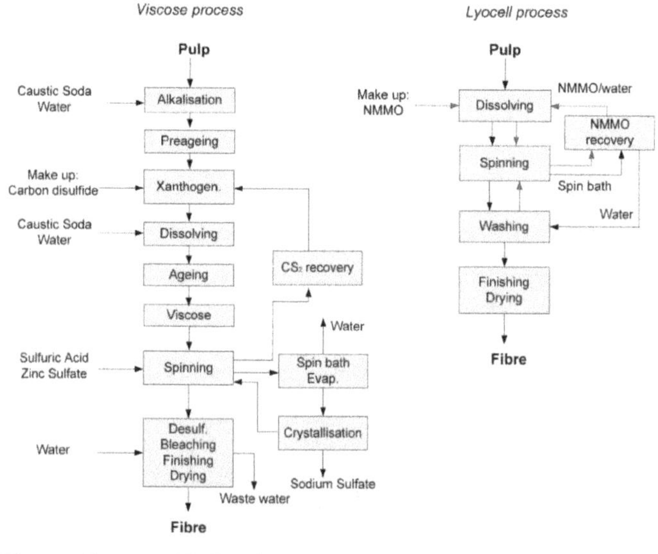

Figure 3. The viscose MMCF and L-MMCF processes (reproduced from [46] with permission from Elsevier

The feedstock for L-MMCF has traditionally been beech and eucalyptus trees, which are fast growing and can be harvested using a technique called coppicing [47,48]. Coppicing involves leaving viable stumps for regeneration when the trees are cut down within a thicket, or copse [49]. Each copse is sectioned into coups, which are harvested on an annual rotation, allowing for complete regeneration of the first coups by the time the last coup is harvested. However, growing and harvesting eucalyptus trees consumes 3200 litres of water per tree; additionally, eucalyptus trees only grow in certain climactic regions [47]. Other cellulosic sources, including bamboo, kudzu, cereal crop agricultural residues (e.g., wheat and barley straw) and sugarcane bagasse [50–54], have been utilized as feedstock; however, all of these cellulosic sources have similar drawbacks. Agricultural crops, including sustainable tree farming, lead to deforestation, excess water consumption (i.e., irrigation), changes to ecological diversity (e.g., use of herbicides and pesticides), freshwater pollution, and soil erosion [9,55]. Bamboo stands in Amazonian forests have resulted in anthropogenically induced climate change [40]. Furthermore, these cellulose sources cannot be grown in a multitude of climactic regions. Table 2 outlines the advantages and drawbacks of the various MMCF manufacturing processes.

Table 2. Advantages and drawbacks of MMCF manufacturing processes.

	Solvent Process	Advantages	Drawbacks
Viscose MMCF (viscose rayon, cellulose acetate, modal, triacetate)	Derivitized	α-cellulose feedstock readily available.	Waste chemicals (e.g., carbon disulphide), leading to contamination of water, air, and soil, and health concerns.
L-MMCF	Direct Dissolution	Environmentally friendly (NMMO 99% recoverable). α-cellulose feedstock readily available.	More expensive to produce than viscose MMCF. Longer manufacturing process.

3. Hemp Biomass Feedstock for MMCF

An alternate cellulose source for L-MMCF is hemp. Two parts of the hemp plant may be utilized for fibre production—the bast (bark) and the hurd (shive) (Figure 4) [56]. In the fibre preparation stage, the bast fibre is separated from the hurd through decortication, scutching, or milling [45]. The process of retting then breaks down lignin (i.e., delignification) and pectin to facilitate further processing.

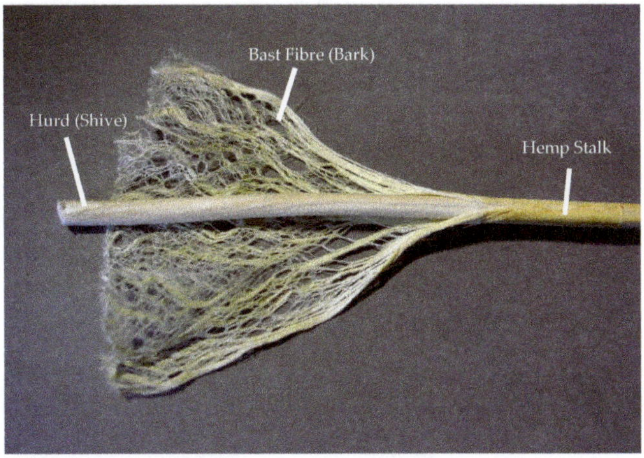

Figure 4. Hemp stalk (Image: Natrij—Public Domain, https://commons.wikimedia.org/w/index.php?curid=2304060, accessed on 1 November 2022).

Both the bast fibres and the hurd may be further dissolved in the NMMO solvent (blended in various ratios, forming a dope solution. The dope solution would then be spu into filaments, which are drawn and washed, mechanically or chemically treated for variou properties (e.g., reduction of fibrillation), and dried [46,57,58]. The Thuringian Institute fo Textiles and Plastics Research (TITK) has demonstrated this novel use of hemp biomass i the development of Lyohemp® [59]. Using high alpha-cellulose pulps from hemp allow for production of high tenacity L-MMCF [44] that would be applicable for use in variou applications such as industrial textiles and apparel (e.g., personal protective clothing dental floss, healthcare products (e.g., disposable masks, gowns, cleaning wipes), airlai and wetlaid nonwoven consumer products (e.g., disposable sanitary products, cleanin wipes), nonwoven industrial products (e.g., biodegradable geotextiles, carpet backings and consumer apparel.

Furthermore, manipulating the bast to hurd blend ratios may alter the viscosity an degree of polymerization of the dissolving pulp, resulting in unique fibre properties fo varying end-use applications [19,60]. Organoleptic, or sensory, properties such as com fort (e.g., cold/warm), hand (e.g., thin/thick, heavy/light), and texture (e.g., smoot rough) [61], may be adjusted based on end-user requirements by varying the amount o bast and hurd within the pulp, resulting in unique fibre and yarn blends. Technical perfo mance, such as tensile strength, abrasion resistance, elongation, moisture management, an antimicrobial properties, may also be adjusted depending on application. Hemp biomas offers a single plant cellulosic source for varying pulp viscosity, which is not possible fro other plant feedstocks such as eucalyptus and beech trees or bamboo.

Research by the authors has determined the alpha-cellulose content of Canadia grown hemp hurd and bast-bleached kraft pulps to be 83.3% and 96.2%, respectively. Th whole plant stem has an alpha-cellulose content of 70%, with the bast at 55–72% and th hurd at 34–44% [62]. In contrast, eucalyptus sawdust has a cellulose content of 41.6% [63 and eucalyptus kraft pulp has an alpha-cellulose content of approximately 92% [64]. Furthe more, bamboo chips have a cellulose content of approximately 48% [65], with a kraft pul alpha-cellulose content of approximately 90% [66]. These results suggest that hemp is viable alternate cellulose source for L-MMCF production with a greater potential alph cellulosic yield than eucalyptus or bamboo. This higher alpha-cellulosic content allow for more versatility for altering the viscosity of the dope solution to potentially modif organoleptic properties and technical fibre performance. Table 3 summarizes the cellulos content of various cellulosic feedstocks.

Table 3. Cellulosic content of various feedstocks (data from authors and [62–66]).

Feedstock	Cellulosic Content (%)	Kraft Pulp, α-Cellulose (%)
Hemp stalk	70	n/a
Hemp hurd	34–44	83.3
Hemp bast	55–72	96.2
Eucalyptus	41.6	92
Bamboo	48	90

4. Agricultural Opportunities and Challenges

Industrial hemp in Canada and Europe is defined as a herbaceous cannabinoid annua plant containing a maximum of 0.3% delta-9-trans-tetrahyrocannabinol (THC) [67]. Whil hemp grows in many different climates, hemp flourishes in temperate climates, with growt rates proportional to the available daytime hours. Hemp can also be grown in tropica and sub-tropical climates, and it is currently being researched for agricultural diversity i these regions [68–70]. As of 2022, there are at least 70 countries growing industrial hem for commercial or research purposes, an increase of 49% from approximately 47 countrie in 2021 [71]. Hemp fibre yields range from 1 to 5 metric ton/hectare (0.4 to 2.2 metri ton/acre) (note—this is not whole plant yield; it is just fibre), whereas cotton fibre yield range from 0.8 to 0.9 metric ton/hectare (0.32 to 0.36 metric ton/acre) [72]; additionall hemp fibre cultivation can lead to a 77% cost savings due to higher yield and reduced water, fertilizer, and pesticide use on one third of the land mass required for cotton cultivation. The estimated global cultivation of industrial hemp in 2021 was 510,000 acres [71], and with a compound annual growth rate (CAGR) of the industrial hemp market estimated at 16.8% (2022–2030) [73], cultivation rates will be increasing annually. Northern temperate climates, as found in central and northern Alberta, Canada, can grow industrial hemp plants that are 20% to 30% taller than in more southern regions due to longer sunlight hours [74]. Furthermore, industrial hemp grows extremely fast, coming second to bamboo in growth rate, and yields four times the biomass of a 25-year tree stand in just 90 days [74].

This "northern advantage", combined with the ability of whole plant utilization for multiple end products, such as fibre, nutraceuticals, food, and bio-fuel, makes industrial hemp a diverse and economically viable agricultural crop [74]. Cultivation and use of hemp for L-MMCF contributes to whole-plant utilization, providing incentives for industrial hemp farmers to grow hemp as an agricultural crop. Agricultural biomass traditionally seen as waste can be diverted to L-MMCF manufacturing, creating value and additional revenue for the hemp industry.

In the 2022 Canadian agricultural growing season, there were 83 approved cultivars of industrial hemp, each grown for specific end uses [75]. Canadian cultivars are typically grown for one of two end products: nutritional seed and oil or nutraceuticals (i.e., cannabidiol oil). Plants grown for nutritional seed and oil are harvested at maturity (e.g., seed stage), resulting in coarser bast fibres that require additional delignification to obtain textile fibres [76]. While these plants are considered dual purpose (e.g., seed and fibre), the coarse fibres are difficult to process into spinnable fibre for flexible textile applications; they require additional processing to "cottonize" or soften the fibre, and field retting results in inconsistent fibres for processing [77]. These fibres are typically utilized towards technical applications (e.g., composites) and paper pulps due to the difficulty in processing to usable textile fibre [78,79]. Nutraceutical plants do not have the stalk growth for quality fibre manufacturing as these plants are harvested at the flowering stage [76,80]; fibres from these plants are often too immature and short to produce quality textile fibre. Other growing conditions influencing fibre quality include rainfall, temperature, and sunlight exposure [80,81]. Producing quality hemp bast fibre for textile applications is additionally challenged in an industry that is growing plants for nutritional and nutraceutical needs, not specifically for textile fibre [62,67,82].

Hemp L-MMCF addresses these fibre concerns since the biomass from both nutritional seed and nutraceutical plants could be used as L-MMCF feedstock, allowing for whole plant utilization. The hemp L-MMCF process avoids additional hemp bast fibre processing requirements, such as retting and decortication, which can damage fibre quality. As a result, the L-MMCF process offers consistent, reliable, and repeatable fibre quality. In summary, hemp biomass may be a preferable feedstock for L-MMCF due to:

- Cultivation in temperate, subtropical, and tropical climates;
- Increased global cultivation (49% increase by country from 2021–2022) at a CAGR of 16.8% (2022–2030);
- Hemp fibre yield 25–500% higher than that of cotton;
- Low pesticide, herbicide, and water consumption requirements;
- High biomass yields in 90 days (4× that of equivalent 25-year tree stand);
- Whole plant utilization.

5. Manufacturing Opportunities and Challenges

Industrial hemp can be grown in a multitude of climates, including temperate, subtropical, and tropical regions, enabling a local-sourced raw feedstock supply for L-MMCF manufacturers [69]. Local supply allows for increased manufacturing traceability, reduced transportation costs, and reduced lead times for product manufacturing. As previously mentioned, both bast and hurd components of the hemp plant may be used for L-MMCF production [83]. These two parts of the plant possess different characteristics [84]; the hurd is comprised of 40–48% cellulose, 18–24% hemicellulose and 21–24% lignin, while the bas includes 57–77% cellulose, 9–14% hemicellulose and 5–9% lignin. Hurd pulp is generall associated with a higher kappa number, which determines the degree of fibrous pulp dige tion and gives an indication of the lignin content [85]. In the case of bast pulp produced b organosolv pulping using ethanol, a large stability of the crystalline structure of cellulos was observed [86]. By blending different ratios of pulp from the alpha-cellulosic-ric bast and the hemi-cellulosic-rich hurd raw materials, different organoleptic and physica properties may be achieved depending on end-use applications.

One of the main concerns with manufacturing L-MMCF from hemp is the presence c transition metals in the hemp fibres. Hemp plants are fast growing and are often used fc contaminated soil remediation due to their quick take-up of nutrients from the soil [87,88 This leads to accumulation of heavy metals in all parts of the hemp plant. Hemp stalks hav been observed to contain high concentrations of heavy metals such as mercury, chromiun nickel, cadmium, lead, and arsenic, as well as high percentages of earth-alkali salts (i.e sodium, magnesium, potassium, and calcium) [87–91]. The percentage and type of meta and salt content are specific to the soil the hemp plant is grown in, and the percentage c heavy metals and salts in hemp plants can be substantially higher than slower-growin plants and trees, such as eucalyptus.

These latent transition metals can pose significant safety concerns during the L-MMC manufacturing process as the oxidation states of transition metals can be a catalyst for e plosive radical decomposition reactions of NMMO [92,93]. Rosenau et al. have extensivel researched these oxidative states, which result in "fast exothermic processes" or radica reactions. To mitigate reactive risks, research has identified stabilizers, such as propy gallate, to incorporate into the L-MMCF manufacturing process. TITK has performe extensive research to further mitigate risks of transition metal content in hemp cellulos during the L-MMCF production process [13,91,94]. They determined that washing th crude pulps prior to dissolution in NMMO with deionized water successfully isolated meta ions (i.e., iron, copper) below hazardous levels. Additionally, calcium and magnesiur ions are addressed within large-scale manufacturing and the use of sequestrants to forr complex compounds [91].

Hemp is a novel feedstock for the L-MMCF manufacturing process. As such, add tional research is required to determine the effects of the pulping parameters, such as th solid content to NMMO solvent ratio and cooking time, to determine the quality of the di solving pulp produced. The carbohydrate component content, kappa value, viscosity, an degree of polymerization of the cellulosic raw material and corresponding dissolving pul solutions need to be characterized to determine the effect of the degree of polymerizatio of the source of cellulose on the quality of the dissolving pulp and its preservation throug the pulping process [95,96].

Additional research is also required to prepare dissolving pulps with the differen blend ratios of hemp bast and hurd to manufacture L-MMCF with varying organolepti and physical properties. Manufacturing processes need to include washing steps fo NMMO removal. Process parameters, including extrusion rate, temperature, and residenc time in the coagulation bath, need to be evaluated and optimal values determined, base on the performance of the filaments, including tensile strength, elongation, moistur content/regain, and residual NMMO content. This research will determine the effect of th processing parameters and type of cellulose produced by the pulping process (type I, I III, and IV depending on the location of hydrogen bonds between and within the cellulos molecules) on the properties of the hemp-based L-MMCF.

To further improve the L-MMCF performance to match the requirements of the di ferent end-use applications, one of the proposed solutions considers the introduction o additives into the regenerated cellulose dope before filament extrusion. The introductio of solid particles, in particular nanoparticles, into polymer fibres has been shown to in prove fibre mechanical performance [97]. For instance, a concentration of 3 wt% cellulos nanocrystals was shown to produce an increase of 24% in the tensile strength of cellulos acetate without negatively affecting its elongation at break [98]. In terms of flammability, additives functionalized with zinc oxide nanoparticles were shown to improve the flame resistance of high-density polyethylene matrix composites [99].

Manufacturing opportunities also exist for the exploration of properties naturally occurring within the cellulose fibers, such as biocidal activity in hemp fibers [100], and to determine whether these properties can be preserved through the L-MMCF process. It has been shown that the activity of bamboo's natural antimicrobial agent, bamboo kun, is maintained through the cellulose dissolution process [53,101]. Hemp hurd contains a high concentration of cannabinoids and similar phenolic compounds as bamboo; its antibacterial activity has been demonstrated against bacteria Escherichia coli [100,102]. Hemp hurd powder exposed to heat treatments of up to 160 °C for up to 3 h did not lower its antibacterial efficiency, but rather increased its antibacterial properties by reducing cross-contamination that had occurred at the retting stage. While challenges exist for hemp-based L-MMCF, the opportunities for fibre manipulation towards end-use applications are endless.

6. Economic Opportunities

Dissolving pulps are produced in NA; however, they are derived primarily from wood and are not considered environmentally sustainable for textile manufacturing. Dissolving pulp from alternate cellulose sources (e.g., hemp, post-consumer cellulosic textiles) would be beneficial from an NA agricultural and environmental vantage. Industrial hemp is a readily accessible commodity already being cultivated for various end usages. Therefore, L-MMCF derived from residual hemp biomass are inherently more sustainable due to the use of waste product as feedstock and aids in whole plant utilization. In addition, hemp biomass harvested from 1 acre produces the same as a 20-year tree stand grown on 4 to 10 acres [103]. The crop is environmentally conscious, requiring less water, fertilizers, pesticides, and herbicides than other cellulosic crops. Manufacturing hemp-based L-MMCF aids in whole plant utilization, and growing agricultural hemp for multiple purposes provides greater economic opportunity and environmental sustainability. In addition, end-users of hemp L-MMCF could have their products dissolved and extruded though the L-MMCF process again to create new L-MMCF, lending to a circular economy. Unfortunately, each time these textiles are recycled via dissolution, the degree of polymerization of the cellulose molecular chain is reduced; however, these recycled L-MMCF textiles could be blended with new raw feedstock (i.e., agricultural biomass) to create unique fibre properties.

Textiles are the second most environmentally damaging commodities following oil and gas production [103], and currently, only 15.2% of textiles are recycled [104]. By utilizing hemp as a feedstock, and by creating a circular economy by post-consumer cellulosic recycling via the L-MMCF process, cellulosic textiles can be diverted from landfills, therefore reducing GHG emissions. The estimated CO2e (CO2 equivalent) emissions per ton of cotton fibre is 2.35 to 4.05 kg [105], which would be reduced substantially when recycled in the L-MMCF process. Replacing cotton cellulosics with industrial hemp and hemp-based L-MMCF would lead to increased carbon sequestration and further reduction in GHG emissions. Industrial hemp is capable of sequestering 22 tonnes of CO2e per acre, making it ideal for regenerative agriculture practices [14]. According to Vosper [12], 0.445 tonnes of carbon are sequestered from the atmosphere for each tonne of industrial hemp grown. Therefore, we can estimate that 44 tonnes of carbon would be sequestered for every 100 tonnes of hemp biomass used for L-MMCF extrusion. Furthermore, manufacturing hemp L-MMCF can offset the use of petroleum-based synthetics, reducing the overall global GHG footprint. In 2021, global market demand for cotton was 22% and MMCF was at 6.4%, with L-MMCF only accounting for 0.28% of the global textile fibre demand [3]. The estimated CAGR for cellulosic fibres, including both natural and MMCF, is 4.2% [106].

Worldwide market size for L-MMCF is estimated at $1.13B [107] with the NA market size estimated at $199.4M [108]. These demands are increasing annually as demands for cellulosic fibres rise (i.e., the Cellulose Gap) [19,23]. The manufacturing cost of L-MMCF compared to other MMCF, such as viscose, is substantially higher due to raw material costs (i.e., NMMO solvent and pulp feedstock) and specialized equipment requirements [4 Based on global fibre production of viscose and L-MMCF in 2021 of 5.8 and 0.3 millio tonnes, respectively [3], and on the global fibre market of viscose and L-MMCF in 2021 c $13.05B [109] and $1.13B [107], respectively, it can be estimated that L-MMCF are approx mately 68% more expensive than viscose. However, demand for more sustainable fibres, a well as the use of less expensive, locally sourced feedstocks such as hemp (due to biomas availability and whole plant utilization), will lead to increase market share of L-MMCF [57 Hemp-based L-MMCF could be a fibre source for the following markets:

1. Personal Protective Equipment (PPE): The current worldwide market for commerci PPE is valued at $52.7 B, with $140 M serviceable towards flame resistant (FR) PP in oil/gas, mining, electrical, and construction applications [110]. The PPE mark is expected to increase to $92.5 B by 2025 [110] and $110.85 B by 2029 [111]. The N market for commercial PPE is $2.9 B, from which $83 M serviceable towards FR PP applications [112]. Targeted market for FR hemp-based L-MMCF at commercializatio is initially estimated at $12.5 M worldwide and $2 M in NA, increasing at scale-u production of hemp-based L-MMCF.
2. Filtration media/Disposable Consumer Products: Single-use disposable non-wover (e.g., facemasks, gowns, flushable wipes, hygiene/sanitation products) are a hig demand commodity. The global non-woven medical textiles market is estimate to exhibit a CAGR of 3.9% with $27.7 B in sales by 2030 [113]. Targeted market fc L-MMCF from hemp and post-consumer cellulosic textiles at commercialization fc use in filtration media/disposable consumer products is estimated at $23 B worl wide and $8.3B in NA [114], increasing at scale-up. Consumption of cotton, othe natural fibres, wood pulp, and rayon fibres for nonwoven applications in NA i 2021 was estimated at 1 million tonnes [115]. Considering a CAGR of 3.9%, the est mated consumption for L-MMCF staple fibres for nonwoven applications in NA a commercialization is 1.2 million tonnes.
3. Apparel: The current global hemp apparel market is estimated to be $23.02 B by 203 (CAGR of 27.1%), with NA and Europe accounting for 65% of global consumptio ($15 B by 2030) [116]. The global L-MMCF market for apparel is approximatel $447 M US (45% of currently L-MMCF market share) [117]. The global ethical fashio market size is estimated to grow from $6.35 B in 2019 to $15.7 B in 2030 with a CAG of 9.1% [118]. Targeted market for L-MMCF from hemp and post-consumer cellulos textiles at commercialization for use in apparel at commercialization is estimate between $6.9 B globally ($4.5 NA) and $10.2 B globally ($6.6 NA).
4. Dental products: The global market for dental floss was valued to $540 M US in 201 with a projected growth to $750 M US in 2025 [119]. There is a growing demand for env ronmentally friendly alternatives to the plastic-based products currently available [12

7. Conclusions

Global demand for cellulosic fibres is increasing annually due to greater awareness c sustainability in textile products. Currently, cotton is the most frequently utilized cellulos fibre; however, due to cultivation restrictions, the maximum global production for this fib is estimated at 26 million metric tons. With the Cellulose Gap increasing, MMCFs off alternatives to cotton cellulosics. With increased awareness of hemp applicability to man different industries, including textile fibre production, industrial hemp as an agricultur crop is making a resurgence internationally. Hemp biomass as L-MMCF feedstock pr motes whole plant utilization, reduces deforestation, reduces water consumption, reduc GHG emissions, promotes carbon sequestration, and can be cultivated via regenerati agricultural processes, lending to an overall reduced environmental impact and reduce environmental burden traditionally seen by the textile industry. While additional researc needs to be completed regionally to assess hemp biomass characteristics (i.e., compositio variance due to remediation properties of industrial hemp growth), the use of industri hemp biomass as a feedstock towards the production of L-MMCF is a promising prospe

Author Contributions: Conceptualization, L.L. and P.I.D.; writing—original draft preparation, L.L.; writing—review and editing, P.I.D., L.M.D., B.B., W.C. and D.K. All authors have read and agreed to the published version of the manuscript.

Funding: Work mentioned in the article has received funding from the National Research Council of Canada Industrial Research Assistance Program (NRC IRAP) (#PN937822).

Institutional Review Board Statement: Not applicable.

Informed Consent Statement: Not applicable.

Data Availability Statement: Not applicable.

Conflicts of Interest: The authors declare no conflict of interest.

eferences

Niinimäki, K.; Peters, G.; Dahlbo, H.; Perry, P.; Rissanen, T.; Gwilt, A. The Environmental Price of Fast Fashion. *Nat. Rev. Earth Environ.* **2020**, *1*, 189–200. [CrossRef]

Bevilacqua, M.; Ciarapica, F.E.; Mazzuto, G.; Paciarotti, C. Environmental Analysis of a Cotton Yarn Supply Chain. *J. Clean. Prod.* **2014**, *82*, 154–165. [CrossRef]

Textile Exchange. Preferred Fiber and Materials Market Report 2022. Available online: https://textileexchange.org/wp-content/uploads/2022/10/Textile-Exchange_PFMR_2022.pdf (accessed on 5 November 2022).

Jiang, X.; Bai, Y.; Chen, X.; Liu, W. A Review on Raw Materials, Commercial Production and Properties of Lyocell Fiber. *J. Bioresour. Bioprod.* **2020**, *5*, 16–25. [CrossRef]

Sayyed, A.J.; Deshmukh, N.A.; Pinjari, D.V. A Critical Review of Manufacturing Processes Used in Regenerated Cellulosic Fibres: Viscose, Cellulose Acetate, Cuprammonium, LiCl/DMAc, Ionic Liquids, and NMMO Based Lyocell. *Cellulose* **2019**, *26*, 2913–2940. [CrossRef]

Briggs, J. Hemp Paper vs. Tree Paper. Available online: https://hempfrontiers.com/hemp-paper-vs-tree-paper/ (accessed on 13 September 2022).

Chua, J.M. Amazon Pledges to Rid Private-Label Brands of Forest-Harming Viscose. *Sourc. J.* **2020**. Available online: https://sourcingjournal.com/topics/raw-materials/amazon-private-label-canopystyle-deforestation-sustainable-viscose-2022-219512/ (accessed on 13 September 2022).

Seisl, S.; Hengstmann, R. Manmade Cellulosic Fibers (MMCF)—A Historical Introduction and Existing Solutions to a More Sustainable Production. In *Sustainable Textile and Fashion Value Chains: Drivers, Concepts, Theories and Solutions*; Matthes, A., Beyer, K., Cebulla, H., Arnold, M.G., Schumann, A., Eds.; Springer International Publishing: Cham, Switzerland, 2021; pp. 3–22, ISBN 978-3-030-22018-1.

Cassman, K.G. Ecological Intensification of Cereal Production Systems: Yield Potential, Soil Quality, and Precision Agriculture. *Proc. Natl. Acad. Sci. USA* **1999**, *96*, 5952–5959. [CrossRef]

10. Ribul, M. Regenerative Textiles: A Framework for Future Materials Circularity in the Textile Value Chain. *Sustainability* **2021**, *13*, 13910. [CrossRef]
11. Janjic, S.; Kostic, M.; Skundric, P. Direct Hemp Cellulose Dissolution in N-Methylmorpoline-N-Oxide. *J. Nat. Fibers* **2007**, *4*, 23–36. [CrossRef]
12. Vosper, J. The Role of Industrial Hemp in Carbon Farming. Available online: https://www.aph.gov.au/parliamentary_business/committees/house_of_representatives_committees?url=ccea/24march2011/subs/sub035.pdf (accessed on 13 September 2022).
13. Kosan, B.; Meister, F.; Sigmund, I.; Paulitz, J. Innovative Dissolving Pulps for Application in Cellulose MMF Production. *Lenzing. Ber.* **2019**, *95*, 9–14.
14. Speicher, S. Hemp Is a Strong Option for Sequestering Carbon. Available online: https://www.lancasterfarming.com/news/main_edition/hemp-is-a-strong-option-for-sequestering-carbon/article_29e4d6d0-bcae-5ca6-bb76-0963c470ffc8.html (accessed on 27 October 2022).
15. Legal History of Cannabis in Canada. Available online: https://en.wikipedia.org/w/index.php?title=Legal_history_of_cannabis_in_Canada&oldid=1109411917 (accessed on 11 October 2022).
16. Lynch, P. Why Was Hemp Banned After World War II? Available online: https://wayofleaf.com/hemp/why-was-hemp-banned-after-world-war-ii (accessed on 11 October 2022).
17. Business Wire. Global Lyocell Fabric Market to Reach 745.9 KT by 2027—Coherent Market Insights. Available online: https://www.businesswire.com/news/home/20191125005611/en/Global-Lyocell-Fabric-Market-to-Reach-745.9-KT-by-2027-%E2%80%93-Coherent-Market-Insights (accessed on 12 October 2022).
18. Fastmarkets RISI. Synthetics Continue to Lead Fiber Supply in 2021, While Cellulosics Reach an All-Time High in Volume. *Sourc. J.* **2020**. Available online: https://thefiberyear.com/wp-content/uploads/2022/07/NWM_14_2022.pdf (accessed on 11 October 2022).
19. El Seoud, O.A.; Kostag, M.; Jedvert, K.; Malek, N.I. Cellulose Regeneration and Chemical Recycling: Closing the "Cellulose Gap" Using Environmentally Benign Solvents. *Macromol. Mater. Eng.* **2020**, *305*, 1900832. [CrossRef]

20. EEA's European Topic Centre on Waste and Materials in a Green Economy. Plastic in Textiles: Towards a Circular Economy fo Synthetic Textiles in Europe—European Environment Agency. Available online: https://www.eea.europa.eu/themes/waste resource-efficiency/plastic-in-textiles-towards-a (accessed on 11 October 2022).
21. Jambeck, J.R.; Geyer, R.; Wilcox, C.; Siegler, T.R.; Perryman, M.; Andrady, A.; Narayan, R.; Law, K.L. Plastic Waste Inputs from Land into the Ocean. *Science* **2015**, *347*, 768–771. [CrossRef]
22. Barnes, D.K.A.; Galgani, F.; Thompson, R.C.; Barlaz, M. Accumulation and Fragmentation of Plastic Debris in Global Environments. *Philos. Trans. R. Soc. B Biol. Sci.* **2009**, *364*, 1985–1998. [CrossRef] [PubMed]
23. Haemmerle, F.M. The Cellulose Gap (the Future of Cellulose Fibres). *Lenzing. Ber.* **2011**, *89*, 12–21.
24. Cherrett, N.; Barrett, J.; Clemett, A.; Chadwick, M.; Chadwick, M.J. *Ecological Footprint and Water Analysis of Cotton, Hemp and Polyester*; Stockholm Environmental Institute: Stockholm, Sweden, 2005; ISBN 978-91-975238-2-0.
25. Zhang, X.; Cai, X. Climate Change Impacts on Global Agricultural Land Availability. *Environ. Res. Lett.* **2011**, *6*, 014014. [CrossRef]
26. United Nations. Population. Available online: https://www.un.org/en/global-issues/population (accessed on 30 October 2022)
27. World Wildlife Fund. Cotton | Industries | WWF. Available online: https://www.worldwildlife.org/industries/cotton (accessed on 21 October 2022).
28. Nyakatawa, E.; Lemunyon, J.L. Predicting Soil Erosion in Conservation Tillage Cotton Production Systems Using the Revised Universal Soil Loss Equation (RUSLE). *Soil Tillage Res.* **2001**, *57*, 213–224. [CrossRef]
29. Nachimuthu, G.; Webb, A. On-Farm Soil Conservation Measures in Cotton Farming Systems of Australia: A Sustainability Analysis. *J. Soil Water Conserv.* **2016**, *71*, 75A–80A. [CrossRef]
30. Food and Agriculture Organization of the United Nations. Water for Sustainable Food and Agriculture: A Report Produced for the G20 Presidency of Germany. Food & Agriculture Org.: Rome, Italy, 2017; ISBN 978-92-5-109977-3.
31. Sustainable Fashion. The Water Consumption Attributable to Cotton Production. Available online: https://sustainfashion.info/the-water-consumption-attributable-to-cotton-production/ (accessed on 21 October 2022).
32. Averink, J. Global Water Footprint of Industrial Hemp Textile. Available online: https://essay.utwente.nl/68219/1/Averink,%20J.%200198501%20openbaar.pdf (accessed on 5 November 2022).
33. Shen, L.; Patel, M. Life Cycle Assessment of Man-Made Cellulose Fibres. *Lenzing. Ber.* **2010**, *88*, 1–59.
34. Changing Markets. Dirty Fashion: How Pollution in the Global Textiles Supply Chain Is Making Viscose Toxic. Available online: http://changingmarkets.org/wp-content/uploads/2017/06/CHANGING_MARKETS_DIRTY_FASHION_REPORT SPREAD_WEB.pdf (accessed on 20 October 2022).
35. Common Objective. Fibre Briefing: Viscose. Available online: http://www.commonobjective.co/article/fibre-briefing-viscose (accessed on 31 October 2022).
36. Sengupta, A.K. Rayon Fibres. In *Manufactured Fibre Technology*; Gupta, V.B., Kothari, V.K., Eds.; Springer: Dordrecht, The Netherlands, 1997; pp. 480–513, ISBN 978-94-011-5854-1.
37. Textile Exchange. Preferred Manmade Cellulosics—About Preferred Manmade Cellulosics. Available online: https://hub.textileexchange.org/textileexchange/learning-center/mmcellulosics (accessed on 14 September 2022).
38. Iroegbu, A.O.C.; Ray, S.S. Bamboos: From Bioresource to Sustainable Materials and Chemicals. *Sustainability* **2021**, *13*, 12200. [CrossRef]
39. Nayak, L.; Mishra, S.P. Prospect of Bamboo as a Renewable Textile Fiber, Historical Overview, Labeling, Controversies and Regulation. *Fash. Text.* **2016**, *3*, 2. [CrossRef]
40. Ferreira, E.; Kalliola, R.; Ruokolainen, K. Bamboo, Climate Change and Forest Use: A Critical Combination for Southwestern Amazonian Forests? *Ambio* **2020**, *49*, 1353–1363. [CrossRef]
41. Yiping, L.; Yanxia, L.; Buckingham, K.; Henley, G.; Guomo, Z. Bamboo and Climate Change Mitigation: A Comparative Analysis of Carbon Sequestration. *Int. Netw. Bamboo Ratt.* **2010**, *30*. Available online: https://www.inbar.int/wp-content/uploads/2020/05/1489457789.pdf (accessed on 6 November 2022).
42. Kauffman, G.B.; Karbassi, M. A Demonstration of the Cuprammonium Rayon Process. *J. Chem. Educ.* **1985**, *62*, 878. [CrossRef]
43. Hong, Y.-K.; Chung, K.-H.; Lee, W.-S. Structure of Regenerated Cellulose Fibers from DMAc/LiCl Solution. *Text. Res. J.* **1998**, *68*, 65–69. [CrossRef]
44. Moriam, K.; Sawada, D.; Nieminen, K.; Hummel, M.; Ma, Y.; Rissanen, M.; Sixta, H. Towards Regenerated Cellulose Fibers with High Toughness. *Cellulose* **2021**, *28*, 9547–9566. [CrossRef]
45. White, P. 4—Lyocell: The Production Process and Market Development. In *Regenerated Cellulose Fibres*; Woodings, C., Ed. Woodhead Publishing Series in Textiles; Woodhead Publishing: Sawston, UK, 2001; pp. 62–87, ISBN 978-1-85573-459-3.
46. Shen, L.; Worrell, E.; Patel, M.K. Environmental Impact Assessment of Man-Made Cellulose Fibres. *Resour. Conserv. Recycl.* **2010**, *55*, 260–274. [CrossRef]
47. Ouyang, Y.; Leininger, T.D.; SRS Center for Bottomland Hardwoods Research. How Sustainable Are Eucalyptus Plantations. Available online: https://www.srs.fs.usda.gov/compass/2016/05/05/how-sustainable-are-eucalyptus-plantations/ (accessed on 21 October 2022).
48. Ouyang, Y.; Xu, D.; Leininger, T.D.; Zhang, N. A System Dynamic Model to Estimate Hydrological Processes and Water Use in a Eucalypt Plantation. *Ecol. Eng.* **2016**, *86*, 290–299. [CrossRef]
49. Williams, D. What Is Coppicing? (And Why It's Amazing). Available online: https://www.growingwithnature.org/what-is-coppicing/ (accessed on 31 October 2022).

Costa, S.A.; Cerón, A.A.; Petreca, B.B.; Costa, S.M. Fibers of Cellulose Sugarcane Bagasse with Bromelain Enzyme Immobilized to Application in Dressing. *SN Appl. Sci.* **2020**, *2*, 285. [CrossRef]

Sirlene, M.; Costa, S.; Costa, S.; Pahl, R.; Mazzola, P. Textile Fiber Produced from Sugarcane Bagasse Cellulose: An Agro-Industrial Residue. *Int. J. Text. Fash. Technol. (IJTFT)* **2013**, *3*, 15–28.

Dever, M.; Collier, B.J.; Petrovan, S.; Collier, J.R. Lyocell Solutions from Alternative Cellulose Sources. *Cloth. Text. Res. J.* **2003**, *21*, 167–173. [CrossRef]

Gericke, A.; Van der Pol, J. A Comparative Study of Regenerated Bamboo, Cotton and Viscose Rayon Fabrics. Part 2: Antimicrobial Properties. *J. Consum. Sci.* **2011**, *39*. [CrossRef]

Millati, R.; Wikandari, R.; Ariyanto, T.; Putri, R.U.; Taherzadeh, M.J. Pretreatment Technologies for Anaerobic Digestion of Lignocelluloses and Toxic Feedstocks. *Bioresour. Technol.* **2020**, *304*, 122998. [CrossRef]

World Wildlife Fund. Sugarcane Farming's Toll on the Environment | Magazine Articles | WWF. Available online: https://www.worldwildlife.org/magazine/issues/summer-2015/articles/sugarcane-farming-s-toll-on-the-environment (accessed on 31 October 2022).

Zhao, X.; Wei, X.; Guo, Y.; Qiu, C.; Long, S.; Wang, Y.; Qiu, H. Industrial Hemp—An Old but Versatile Bast Fiber Crop. *J. Nat. Fibers* **2021**, *19*, 1–14. [CrossRef]

Zhang, S.; Chen, C.; Duan, C.; Hu, H.; Li, H.; Li, J.; Liu, Y.; Ma, X.; Stavik, J.; Ni, Y. Regenerated Cellulose by the Lyocell Process, a Brief Review of the Process and Properties. *BioRes* **2018**, *13*, 4577–4592. [CrossRef]

Periyasamy, A.P.; Khanum, R.; Effect of Fibrillation on Pilling Tendency of Lyocell Fiber. Bangladesh Text. Today 2012. Available online: https://www.textiletoday.com.bd/effect-of-fibrillation-on-pilling-tendency-of-lyocell-fiber/ (accessed on 31 October 2022).

TITK. Cellulose Research: Lyohemp®—Latest Generation Sustainable Fiber. Available online: https://www.titk.de/en/innovations/lyocell-/-lyohemp/ (accessed on 5 November 2022).

Zhou, Y.; Zhang, X.; Zhang, J.; Cheng, Y.; Wu, J.; Yu, J.; Zhang, J. Molecular Weight Characterization of Cellulose Using Ionic Liquids. *Polym. Test.* **2021**, *93*, 106985. [CrossRef]

Nogueira, C.; Silva, M.E.C. Sensory Analysis of Textiles Products. Presented at the Semana de Engenharia 2010, Guimarães, Dubrovik, Croatia, 3 October 2010.

Thomsen, A.B.; Rasmussen, S.; Bohn, V.; Nielsen, K.V.; Thygesen, A. *Hemp Raw Materials: The Effect of Cultivar, Growth Conditions and Pretreatment on the Chemical Composition of the Fibres*; Risø DTU-National Laboratory for Sustainable Energy: Roskilde, Denmark, 2005; p. 31.

Chen, X.; Zhang, K.; Xiao, L.-P.; Sun, R.-C.; Song, G. Total Utilization of Lignin and Carbohydrates in Eucalyptus Grandis: An Integrated Biorefinery Strategy towards Phenolics, Levulinic Acid, and Furfural. *Biotechnol. Biofuels* **2020**, *13*, 2. [CrossRef] [PubMed]

Tonoli, G.H.D.; Teixeira, E.M.; Corrêa, A.C.; Marconcini, J.M.; Caixeta, L.A.; Pereira-da-Silva, M.A.; Mattoso, L.H.C. Cellulose Micro/Nanofibres from Eucalyptus Kraft Pulp: Preparation and Properties. *Carbohydr. Polym.* **2012**, *89*, 80–88. [CrossRef]

Moradbak, A.; Tahir, P.M.; Mohamed, A.Z.; Halis, R. Alkaline Sulfite Anthraquinone and Methanol Pulping of Bamboo (*Gigantochloa Scortechinii*). *Bioresources* **2016**, *11*, 235–248. [CrossRef]

Chen, H.; Wang, Y.; Qiu, J.; Song, J.; Wang, J.; Liu, M.; Wo, Q.; Jiang, L.; Yang, T. Properties and Application of Kraft Pulp Prepared from Waste Bamboo Powder. *BioRes* **2022**, *17*, 6262–6276. [CrossRef]

Blade, S.; Gaudiel, R.; Kerr, N. Low-THC Hemp Research in the Black and Brown Soil Zones of Alberta, Canada. In *Perspectives on New Crops and New Uses*; ASHS Press: Alexandria, VA, USA, 1999; pp. 306–310.

Wimalasiri, E.M.; Jahanshiri, E.; Chimonyo, V.G.P.; Kuruppuarachchi, N.; Suhairi, T.A.S.T.M.; Azam-Ali, S.N.; Gregory, P.J. A Framework for the Development of Hemp (*Cannabis sativa* L.) as a Crop for the Future in Tropical Environments. *Ind. Crops Prod.* **2021**, *172*, 113999. [CrossRef]

De Prato, L.; Ansari, O.; Hardy, G.E.S.J.; Howieson, J.; O'Hara, G.; Ruthrof, K.X. The Cannabinoid Profile and Growth of Hemp (*Cannabis sativa* L.) Is Influenced by Tropical Daylengths and Temperatures, Genotype and Nitrogen Nutrition. *Ind. Crops Prod.* **2022**, *178*, 114605. [CrossRef]

Wimalasiri, E.M.; Jahanshiri, E.; Syaherah, T.A.; Kuruppuarachchi, N.; Chimonyo, V.G.P.; Azam-Ali, S.N.; Gregory, P.J. Datasets for the Development of Hemp (*Cannabis sativa* L.) as a Crop for the Future in Tropical Environments (Malaysia). *Data Brief* **2022**, *40*, 107807. [CrossRef]

Cannabis Business Plan. Global Hemp Market. Available online: https://cannabusinessplans.com/global-hemp-market/ (accessed on 6 November 2022).

Duque Schumacher, A.G.; Pequito, S.; Pazour, J. Industrial Hemp Fiber: A Sustainable and Economical Alternative to Cotton. *J. Clean. Prod.* **2020**, *268*, 122180. [CrossRef]

Grand View Research. Industrial Hemp Market Worth $16.75 Billion By 2030. Available online: https://www.grandviewresearch.com/press-release/global-industrial-hemp-market (accessed on 7 November 2022).

Alary, B. How Alberta Is Poised to Become a Hub of the Hemp Industry. Available online: https://www.techlifetoday.ca/articles/2019/alberta-poised-to-become-a-hemp-hub (accessed on 25 October 2022).

Health Canada. List of Approved Cultivars for the 2022 Growing Season: Industrial Hemp Varieties Approved for Commercial Production. Available online: https://www.canada.ca/en/health-canada/services/drugs-medication/cannabis/producing-selling-hemp/commercial-licence/list-approved-cultivars-cannabis-sativa.html (accessed on 25 October 2022).

76. Government of Alberta—Agriculture and Forestry. Growing Hemp in Alberta—Growing Hemp in Alberta—Open Governme Available online: https://open.alberta.ca/dataset/growing-hemp-in-alberta/resource/9babce43-b6f6-422c-85cd-283ea1a561 (accessed on 14 September 2022).
77. Musio, S.; Müssig, J.; Amaducci, S. Optimizing Hemp Fiber Production for High Performance Composite Applications. *Fro Plant Sci.* **2018**, *9*, 1702. [CrossRef] [PubMed]
78. Horne, M.R.L. 5B—Bast Fibres: Hemp Cultivation and Production. In *Handbook of Natural Fibres*, 2nd ed.; Kozłowski, R.M Mackiewicz-Talarczyk, M., Eds.; Woodhead Publishing Series in Textiles; Woodhead Publishing: Sawston, UK, 2020; pp. 163–19 ISBN 978-0-12-818398-4.
79. Manian, A.P.; Cordin, M.; Pham, T. Extraction of Cellulose Fibers from Flax and Hemp: A Review. *Cellulose* **2021**, *28*, 8275–829 [CrossRef]
80. Zimniewska, M. Hemp Fibre Properties and Processing Target Textile: A Review. *Materials* **2022**, *15*, 1901. [CrossRef] [PubMe
81. Żuk-Gołaszewska, K.; Gołaszewski, J. *Cannabis sativa* L.—Cultivation and Quality of Raw Material. *J. Elem.* **2018**, *23*, 971–98 [CrossRef]
82. Ahmed, A.T.M.F.; Islam, M.Z.; Mahmud, M.S.; Sarker, M.E.; Islam, M.R. Hemp as a Potential Raw Material toward a Sustainab World: A Review. *Heliyon* **2022**, *8*, e08753. [CrossRef] [PubMed]
83. Meister, F.; Sigmund, I.; Paulitz, J.; Kosan, B.; Kreig, M.; Mooz, M. Novel Dissolving Pulps Based on Annual Plants for th Application in the Lyocell Process. Presented at the 56th Dornbirn MFC, Dornbirn, Austria, 13–15 September 2017.
84. Han, J.S. *Properties of Nonwood Fibers*; The Korean Society of Science and Technology: Seoul, Korea, 1998; pp. 3–12.
85. Krotov, V.S. Use of AAS Pulping for Flax and Hemp Shives. *J. Int. Hemp Assoc.* **1995**, *3*, 16–18.
86. Gümüşkaya, E.; Usta, M.; Balaban, M. Carbohydrate Components and Crystalline Structure of Organosolv Hemp (*Cannabis sati* L.) Bast Fibers Pulp. *Bioresour. Technol.* **2007**, *98*, 491–497. [CrossRef]
87. Ahmad, R.; Tehsin, Z.; Malik, S.T.; Asad, S.A.; Shahzad, M.; Bilal, M.; Shah, M.M.; Khan, S.A. Phytoremediation Potential Hemp (*Cannabis sativa* L.): Identification and Characterization of Heavy Metals Responsive Genes. *CLEAN—Soil Air Water* **201** *44*, 195–201. [CrossRef]
88. Griga, M.; Bjelková, M. Flax (*Linum usitatissimum* L.) and Hemp (*Cannabis sativa* L.) as Fibre Crops for Phytoextraction of Heav Metals: Biological, Agro-Technological and Economical Point of View. In *Plant-Based Remediation Processes*; Gupta, D.K., Ed.; Sc Biology; Springer: Berlin/Heidelberg, Germany, 2013; pp. 199–237, ISBN 978-3-642-35564-6.
89. Linger, P.; Ostwald, A.; Haensler, J. *Cannabis sativa* L. Growing on Heavy Metal Contaminated Soil: Growth, Cadmium Uptak and Photosynthesis. *Biol. Plant.* **2005**, *49*, 567–576. [CrossRef]
90. Shi, G.; Liu, C.; Cui, M.; Ma, Y.; Cai, Q. Cadmium Tolerance and Bioaccumulation of 18 Hemp Accessions. *Appl. Bioche Biotechnol.* **2012**, *168*, 163–173. [CrossRef]
91. Paulitz, J.; Sigmund, I.; Kosan, B.; Meister, F. Lyocell Fibers for Textile Processing Derived from Organically Grown Hem *Procedia Eng.* **2017**, *200*, 260–268. [CrossRef]
92. Rosenau, T.; Potthast, A.; Adorjan, I.; Hofinger, A.; Sixta, H.; Firgo, H.; Kosma, P. Cellulose Solutions in N-Methylmorpholine-N Oxide (NMMO)—Degradation Processes and Stabilizers. *Cellulose* **2002**, *9*, 283–291. [CrossRef]
93. Rosenau, T.; Potthast, A.; Kosma, P.; Chen; Gratzl, J.S. Autocatalytic Decomposition of N-Methylmorpholine N-Oxide Induced b Mannich Intermediates. *J. Org. Chem.* **1999**, *64*, 2166–2167. [CrossRef]
94. Kosan, B.; Michels, C.; Meister, F. Dissolution and Forming of Cellulose with Ionic Liquids. *Cellulose* **2008**, *15*, 59–66. [CrossRe
95. Fang, Z.; Li, B.; Liu, Y.; Zhu, J.; Li, G.; Hou, G.; Zhou, J.; Qiu, X. Critical Role of Degree of Polymerization of Cellulose i Super-Strong Nanocellulose Films. *Matter* **2020**, *2*, 1000–1014. [CrossRef]
96. Krässig, H.; Kitchen, W. Factors Influencing Tensile Properties of Cellulose Fibers. *J. Polym. Sci.* **1961**, *51*, 123–172. [CrossRef]
97. Mechanical Properties of Polymer Nanocomposites. In *Fundamentals, Properties, and Applications of Polymer Nanocomposite* Koo, J.H. (Ed.) Cambridge University Press: Cambridge, UK, 2016; pp. 273–331, ISBN 978-1-139-34276-6.
98. Gwon, J.-G.; Lee, D.-B.; Cho, H.-J.; Lee, S.-Y. Preparation and Characteristics of Cellulose Acetate Based Nanocomposit Reinforced with Cellulose Nanocrystals (CNCs). *J. Korean Wood Sci. Technol.* **2018**, *46*, 565–576. [CrossRef]
99. Bajwa, D.S.; Rehovsky, C.; Shojaeiarani, J.; Stark, N.; Bajwa, S.; Dietenberger, M.A. Functionalized Cellulose Nanocrystals: Potential Fire Retardant for Polymer Composites. *Polymers* **2019**, *11*, 1361. [CrossRef]
100. Khan, B.A.; Warner, P.; Wang, H. Antibacterial Properties of Hemp and Other Natural Fibre Plants: A Review. *BioResources* **201** *9*, 3642–3659. [CrossRef]
101. Arafa Badr, A. Anti-Microbial and Durability Characteristics of Socks Made of Cotton and Regenerated Cellulosic Fibers. *Ale Eng. J.* **2018**, *57*, 3367–3373. [CrossRef]
102. Khan, B.A.; Wang, J.; Warner, P.; Wang, H. Antibacterial Properties of Hemp Hurd Powder against *E. Coli*. *J. Appl. Polym. S* **2015**, *132*. [CrossRef]
103. O'Connell, K. Hemp vs Paper | Hemp Paper Benefits. Available online: https://ministryofhemp.com/blog/hemp-pape (accessed on 13 September 2022).
104. Brown, R. The Environmental Crisis Caused by Textile Waste. Available online: https://www.roadrunnerwm.com/blog/textil waste-environmental-crisis (accessed on 13 September 2022).
105. Ecotextiles (and Two Sisters Ecotextiles). Estimating the Carbon Footprint of a Fabric. Available online: https://embe tumblr.com/widgets/share/button?canonicalUrl=https%3A%2F%2Foecotextiles.blog%2F2011%2F01%2F19%2Festimating

the-carbon-footprint-of-a-fabric%2F&postcontent%5Btitle%5D=Estimating%20the%20carbon%20footprint%20of%20a%20 fabric&postcontent%5Bcontent%5D=https%3A%2F%2Foecotextiles.blog%2F2011%2F01%2F19%2Festimating-the-carbon-footprint-of-a-fabric%2F (accessed on 25 October 2022).

6. Future Market Insights Cellulose Fiber Market: Cellulose Fiber Market by Application, Process of Manufacturing & Region—Forecast 2022—2032. Available online: https://www.futuremarketinsights.com/reports/cellulose-fiber-market (accessed on 6 November 2022).

7. Research and Markets ltd. Lyocell Fiber Market: Global Industry Trends, Share, Size, Growth, Opportunity and Forecast 2022–2027. Available online: https://www.researchandmarkets.com/reports/5530547/lyocell-fiber-market-global-industry-trends (accessed on 13 September 2022).

8. IMARC. North America Lyocell Fiber Market Report: Industry Trends and Forecast 2022–2027. Available online: https://www.imarcgroup.com/north-america-lyocell-fiber-market (accessed on 24 October 2022).

9. Proficient Market Insights. Viscose Staple Fiber Market Projected to Surpass 16924.39 Million USD and Grow at a 4.43% CAGR During the 2022–2027 Forecast Timeframe. Available online: https://www.globenewswire.com/news-release/2022/10/20/2538221/0/en/Viscose-Staple-Fiber-Market-Projected-to-Surpass-16924-39-million-USD-and-Grow-at-a-4-43-CAGR-During-the-2022-2027-Forecast-Timeframe-104-Pages-Report.html (accessed on 7 November 2022).

10. VynZ Research USA. Global Personal Protective Equipment (PPE) Market Was Valued at USD 52.7 Billion in 2019 and Is Expected to Reach USD 92.5 Billion by 2025, Observing a CAGR of 8.7% during 2020–2025. Available online: https://www.globenewswire.com/news-release/2020/04/14/2015737/0/en/Global-Personal-Protective-Equipment-PPE-Market-was-Valued-at-USD-52-7-billion-in-2019-and-is-Expected-to-Reach-USD-92-5-billion-by-2025-Observing-a-CAGR-of-8-7-during-2020-2025-Vy.html (accessed on 13 September 2022).

11. Fortune Business Insights. *Personal Protective Equipment [PPE] Market Size | Trends, 2029*. Available online: https://www.fortunebusinessinsights.com/personal-protective-equipment-ppe-market-102015 (accessed on 11 October 2022).

12. Available online: https://www.alliedmarketresearch.com/north-america-protective-clothing-market-A06968 (accessed on 11 October 2022).

13. Growman Research and Consulting Pvt Ltd. Medical Nonwoven Disposables Market Size to Surpass US$ 27.70 Billion by 2030—Growth Plus Reports. Available online: https://www.globenewswire.com/news-release/2022/09/26/2522543/0/en/Medical-Nonwoven-Disposables-Market-Size-to-Surpass-US-27-70-Billion-by-2030-Growth-Plus-Reports.html (accessed on 11 October 2022).

14. Fact.MR. Medical Nonwoven Disposables Market Analysis By Product (Incontinence Hygiene Products, Sterile Nonwoven Products & Wound Dressing) By Material (Polypropylene, Polyethylene, Acetate, Rayon, Acrylic, Polyamides & Polyesters) By End-User & Region—Global Market Insights 2022 to 2032. Available online: https://www.factmr.com/report/medical-nonwoven-disposables-market (accessed on 11 October 2022).

15. Kalil, B. A View of the North American Nonwovens Market and Material Consumption. Available online: https://www.fiberjournal.com/a-view-of-the-north-american-nonwovens-market-and-material-consumption/ (accessed on 11 October 2022).

16. Allied Market Research. Hemp Clothing Market Size, Share And Trends | Forecast By 2031. Available online: https://www.alliedmarketresearch.com/hemp-clothing-market-A16939 (accessed on 11 October 2022).

17. Business Wire. Global Lyocell Fiber Market (2021 to 2029)—Increased Product Demand in the Apparel Manufacturing Industry Is Driving Growth. Available online: https://www.businesswire.com/news/home/20210709005355/en/Global-Lyocell-Fiber-Market-2021-to-2029---Increased-Product-Demand-in-the-Apparel-Manufacturing-Industry-is-Driving-Growth---ResearchAndMarkets.com (accessed on 11 October 2022).

18. TBRC Business Research PVT LTD. Sustainable Fashion Market Analysis Shows The Market Progress In Attempt To Decrease Pollution In The Global Ethicalfashion Market 2020. Available online: https://www.globenewswire.com/news-release/2020/10/28/2116073/0/en/Sustainable-Fashion-Market-Analysis-Shows-The-Market-Progress-In-Attempt-To-Decrease-Pollution-In-The-Global-Ethicalfashion-Market-2020.html (accessed on 11 October 2022).

19. Available online: https://www.grandviewresearch.com/industry-analysis/dental-floss-market (accessed on 13 September 2022).

20. Business Wire. Global Dental Floss Market to Grow by USD 1.03 Billion during 2019-2023 | Technavio. Available online: https://www.businesswire.com/news/home/20190530005031/en/Global-Dental-Floss-Market-to-Grow-by-USD-1.03-Billion-during-2019-2023-Technavio (accessed on 31 October 2022).

sustainability

MDPI

Article

Fabrication of Novel Polymer Composites from Leather Waste Fibers and Recycled Poly(Ethylene-Vinyl-Acetate) for Value-Added Products

Shubham Sharma [1,2,3], P. Sudhakara [1], Jujhar Singh [2], Sanjay M. R. [4,*] and S. Siengchin [4]

1 CSIR—Central Leather Research Institute, Regional Centre, Jalandhar 144021, Punjab, India
2 Department of Mechanical Engineering, IK Gujral Punjab Technical University, Jalandhar-Kapurthala Roa Kapurthala 144603, Punjab, India
3 Mechanical Engineering Department, University Centre for Research and Development, Chandigarh University, Mohali 140413, Punjab, India
4 Natural Composites Research Group Lab., Department of Materials and Production Engineering, The Sirindhorn International Thai-German Graduate School of Engineering (TGGS), King Mongkut's University of Technology North Bangkok (KMUTNB), Bangkok 10800, Thailand
* Correspondence: mavinkere.r.s@op.kmutnb.ac.th

Abstract: This investigation was focused on evaluating the utilization of Leather-waste, i.e., "Leathe Shavings", to develop "Poly(ethylene-vinyl-acetate)" (EVA) based "polymer matrix composites Composites with the highest ratio of 1:1 were developed using a rolling-mill, which was then sub jected to hot-press molding for value-added applications, notably in the "floor-covering", "structural "footwear", and "transportation domain". The specimens were examined for evaluating the "physic mechanical characteristics" such as, "Compressive and Tensile, strength, Abrasion-resistance, Densit tear-resistance, hardness, adhesion-strength, compression, and resilience, damping, and water ab sorption" as per standard advanced testing techniques. Raising the leather-fiber fraction in th composites culminated in considerable enhancement in "physico-mechanical characteristics" incluc ing "modulus", and a decline in "tensile-strain" at "fracture-breakage". The thermo-analytic method viz. TGA and DSC studies have evidenced that substantial enhancement of thermo-stability (up t 211.1–213.81 °C) has been observed in the newly developed PMCs. Additionally, the DSC stud showed that solid leather fibers lose water at an endothermic transition temperature of around 100 °C are thermo-stable at around 211 degrees centigrade, and begin to degrade at 332.56-degree centigrad for neat recycled EVA samples and begin to degrade collagen at 318.47-degree centigrade for "leathe shavings/recycled EVA polymer composite samples", respectively. Additionally, the "glass transitio temperature" (T_g) of the manufactured composites was determined to be between −16 and 30 °C Furthermore, SEM and EDAX analysis have been used to investigate the morphological characteris tics of the developed composites. Micrograph outcomes have confirmed the excellent "uniformit compatibility, stability and better-bonding" of leather-fibers within the base matrix. Additionall the "Attenuated-total-reflection" (ATR-FTIR) was carried out to test the "physicochemical chemica bonding", "molecular-structure", and "functional-groups" of the "base matrix", and its "composites further affirm the "recycled EVA matrix" contained additives remain within the polymeric-matri An "X-ray diffraction study" was also conducted to identify the "chemical-constituents" or "phases involved throughout the "crystal-structures" of the base matrix and PMCs. Additionally, AFM analysis has also been utilized to explore the "interfacial adhesion properties" of mechanically teste specimens of fabricated polymeric composite surfaces, their "surface topography mapping", and "phase-imaging analysis" of polymer composites that have leather-shavings fibers.

Keywords: leather shavings; recycled EVA; compression molding; rolling; footwear and leathe ancillaries(s) applications

check for updates

Citation: Sharma, S.; Sudhakara, P.; Singh, J.; M. R., S.; Siengchin, S. Fabrication of Novel Polymer Composites from Leather Waste Fibers and Recycled Poly(Ethylene-Vinyl-Acetate) for Value-Added Products. *Sustainability* **2023**, *15*, 4333. https://doi.org/10.3390/su15054333

Academic Editor: Jorge de Brito

Received: 17 December 2022
Revised: 11 February 2023
Accepted: 17 February 2023
Published: 28 February 2023

1. Introduction

A crucial worldwide commodity is leather. With an estimated USD 200 billion in annual global trade value for leather and leather goods, the leather industry is a significant contributor to the economy. However, the wastes ('solid and liquid') of such industries, such as buffing dust sludge etc., regularly pose a threat to the industry [1]. Land disposal is the most common method of treating solid waste [1,2]. Numerous waste-management options work to create value from trash by creating new materials and technology [3–5]. The fibrous structure of the leather is one of its key characteristics. It also offers excellent characteristics such as "high tear resistance, flexibility, thermal insulation, water resistance, and form maintenance". Therefore, adding this leather waste (specifically leather shavings) to any appropriate polymer matrix may result in composites with improved qualities for a variety of applications, including 'footwear, leather ancillaries, automotive, transportation, packaging, etc. Making valuable goods out of unused material would have a synergistic influence on the atmosphere and add value to the wastes produced by the polymer and leather industries, respectively. By using environmentally friendly polymers as matrix materials, numerous researchers have attempted to use leather wastage to make bio-composite materials, including 'films, sheets, fibers, etc.' [6–25].

The development of footwear and clothing made of thermoplastic elastomeric (TPE) material is rapidly accelerating. Numerous factors and criteria have been made possible by TPEs in order to meet consumer expectations [26–29]. EVA is used to make 'inserts/soles and midsoles' throughout the industrial footwear sector, accounting for 18% to 28% of the total. Prior to adding a mixture of Natural and Butadiene, rubbers possess better mechanical characteristics such as "high capabilities, minimum slippage, shrinkage, slipping, and delivers a ferocious compression-set," EVA has "low compression-set, abrasion, and tearing characteristics". These materials have exceptional 'physical and compression' characteristics together with 'least slippage' when compared to EVA/BR and EVA/NR. In the production of "hiking boots, basketball shoes, and virtually every other type of athletic footwear", the EVA midsole is an excellent 'vibration-damping' based cushioning material, which is 'light in weight, resists compression set, easily available in any color, and is made conveniently' [27,28,30].

For the current study, recycled EVA polymer matrix and leather shavings, two substantial leather solid wastes that are produced in huge quantities globally, have been discovered. Leather shavings are considered hazardous solid waste because they contain trivalent chromium, which, in some circumstances, can transform into the poisonous form known as hexavalent chromium. Traditional disposal techniques such as landfilling and incineration can easily produce secondary contamination and can't get rid of the hazardousness. In order to significantly reduce the hazardousness, it may be best to impregnate this leather waste into a recycled EVA polymer matrix. Additionally, the collagenous protein of the skins and hides creates a 3D cross-linking structure with the chromium, resulting in so-called durable leather. When this chrome-tanned leather waste is incorporated as reinforcement in the recycled EVA matrix, the physicochemical strength and thermal resistance of the composite blends will be enhanced. For instance, in the case of vegetable-tanned leather, where the chromium is absent, the degradation temperature and physico-mechanical properties of the composite blend will be inferior to that of chrome-tanned leather.

Finally, the literature review revealed that combining leather solid waste with recycled thermoplastic elastomeric polymers should result in composites with outstanding qualities [3,5,16,18,21,31]. No work related to leather-shaving fibers with recycled EVA polymer matrix composites with a 1:1 ratio has been reported by any researcher. The method of impregnating leather-shaving into a polymer in max. percent-wt. reinforcement of 1:1 is undoubtedly difficult. To enhance the 'homogeneity'/'proper blending' of the polymer and leather, however, in this work, appropriate additives, including lubricants/oils, paraffin and naphthalenes, have been used, resulting in composites with superior qualities for the desired multifunctional "value-addition" applications. From in-depth literary studies, it has been unveiled that no work has been made accessible on carrying-out experiments regarding the efficient utilization of leather waste (Leath shavings) with max leather fiber-loading of 1:1 in combination with recycled EVA polym matrix for footwear, transportation applications, etc., ref. [16,32–44]. As a consequence, th recommended suggestions of this research are valuable for the leather as well as polymer industries in order to significantly alleviate a load of disposal of solid waste to the greate extent possible. To date, no work has been reported on the utilization of solid leather wast as a reinforcement in the reused "EVA thermoplastic" matrix for the fabrication of leath waste-reinforced polymeric composites. Thus, the suggested concept of such research coul be substantially effective and compelling for the leather and polymeric sectors to eradica the strain of solid-waste management to the greatest extent practicable.

In addition, no literature has been made available in which researchers have tried modify the processing method of solid leather waste/recycled EVA thermoplastics matri based polymeric composites by adding preferable additives, namely paraffin lubricants/oi and naphthalene to enhance the appropriate blending of leather and polymer thus to produ composites with better properties for proposed applications. Limited work is available on th study of mechanical, thermal, structural, morphological, CRT, and cushioning characterist of leather waste-reinforced recycled/scrap thermoplastic polymer composites. Therefore, o portunities exist to investigate the characteristics of these leather waste reinforcement polym composites so as to explore their usage for multifunctional applications. It is also intende to see the effect of additives and plasticizers on the physicochemical, mechanical, therma volumetric wear-loss, damping, cushioning/shock absorption, compression, resilience, an Morphology analysis on the fabricated polymer composites.

Thus, the study examines the potential use of "leather shaving waste" as reinforci fibers in "recycled EVA polymeric-matrix" in light of their "intrinsic fibrous-nature", an "renewability characteristics". The intent was to explore the implications of "leather sha ing fibers" on the "recycled EVA matrix" based on its "physico-mechanical", "thermal "morphological", "chemical", "structural crystallization", "elemental mapping", and "t pographic image-mapping properties". From an industrial standpoint, "EVA" is wide employed in the production of "inserts", and "soles" in the "footwear sector", accounti for 18% to 28% of total usage. "EVA" is known for its "low compression-set", "abrasion and "tearing resistance" and its exceptional "physical" and "compression characteristics EVA is also used as a "cushioning material" in "athletic footwear" owing to its "lig weight", "resistance to compression-set", and "vibration-damping characteristics". A com posite composed of "recycled-EVA polymer" combined with "leather shaving waste" undoubtedly an attractive option, as it has the significant merit of being in accordance wi the underlying concept of "circularity" throughout the underlying "production chain" f "value-addition applications".

2. Experimental

2.1. Materials

As far as the "sample location" is concerned, Laxmi Polymers, a company in B hadurgarh, India, has supplied recycled EVA granules from the footwear industries. Aft confirming with the Gel Permeation Chromatography (GPC) analysis with the suppli for the "recycled EVA polymer", the average molecular weight of "Recycled polyethylen vinyl acetate polymer" (RPEVA) is 170,000 g/mol, and the "Melt flow index" (MFI) RPEVA can range from 0.1 to 45 g/10 min [45]. As shown in (Figure S1a,b), the solid leath wastes (Leather-shaving fibers) were gathered from the 'tanneries' of "Leather Complex "Jalandhar Leather India Pvt. Limited", 'Punjab'. Solid leather fibers have blended proper with the recycled polymer matrix with the usage of 'fillers' and 'plasticizers' such as "zi octadecenoate" and "octadecanoic acid" that have been procured from a local source, In pact Agencies, Industrial Estate, Jalandhar, Punjab. In the hot -press, "Poly-vinyl Alcoh (P.V.A.) and "Teflon" (P.T.F.E.) sheets wrapped around 'steel-molds' are used as "mol release agents" provided by "Sigma Aldrich chemicals" as part of the 'post-processing' an 'finishing operations'.

2.2. Fabrication of Composites

Fabrication of Thermoplastic Elastomer Polymer Composites, including Leather Flakes and Recycled EVA

The ratio of recycled EVA polymer and leather shavings in the compositions of leather polymer composites is 1:1. (wt. percent). The current study's methodology is shown in sequential sequence in Figure S2. The following Table S1 presents the formulations of "polymer composite" specimens in weight -ratios based on the maximal "leather-shaving" fiber loading of 1:1.

As per a particle size analyzer, the leather shavings had an average particle size of 500 microns. In accordance with the sampling process is concerned, the two-roll mill (manufactured/exported from the Ravi Engg. Works, New Delhi, India), normally employed for rubber compounding, has been purposely used to mix the components, and the experimentation has been carried -out at local manufacturer, Impact Agencies, Industrial Estate, Jalandhar, Punjab. Optimal huge-shear blending can be achieved by orienting two rolls facing one another at 368.15 K with a user-specified preset, configurable, or customizable gap to facilitate material permeation. The rollers' spacing/gap has been kept at 1 to 1.5 mm [16,39,40,46–48]. Usually, a rear-roll rotates more quickly than a front-roll. In this instance, the front roll rotates at 10 rpm while the back roll rotates at 15 rpm. To develop the necessary flexible -sheet with specifications of 185 mm × 185 mm × 3.5 mm, dies have been fabricated. After compounding, materials were introduced in lower-dies. The heat was applied to the dies, reaching 368.15 K. After cooling in the die for 30 min, the upper die was squeezed against the lower die with an applied -load of 20 tonnes. With the use of the ejector pin, the composite material that had been compounded was removed from the die.

Two roll mills were used to mix the materials for the creation of flexible composites under specific conditions. The uniform blend was transformed to a hot-press molding, Manufacturer: Techno Search Instruments, Model PF M-15 containing preheated molds with 20 cm × 20 cm with variation in thickness of 1, 2, 3, and 5 mm. Fillers and Plasticizers such as "zinc octadecenoate" and "Octadecanoic acid" were also utilized to optimize the processing variables. The necessary flexible visco-elastic damping composite sheets will be produced by hot-press, die-molding the fine particulate-matter blended mix. To squeeze out more water, a hydraulic press was used to push the moist sheet that had developed. As shown in Figure S3a,b, the "leather-fibrous composites" were "dried and hot-pressed" under the 90 to 120-degree centigrade temperature and 77 to 108 psi of pressure applied for 20 to 40 min.

2.3. Sample Preparation

The standards for the performance of leather-fiber composite materials are provided by ASTM/ISO/SATRA. It is imperative to understand that 'material characteristics' differs in relation to "specimen preparation", "specimen dimension", "speed", and "environment of testing". A series of experiments are necessary for this research project. Hence, the 3–5 samples from each category were examined. The equipment utilized, the different test techniques, and the specific standards employed in this investigation are all shown in Table S2. As shown in Table S2, the specimens for each test were prepared in accordance with SATRA/ISO/ASTM criteria.

To determine how much force it can withstand the fabricated material using the tensile test according to ASTM/SATRA/ISO standards. By using Instron equipment, the maximum specimen elongation can be determined. The variations of Stress-displacement and load-displacement can be calculated by using tensile test, which determines the tensile modulus. From the tensile test readings, we can come to which material can be used to withstand the load and design it so that it can pass the quality check control of materials. High aspect ratio and young's modulus with Tensile strength, in mixture with thermostability and other physico-mechanical characteristics, make them smart materials and open new thoughts for Smart materials.

The aforementioned artificial SATRA testing standards are preserved in a compute ized UTM machine. To determine the material's elongation and strength, the specimen i kept in the UTM at the right spacing and pulled until it breaks at a 60 mm/min speed.

2.4. Characterizations

To meet the specimen specifications set out by ASTM, SATRA, and ISO, strips of the bas matrix and its composites were prepared into different sizes. The samples were then measure for thickness at a minimum of four randomly chosen places, with the findings average Table S2 lists the specimen dimensions for various mechanical testing in accordance wit ASTM/SATRA/ISO standards. The details of the specimen, test technique, test significanc and testing equipment parameters are covered subsequently in the corresponding sections.

Moisture content, PH, chromium trivalent, and chromium content of leather shaving that were chosen specifically for the creation of composites were measured in order t ascertain their physicochemical properties.

2.4.1. pH of Leather-Shavings

To find the pH level, 5 g of the specimen was soaked in 100 mL "distilled water" fo about 16–24 h, and then the pH level was directly measured using the SLC 13 standar procedures (SLC, 1996) [28].

2.4.2. Moisture Content of Leather-Shavings

A "physical sieve-shaker" is used to produce the "fine-grade micron-sized particles" o the leather-fibers of less than 650 μm. The materials were heated for about 30 min at 100 °C in an 'oven' to remove the 'moisture'. These particles were initially dried to a moistur level of % in a small dry kiln operating between 95 and 100 °C. At 23 degrees centigrad and 0.5 "relative-humidity", properties were then measured. By measuring "3 mg of th sample in a crucible and then placing it in an oven for 5 to 6 h", the moisture content of th samples was then evaluated using the gravimetric method, SLC-3 (SLC, 1996). The result showed that the moisture content of the samples was 9 ± 2% [28].

2.4.3. Chromium Trivalent of Leather-Shavings

Chromium trivalent was identified using "SLC 8 in SLC 1996" [28]. A "5 mL of H_2SO 10 mL of $HClO_4$, or 15 mL of the blend and 15 mL of HNO_3 were added to 2 g of grin chrome shaving". Using a "hot plate heated to a moderate boil, a funnel was positioned o the flask in anticipation of the reaction blend starting to convert into orange". The 'flame' wa then reduced until the color change was complete. The mixture was then gently heated for a additional two minutes, quickly cooled in air, and then diluted to about 200 mL. To remov the chlorine interference, the mixture was further heated for about 10 min, allowed to coo and then 20 mL of 10% potassium iodide and 15 mL of orthophosphoric acid were adde to conceal the presence of any iron. This mixture was then allowed to stand for 10 min in dark area. Using 5 mL of a 1% starch indicator, "the sample was titrated against 0.1 N Sodiun thiosulphate until the solution in the flask turned light green (SLC, 1996)" [28].

$$\text{"}Cr_2O_3(\%)\text{"} = \frac{\text{"titration volume} \times 0.00253 \times 100 \times \text{correction factor"}}{\text{"Sample weight"}} \quad (1$$

where 0.1 N "Sodium thiosulphate ($Na_2S_2O_3.xH_2O$)" = 0.00253 g Cr_2O_3

3. Chemical Properties EVA-Based Composites

As per "IUC 8 and 18 test standards", the UV-Visible spectrophotometry instrumen (Inkarp, Sican 2600 Series model) provides the most appropriate, efficient, and precis methodology for analyzing "the chromium levels in leather savings and "Polymeric matri composites" (PMC's) [29,30]. Therefore, "the chromium content of the control sample an the leather fibrous composites was determined" as per "IUC 8 and 18 test-standards/ISC

17072-1:2011(E): IULTCS/IUC 27-1:2011(E) test procedure" [29,30]. A 100 mL of the "acid artificial-perspiration acid" (APA) solution was mixed with the precisely weighed 2 g of ground leather board sample. The specimen was then slowly shaken in water at 37 °C ± 2 °C for 4 h ± 5 min. This APA solution was then made by blending 5 g of NaCl, 2.2 g of H_6NaO_6P, and 0.5 g of L-histidine—monohydrochloride monohydrate. The 'extracted mixture' was filtered with a membrane filter before being filtered initially with filter paper. An appropriate amount of extract was obtained for analysis, and nitric acid at a concentration of 5% (by volume) was added for the direct measurement of the elements. The dilution factor took this contribution into account. To control impurities using the same process, blank samples were run concurrently with the sample. A sample container is used to collect an aliquot of acid sweat, which is then treated as a sample for all tests. The amount of chromium was measured as per the 'IUC 8 test method' with the help of a 'UV-visible spectrophotometry device (Agilent 700 Series)' [30]. Nitric acid was added to a 2 g sample, then 20 mL of a blend of perchloric acid and H_2SO_4 in a 70:30 ratio was added. Heat was applied to the mixture until an orange color appeared. After cooling the digested sample, 100 mL of distilled water was introduced and boiled for 10 min to eliminate excess chlorine. An additional 15 mL of orthophosphoric acid was added to prevent iron interference. To avoid "light interference" to the 'iodine', 20 mL of 10% "Potassium iodide" was mixed and preserved in a "dark area for 10 min". Using starch as an indicator, "the sample was titrated against 0.1 N Sodium Thiosulphate until a sky blue color was seen (IUC/SLC, 1996)" [28–30]". By immersing "5 g sample in 100 mL of distilled water and keeping it in an orbital shaking device for 16 to 24 h", the pH of the reinforcing agent and its composites were measured directly using the SLC 13 standard methods (SLC, 1996) [28–30].

4. Physico-Mechanical Characteristics

4.1. Tensile Strength

Tensile strength is one of the factors that can be evaluated as a generic feature even though it has no direct impact on the properties and expectations of the soling materials. The fracture force/cross sectional-area is referred to as tensile strength. After cross sectional-area test specimen, the tensile tester is filled with both jaws, as shown in Figure S4. The specimen is pulled with force to ascertain. The "percent-age of elongation at break" is also calculated as it is an important parameter to determine the flexibility of the material. The tensile test on composite sheets was conducted as per SATRA TM-137 1995, using a "computer-controlled INSTRON apparatus, model 3369J7257, with an optical-extensometer gauge-length of 50 mm" [49–53]. After that, the tensile tester was fastened to the leather composite sample. Jaw separation was limited to 100 ± 20 mm per min. The leather composite sample breaking point was recorded. Before beginning the instrument (L_1) and when the test specimen broke (L_2), the tensile tester's two separating jaws' distances were measured. The %age elongation at break and tensile strength were computed. These properties are crucially used to gauge the durability of "leather and footwear-related items". Four tests were conducted for every developed composite, and their average was taken for analysis.

$$\text{Tensile Strength}\left(\frac{\text{N}}{\text{mm}^2}\right) = \frac{\text{Fracture Force}}{\text{cross} - \text{sectional area}} \tag{2}$$

$$\text{"Elongation at break (\%)} = \frac{L_2 - L_1}{L_1} \times 100\text{"} \tag{3}$$

"where, L_2 = final length after fracture", and L_1 = initial length of the specimen.

4.2. Compressive Strength

The compressive test on composites was carried out in accordance with ASTM D-3410 [54]. The experiments were conducted on "computer-controlled INSTRON equipment, model 3369J7257, with an optical-extensometer gauge-length of 50 mm and a test-speed of 50 mm/min". Two cross heads are used to fix the test specimen in the loading unit, and the 'drive unit' supplies the force at the desired rate. Depending on the kind of test, replaceable load cells with capacities of 5 KN and higher are positioned above or below the cross-heads. There are several ways to adjust, control, or vary the 'speed' of the moving cross-head. The final physico-mechanical properties of the composites were calculated by averaging results from several observational and measurement methods.

4.3. Tearing Resistance

Tearing resistance reflects the ability of polymer material to resist tearing. For this tearing test samples were made according to "SATRA TM-218 1999/ISO 20344:2011" [55]. Tests were performed at 50 mm/min speed, and four tests per composite were conducted.

4.4. Hardness

"The penetration of a rigid ball into the test pieces under particular conditions" is measured using a device called a hardness tester to determine the hardness of a material. The Shore Using a shore test, the composites' hardness was evaluated by employing "Shore-A hardness tester model Digitest" using the "SATRA TM-205:2016/ISO 868:2013 test" standards [56].

4.5. Density

The density of the material is proportionate to its weight and is represented as "mass per unit of volume". This illustrates "how a material becomes lighter or heavier meaning denser the material resulting larger its density" and vice versa. To estimate the 'density' of flat materials, uniformly thick test specimens have been placed in a 'circular' or 'square' container. By measuring both the 'thicknesses' and the 'diameter/length', the volume may be computed. The mass of a sample is furthermore evaluated, and "the density is derived by dividing mass by volume". It is hard to ascertain "the volume of a moulded unit-sole dye to its uneven-surface". Nevertheless, the volume of such specimen soles is measured in both 'normal' and 'immersed' conditions. "As per the ASTM-D-792-00, the weight-loss in water indicates the volume of water-replaced, which is identical to 'sole-unit', and the 'density' may then be computed" as per equation given as under:

$$\rho = \frac{w_a \times 0.9975}{w_a - (w_w - w_b)} \quad (\cdot$$

where ρ = Density of sample, w_a = Wt. of the specimen while suspended in the atmosphere, w_w = Wt. of partially-immersed 'wire-holding' specimen, w_b = Wt. of the specimen when completely immersed in 'distilled-water', as well as the partially-immersed 'wire-holding' the sample, and the 'density' of the 'distilled-water' at 23 °C in gm/cm^3.

Samples were acquired in any form with a volume to calculate a density using the equation for volume, mass, and density.

4.6. Adhesion Strength

"The adhesion strength of the leather shaving and its PMCs was assessed Using SATRA TM 401 2000 at room temperature" with an aim to determine how well the lining adhered [57].

4.7. Compression and Resilience Test (CRT)

The Compression and Resilience Test (CRT) of the PMCs and leather savings was performed on PMCs to evaluate their compressive characteristics. The tests were conducted using "SATRA TM 64:1996 at room temperature" [46].

4.8. Abrasion Resistance

"The wear resistance of the leather shavings and its PMCs was evaluated using DuPont based unidirectional rotary drum abrasion model STM-140 UK tester" in accordance with ISO20871:2001 requirements [58]. At 30 revolutions per minute, composites with the prope specifications were attached to the coarse wheel. There were 1000 cycles of testing. A rotating cylinder filled with "50-mesh emery abrasive paper" was used to measure the weight loss caused by the abrasive paper. "The specimen-holding limb was propelled by a 1500 gm weight that provided a 2.03943 Kgf contact normal force and moved parallel to the (horizontal) cylinder axes". An alternative method involved cutting leather composite samples with a conventional die, weighing them, and clamping them firmly in an "abrasive disc against emery paper". The machine was now turned on, and measurements of weight loss were made after 500, 1000, 1500, and 2000 rotations. Volumetric wear-loss is therefore calculated based on the reduction in weight of the specimen prior to and after the testing.

4.9. Water-Absorption Test

The specimen used in this study is a sheet of composite plate measuring 25 mm by 25 mm. After thoroughly cleaning the composite plate, the sample of the specimen being tested was measured, and the initial weight was retained for reference. The sample was placed in a jar with one liter of fresh water, kept inside, and submerged entirely in water for 12 h. After that, the water sample was taken, and the percentage of water absorption was determined by monitoring the change in sample weight over time.

$$\text{"Percentage water absorption"} = \frac{\text{"Volume (mL) of water absorbed"}}{\text{"Weight (gm)of leather specimens"}} \times 100 \quad (5)$$

5. Thermal Properties

"The thermal properties of leather and its composites reflect its stability under high temperature". Given the high polarity of both the leather fibers and the matrix, interactions between the two may influence how the leather fibers melt. Therefore, in this section, thermogravimetry and DSC techniques are employed to investigate "the temperature-dependent degradation of the leather-filled composites as well as the melting behaviour of both neat, base polymer with/without leather fiber".

5.1. Thermo-Gravimetric Analysis (TGA)

The thermo-stability testing of base material and their composites were performed with the help of "TGA/DSC analyzer TA instrument, Waters Austria type Q50". "In a N_2 atmosphere, a sample of 5 mg was treated in an aluminium pan and heated from 30–800 °C at a rate of 20 °C per minute. As a minimum weight loss of the sample, the temperature in the TGA thermogram that corresponds to a 5% weight loss is used". The quantitative information on "reduction in weight decomposition, and the products created on decomposition" is provided by the TGA thermogram of leather shaving and PMCs.

5.2. Differential Scanning Calorimetry (DSC)

The samples were conditioned for 24 h at room temperature after being pre-dried for 10 min at 100 °C in an air-circulating oven. With a thermo-gravimetric analyzer, the thermal stability of leather shavings/recycled EVA composites and base matrix with/without any leather fiber was monitored (TA instruments, Waters Austria, model Q50).

Utilizing differential scanning calorimetry, the thermal characteristics of leather-shavings/recycled EVA polymer composites have been studied (DSC). The DSC2A-00837 (192.168.1.11) was used for the tests, along with standardized aluminum crucible pans and lid coverings. The cell environment has been injected or pumped with nitrogen. The first samples were evaluated at room temperature and then "quickly chilled to a temperature of about −25 °C and then raised up to 400 °C at a rate of 10 °C per minute". They were maintained at this temperature for two minutes. The specimens were then cooled to −20 °C and heated to 400 °C at a rate of 10 °C per minute. Re-crystallization and "melting temperatures" were calculated using "heat-flow indicators", which were utilized to determine the glass-transition temperature. The initial heating diminished the material's manufacturing background (T_g).

DSC provides details on the thermal transitions that occur in polymers, includin melting, the transition to glass, oxidation, etc. Additionally, shifts or modifications in th crystallization peaks can be used to analyze "the miscibility of the polymeric blend system

A Q200 from A TA Instruments, Waters Austria, DSC2A-00837 (192.168.1.11) Th melting behavior and thermal transitions of "neat recycled-EVA and leather fibers" whe blended with recycled EVA polymer were monitored using a "differential scanning calorim ter connected to a 990 thermal analyzer". The material was put into the DSC cell in a aluminum pan that contained around 20 mg of it. As a point of reference, an empty alu minum pan was used. "DSC examination was conducted under nitrogen environment wit a heating rate of 10 °C min^{-1} and temperature ranges of −25 °C to 400 °C".

6. Fourier-Transform Infra-Red Spectroscopy (FT-IR)

The samples' interfacial contact has been verified using FTIR (FT/IR-4700typeA Seria Number C016661788). In transmission mode, FTIR has been performed in a humidity-fre environment at ambient temperature. For the powdered material, spectra between 4000 an 600 cm^{-1} were recorded. By using ATR, solid samples have been examined (Attenuate Transmission Reflectance).

The chemical properties of the neat and leather shavings/recycled EVA polyme samples were analyzed using the FT/IR-JASCO, model 4700typeA. Using a SHIMAZU I affinity -IS spectrometer, "the FTIR spectra for collagen hydrolysate were captured wit KBr pellet". The samples were reduced to a fine powder, combined with KBr, and then pu in an IR cell that was stored in a container with a regular slit. As the IR and air spectr were being captured, the reference was used. "With a resolution of 4 per cm and a sca speed of 2 mm/s in the wavenumber range of 600–4000 cm^{-1} at room temperature, it wa captured in the transmittance-vibrational modes". This method generates a constraine compositional profile while measuring the surface composition, bonding, and structure.

7. Morphological and Elemental Analysis

Understanding the potential morphological changes in leather fibers during moldin at high temperatures, as well as the potential alterations at the interface between the matri and the fiber, was carried out using a "Scanning Electron Microscopy" (SEM) equippe with "Energy-dispersive analysis of X-ray" (EDAX), "a Phenom World PhenomPro mode and an accelerating voltage of 10 kV". Gold ions were used to sputter-coat samples (1 cm^2 acting as a "conducting medium" when they were scanned with a "scanning microscope made by Phenom World, model PhenomPro. The surface morphology of leather fibers, th x-section morphology of "leather fiber reinforced recycled EVA polymer composites", th "degree of fiber alignment, the degree of adhesion, the fusion of leather fibers with recycle EVA polymer, the uniformity of fiber dispersion, and the extent of fiber polymer adhesion have all been studied using SEM analysis. To determine the occurrence or existence o chemical-elemental compositions, E-DAX is examining the stoichiometric compounds an percent chemical purity of the specimens.

Without any further additives, untreated leather fibers (Shaving wastes) were include into "the recycled EVA polymer matrix and mixed either at ambient temperature or a 120 °C on a mixing mill that also served as a hot-press moulding machine". The "nea recycled-EVA" polymer with no leather fiber content and the composites made of leathe shavings and recycled EVA polymer was utilized for SEM morphological analyses after th mixed samples were kept for a period of 24 h.

8. X-ray Diffraction Analysis

An "X-ray Diffraction (XRD) analysis" was conducted to identify the chemical constituent or phases involved throughout the crystal structures of recycled EVA polymer composite The specimens were powdered with neat Recycled EVA material of 60 shore-hardness an leather-shavings/recycled EVA polymer material of 90.5 shore-hardness on a wide-angle X ray diffractometer (XRD), made with model name (Bruker AXS, ECO D8 Advance) at th

Chandigarh University, Mohali, India. Such diffractometer makes utilization of the copper (Cu-K) radiation sources that have been driven by 25 mA, and 40 kV. For attenuating an interference peak, K-beta filtering is employed. Divergence and scattering-slits of 0.4-degree were employed, including a receivers-lit of 0.2 mm. The composite components were positioned in a sample container, and the analysis was continued perpetually. The trial evaluation was performed by observing the diffraction-patterns between in the 2-theta ranges from 5-degrees to 90-degrees at a scanning-rate of 2-degrees/min and a scan-step of 0.05-degrees.

9. Atomic Force Microscopy (AFM) Analysis

An optical-microscopy, and an intermittent-contact atomic force microscopy (AFM, Make NT-MDT, Ireland, Model name, NTEGRA PRIMA) were used to investigate its molecular structure, leather-shavings/PMC phases; to analyze their cross sectioned area.

AFM has been utilized to study the adsorption of "neat recycled EVA and recycled EVA polymer composites on the leather fibers". The most powerful objective lenses and phase contrast microscope (Leica DM 750) were used for the microscopic morphological research (i.e., 100× lens).

A Cryomicrotome equipped with a diamonds-knife is being used to create test specimens for AFM experiments, and the samples were kept at −80 °C. A microscope was used to record "the AFM pictures in tapping mode (TM) under room-temperature". Ambient-air (AFM, Make NT-MDT, Ireland, Model name, NTEGRA PRIMA). "Commercial silicon-tips with a spring-constant of 24–52 N/m, a resonating frequency of 264–339 kHz, and an average radius of the curve throughout the 10–15 nm region was used as probing for the test". This study used the greatest sampled resolution accessible to generate phase-detection images. The recycled EVA thermoplastic polymer matrix will be enhanced by incorporating fillers, plasticizers such as stearic acid, zinc stearate, and other lubricants such as paraffins and naphthalene oils.

AFM has been utilized to determine the size (thickness, width etc.,) of the Interphases and their stiffness relative to the bulk phase of PMCs and their constituents. AFM has been utilized to examine the AFM Roughness Analysis and AFM Grain Analysis of the "neat recycled EVA polymer composites, and leather-shaving fibers reinforced with the recycled EVA polymer composites".

10. Results and Discussions

10.1. Moisture Content, Chromium Trivalent and Chemical Composition

As a result of the uniform-ragged powdery form-structure, the leather-shavings exhibit a "moisture content" of 5.12%. As a result, drying time for leather shavings waste must be increased, or another "drying method" must be selected further.

Additionally, UV-Visible spectroscopy was used as a quick method of detecting Cr(VI) in the "solid leather wastes/recycled EVA thermo-plastic polymer composites". The absence of Cr(VI) in thermoplastic elastomer-based polymer composites made from recycled materials or leather shavings is demonstrated by the results. Results from Tables S3 and S4 indicate that the tested leather shaving waste-reinforced recycled EVA thermoplastic elastomeric polymer composite samples did not contain any hexavalent chromium.

10.2. Physicomechanical Characteristics

The tensile slabs were molded in a "hydraulic press with electrically heated platens operating at 110 °C and 6 tonnes of pressure" to evaluate physical and mechanical properties. The mold was filled with the compounded sheet, which had a thickness of about 3.54 mm, and the platens were compressed to seal.

The percent leather shavings as fiber concentrations had an indispensable effect on the "physico-mechanical characteristics of the solid leather wastes/recycled polymer composites". The "neat recycled-EVA polymer" tensile strength was discovered to be 12.295 MPa. However, it was discovered that recycled EVA polymer composites using leather shavings as reinforcing fiber had an average tensile strength of 10.615 MPa. This behavior was attributed to compelling interface bonding or adhesion strength and the fact that the fiber agglomerations are closer together with a 1:1 ratio of reinforcement and bas matrix. Th stress may have been transmitted from one aggregate to another as a result of the recycle EVA polymer acting as an "adhesive-agent" between the "coalescence agglomerations Thus, it is abundantly obvious that the addition of leather shavings as fiber causes enhance ductility and a decrease in the "elongation at break" of the base matrix.

The elasticity of the composite materials can be shown in the % elongation at brea As shown in Figure 1, the "leather shavings/recycled EVA polymer composites" showe lower "percentage elongation at break" of 67.91 percent as compared to the control sampl i.e., "neat recycled-EVA polymer", which is discovered to be much higher at 192.16 percer This low value of "elongation at break" demonstrate how stiffer the leather fibers becom when combined with the recycled EVA polymer matrix. As the "leather loading in th polymer" grows, the failure process becomes more brittle [16,18,35,37,46].

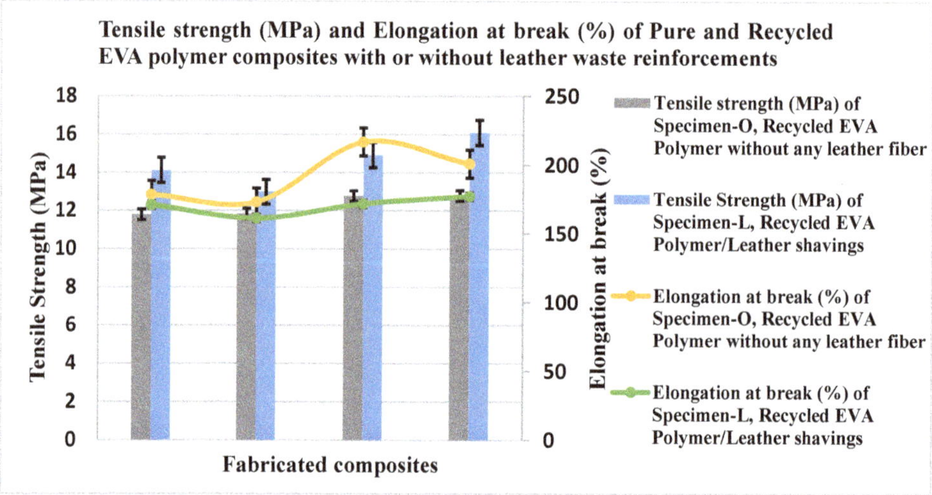

Figure 1. Comparison of the "tensile strength" of "neat recycled EVA polymer composites" an "Leather shavings/recycled EVA polymer composites".

The production of "crevasses, deformations, perforations, microscopic cracks enclosin the fillers, and the emergence of void cavities" due to a localized "detachment of the recycle EVA matrices from the leather fabrics" could be blamed for the poor elongation at brea values [35,37,39,40,47,48]. As the composites get closer to brittle failure, there is a significar reduction in the elongation at break values. When filler loading is large (1:1), voids develop th compromise the structural integrity of the composites. The matrix and fiber become separate when the voids expand and interact with one another. The reinforcement provided by th leather fibers is outweighed by the increased contact between the nearby spaces. Additionall each filler particle experiences normal stresses while under a tensile load. The most vulnerab locations are those where these pressures are greatest since that is where "the separation c the recycled EVA matrix from the leather fiber is most likely to begin". The leather fibers ma also selectively form a cohesive network between themselves as the loading increases an reach up to 1:1, which causes a weak interfacial bonding with the "recycled EVA matrix" an poor mechanical qualities. Even so, "the leather shavings reinforced recycled EVA polyme composites" still display an exceptional outstanding mechanical strength when compared t the features of these composites with the prior discoveries [16,18,35,37,39,40,46–48].

The related results in context with the distribution of stress from one aggregate t another one have fervently been reported by Ambrósio et al. (2011), where, as the percent c leather-particulates in the composite materials keeps increasing, the tensile-strength gra ually increases [37]. The physico-mechanical characteristics of leather-fibres/polymeri composites have been ascertained not only from the properties of the polymer-matrix and reinforcing leather-fibers, as well as by the interfacial seen among the matrix and fiber-reinforcement, which would be absolutely crucial for the transmission of inter-facial stress [37]. A sudden drop in the tensile-strength is a widespread occurrence in thermoplastic materials loaded with natural-filler particles; as the filler-content rises, thus also continues to increase the interfacial contact-area, the interfacial-adhesion, and bonding-strength between both the polar-solvents based-filling materials and the matrix-polymer deteriorates/decomposes [59–61].

The coupling inhibitors and compatibilizing-agents enhance the tensile strength of composites by lowering the gap between the tensile strength of "neat polymer-composite matrices" and that of the composites with "unprocessed natural fabrics" [37]. An enhancement in the tensile-strength implies that stress is transmitted from the matrices to the filling additive materials [37].

According to the SEM-fractography analysis, the findings have indicated that only certain cryogenic-fracturing cracks, fissures, and deformations of the leather/poly(vinyl butyral) (PVB) specimens happened via aggregates or coalescence or agglomerations, revealing that the interfacial-interaction transmitted the stresses of poly(vinyl butyral) (PVB)-matrices to the leather-fibers [37]. Even though enhancing "the wt.% of leather-reinforcing in the composite materials diminishes tensile-strength, excellent interfacial-bonding interaction among the PVB-matrix with the vinyl-alcohol-hydroxyl as well as R1-C(=O)-C-R2 functional-groups and the leather-fiber with their hygroscopic-collagenous fibril molecules might well have improved strength properties" [37]. This could lead to a favorable equilibrium and, as a result, an enhancement in physico-mechanical characteristics.

As the wt.% of leather-fibers tends to increase, the gap-distances/spacing between the fabric agglomerations reduces, which can be evident in the SEM-fractography analysis for the composites with max. percent wt. of leather-loading [37]. Due to the obvious superior interfacial-bonding and crosslinking-density mechanisms of poly(vinyl butyral) (PVB) with the leather-fiber aggregates, the PVB may very well have behaved/functioned as an adhesives binding-agent among both of the agglomerations, thereby distributing the stresses of one coalescence conglomeration to the other one [37].

The leather-fiber content had a significant effect on the physic-mechanical properties of leather/poly(vinyl butyral) (PVB) composites. The tensile strength of composites was lower than those of base PVB. Whenever the volume of leather-fibers in the composites has been raised, the tensile strength is considerably enhanced to a substantial extent [37]. This mechanism was attributed to strong interfacial interaction with a firm bonding and the notion that perhaps the fabric-agglomerations get nearer to one another as the wt.% of leather-loading rises. As a result, a PVB might well have functioned as a bonding agent between the agglomerations, transmitting stress from one to the next [37]. The tensile-modulus of the composites increased substantially as the quantity of leather-fiber within composites increased. This phenomenon was supported based on the premise that the leather-fiber agglomerations are rather near to one other, posing the possibility of entanglement [37]. Because of the elasticity of the leather-fibers and the matrices of plasticized-PVB, the leather-fiber agglomerations might even get nearer and undergo tangling or entrapment as the wt.% of leather-loading increases [37]. The entanglement agglomerations encompassed by the PVB-matrices have the potential to significantly improve the elasticity, modulus, and strength of the composites and, thus, the young's modulus [37].

If the matrix and the fiber have sufficient adhesion, the interfacial strength is often improved. The internal resistive power to come apart the sample is controlled by the "stress-concentration" across the region of the "nano-additives or nano-filling agent" if the "interfacial strength" is greater than the matrix "cohesion force". In these circumstances, it is common to see an increase in modulus, a sharp decrease in elongation at break, and an increase in tear strength. Since the variation of tensile strength relies on the loading, the kind of fiber, the makeup of the matrix, and other factors, it normally does not show any particular trend. When the fiber concentration is below the threshold volume percentage, composite materials can have tensile strengths that are even lower than those of the unfilled matrix [62]. However, as the fiber concentration rise above the "critical volume fraction", mechanical properties get better.

The comparable findings have been observed by Nanni et al., wherein, Poly(1 aminododecanoic acid lactam) and thermoplastic poly-urethane-predicated composition revealed desired behavior (about +47 percent modulus and +40 percent creep-resistanc even though not-optimum fabric-matrices interfacial-adhesion, as well as inadequate def bration of leather-fiber wastes, substantially decreased strength along with elongation-a break [38]. Eventually, the thermoplastic elastomeric-based specimen behaved the poores leading to a complete lack/absence of compatibility interaction and affinity between bot the polar-molecules-based-leather-waste fibers as well as the non-polar based-polyme matrices [38].

Leather fiber agglomerates could be brought closer together and demonstrate an unbrea able link thanks to the flexibility and toughness of the "leather fibers and the plastic recycle EVA matrix", as depicted in Figure S5a,b. Leather shavings could also be entangled as muc as possible according to these properties. According to Ambrosio et al., the linked collectiv "covered in the recycled EVA matrix could significantly improve the stiffness, elasticity, an stability of the composites", resulting in an elastic module (2011) [38].

Dodwell (1989) claims that "expensive and light footwear could have been made fror "leather-boards having a tensile strength of 4.0 MPa, fashionable, aesthetically pleasin and practical footwear could have been made from leather-boards having tensile strengt of 5.5 MPa" [63].

A wide range of behavior possibilities is also shown in Figure S5a,b, ranging fron "hard, brittle" to "ductile", with a yield point similar to that of a "thermoplastic polyme when subjected to mechanical agitations (stresses) [64].

The matrix had already effectively received the stiffness of the reinforcement. Accor ing to Covington (2009), the higher-modulus at a 10% waste-content could be explained b the HDPE90/Chrome tanned wastes' amorphous form after being chilled from melt [47].

The "neat recycled-EVA" polymer sample's tensile-fractured micrographs are shown i Figure 2a,b. The surface is lined and filled with "bulged, twisted, and elongated materials which appears to be a sign of such "lateral-pressure bearing strength, significant deform tion capacities as might be anticipated for relatively flexible recycled EVA co-polymer It also supports the "enormous strains as measured through experimental studies wit a tensile strength of 12.295 MPa and percent-elongation at break of 192.17 percent as th comparable values by Stael et al. (2005)" [39]. Figure 2a,b has illustrated the fractograph surface-morphology analysis, which evidently unveiled that the surface is encompassed b stretch material, which seems to be an indication of the significant distortion abilities tha might be envisaged with the exceptionally elastic and resilient EVA polymer, so it appear to confirm the huge strains analytically reported. Comparable findings have been reporte in prior studies conducted [16,21,26,35,37,39,46–48,65].

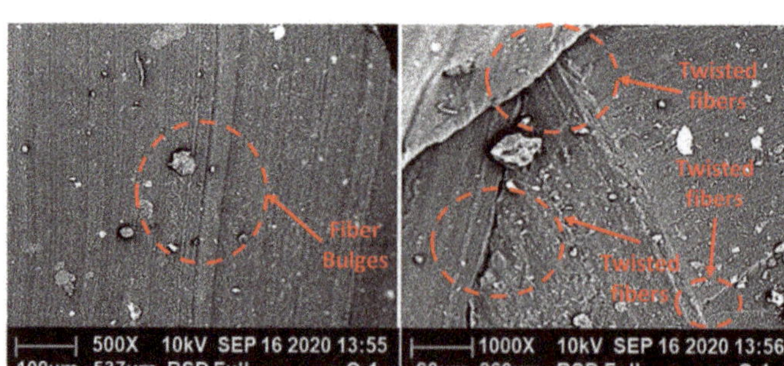

(**a**) (**b**)

Figure 2. "SEM tensile-fractured micrographs of Neat recycled EVA polymer without leather fiber at a magnification of (**a**) 500× and; (**b**) 1000× Adapted from the reference [45].

Figures 3a,b and 4a,b shows certain perforations, fractures, fissures, holes, and voids in the surface as well as the cross-sectional area of the recycled EVA polymer surface in addition to the collagen fibrils-bundles. This may be caused by fibers from leather shaving waste being removed after tensile testing, as shown in Figures 1 and S5a,b. It was simple to identify the cause or position of fiber rupture or deformation close to pull-out apertures, indicating poor adhesion between leather shaving fibers and recycled EVA polymer. However, increasing the number of leather shaving fibers in composites improved their permeability, bonding, and interfacial-adherence to recycled EVA polymer. In fact, as demonstrated in Figures 1 and S5a,b, the distribution of leather shaving fibers in composites made of recycled EVA polymer and leather shaving fibers was fairly intense. The leather shaving fibers appeared to be getting smaller, proving that the bundle of leather fibers had been broken up into individual single fibers. As a result, every fiber in the recycled EVA polymer matrix can effectively shift and reorient, resulting in a reasonable distribution of leather-shaving fibers. However, as leather fiber volume increased, the region of the stress-relaxation gradient curvature was dramatically boosted by a decrease in fiber-matrix interaction. Additionally, the strong bonding between the leather shaving fibers and recycled EVA polymer shows a remarkably large improvement in the physico-mechanical properties of the composite materials made from leather shavings and recycled EVA.

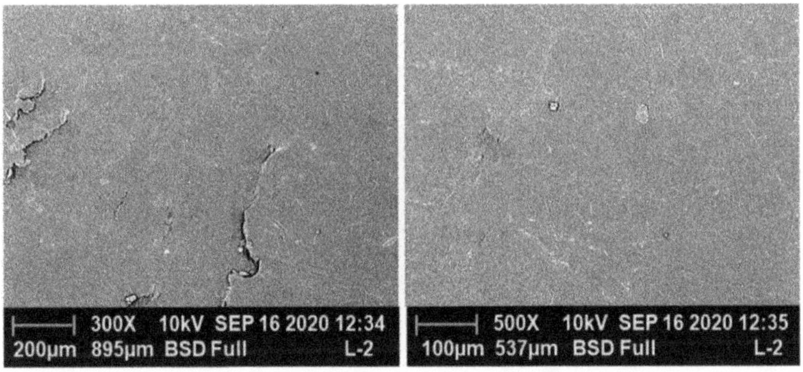

(a) (b)

Figure 3. "SEM tensile-fractured micrographs of the surface of Leather Shavings/recycled EVA polymer composites" at a magnification of (**a**) 300× and; (**b**) 500×.

(a) (b)

Figure 4. SEM tensile-fractured micrographs of the cross-section of "Leather Shavings/recycled EVA polymer composites at magnification of, (**a**) 1000× and; (**b**) 500×".

Leather shavings/recycled EVA polymer composite, a representative specimen, ha a lower average compressive modulus, measuring 63.63 MPa. As shown in Figure 5, th average compressive modulus of "plain recycled EVA polymer" without fiber conte was determined to be 66.695 MPa. Though the compressive modulus is a measure c "the material's stiffness, and higher is the compressive modulus which leads to stiff in the recycled polymer composites" [16,21,26]. The comparable outcomes have bee unveiled by the prior studies wherein the investigators have enumerated that brittleness a characteristic of polymeric materials that have been simply cracked, fractured, distorte ruptured, and/or shattered [16,21,26,46]. As brittleness is characteristic of the materials th fracture once stressed but seem to have minimal ability to degrade/deform prior to fractu or breakage. Brittle materials exhibit minimal deformation/distortion/cracking, a lowe capability to withstand/bear the shock as well as vibrating deformation, a relatively high compressive-strength, and a poor tensile-strength [16,21,26,35,37,39,46,47]. At the sam time, stiffness seems to be a property that indicates resistance to cracking and therefo is considerably easier than physico-mechanical strength, which has been associated wit failure breakdown [16,21,26,35,37,39,46,47]. Also, stiffness is a polymeric characterist represented by the Flexural-modulus as well as bending elastic-modulus. The three-poi bending-test was utilized to measure a material's stiffness/resistance to bending whenev a loading is acting perpendicular towards the longest side of a specimen. The stiffer an denser the materials, the significantly larger the bending modulus of elasticity; the less the bending modulus, therefore the more flexibility it is indeed [16,21,26,35,37,39,46,47]. A reported from the literary sources wherein the difference between stiffness and brittlenes has been reported in such a manner that the stiffness alludes toward the leather polymer composites that are rigid/stiff, inflexible, unyielding, difficult to bend, as well as inflexibl whereas, brittle pertains to that of leather polymeric composites that would be inflexib and likely to fracture, crack, shatter, or rupture quickly underneath applied stress c pressure [35,37,39,46,47].

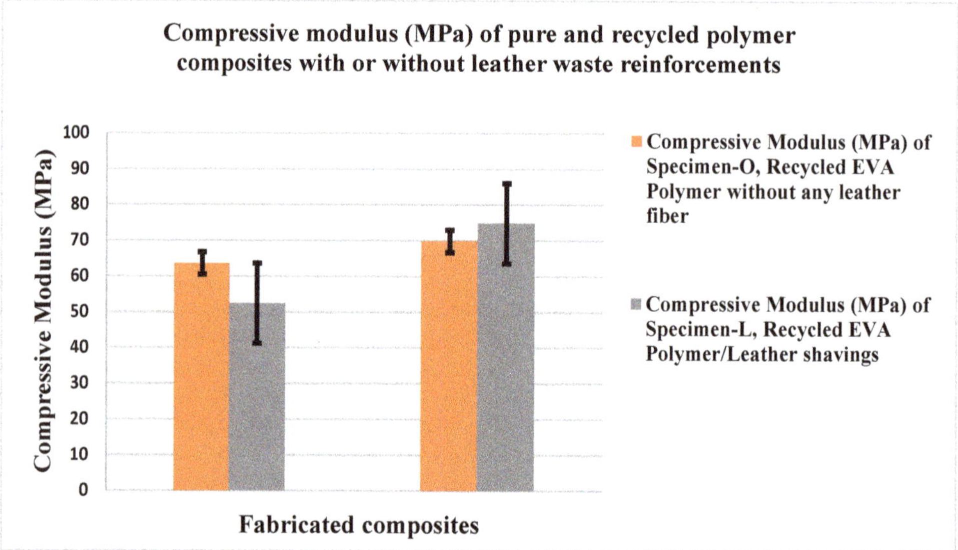

Figure 5. Comparison of "Compressive modulus (MPa)" for "neat-recycled EVA" and "recycled-EV polymer composites".

As shown in Figure S6, the compressive strength for the leather shavings/recycle EVA polymer composite tends to be somewhat higher (9.88 MPa) than the plain recycle EVA polymer (9.87 MPa).

The "stress-strain profile of thermoplastic polymers" is consistent with the nonlinearity of the stress-strain curve. According to Musa et al. (2017) [66], the control (HDPE) had a higher strain than its composites because the filler had not been used to temper the ductility. The stiffness and brittleness of a filler have infused its stringency into some of the "most ductile matrix (HDPE)" that results in reducing strain at minimal waste content (within 10–20 wt.% of fiber). Results also demonstrate that the composite "HDPE90/Chrome-tanned waste at 10% has higher stress and strain than the control (HDPE)". Due to the presence of natural rubber, the strain was further impacted (serving as an extender), and the filling dispersion in the matrix was facilitated. As a result, "HDPE90/Chromium-tanned waste with a 10% loading had become more robust and ductile".

The "solid leather waste/recycled EVA composites" tear strength also shows how dependent this feature is on the leather waste, which has the highest fiber concentration in the composites. The average tear strength, however, was discovered to be around 17.5575 N/mm when leather shavings were used as reinforcement fibers. The average tear strength for "the "neat recycled-EVA" polymer was determined to be 9.48 N/mm", as shown in Figure 6. According to Madera-Santana et al. (2004) [16], the highest leather fiber percentage increases tear resistance. This is due to the collagenous (proteinous) fibrous shredded form of "leather shavings" in the "recycled EVA matrix" being assimilated ("uniformly dispersed/blended with matrix").

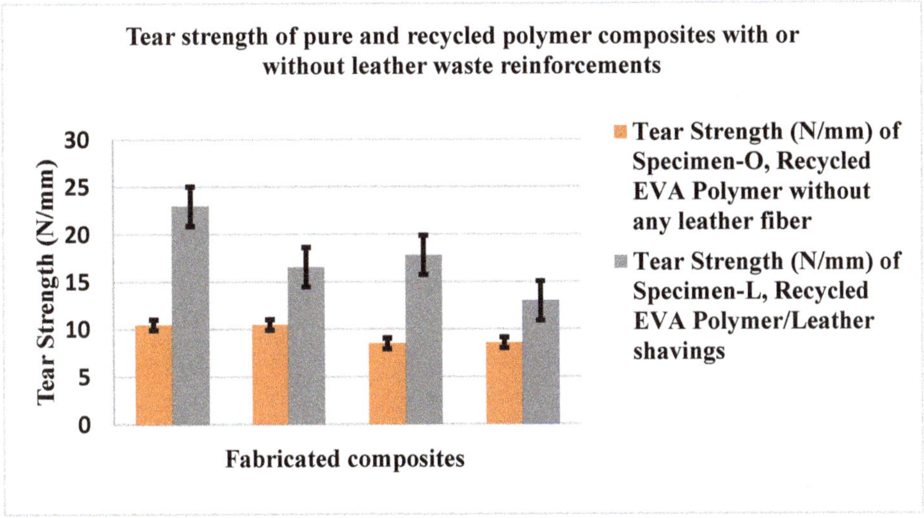

Figure 6. Comparison of "tear strength (N/mm)" for "neat-recycled EVA", and "recycled-EVA polymer composites".

The presence of "additives, according to Ambrósio et al. (2011) and Musa et al. (2017), has increased the fiber dispersion in HDPE-matrix, enabling superior compatibility and interacting between the waste fibers and matrix" [38,66]. The main function of the additives was "to act as a chemical group, creating an interface between the matrix and the fiber surface that allowed for effective stress transfer". A lack of a wetting surface could cause the deterioration of greater waste contents by drastically reducing stress transfer through the interface.

According to Ravichandran and Nmg (2005), the mechanical characteristics of "NBR/PVC composites filled with treated leather fibers" have significantly improved [67]. The elasticity modulus of the mixes is increased when leather fibers are added. Ammonia-treated leather shavings significantly increase tensile strength and tear strength. It is possible to ascribe the strong polymer-leather contact and effective vulcanization properties to the increased modulus and decreased elongation at break values. It is also well known that a composite alway has a larger effective surface fracture energy than an empty polymer. Dispersed leather fibe lengthen or delay the crack propagation process. Additionally, they take in some of the energ which slows down the matrix's deformation. The enhanced "tear strength" of the composite with leather filling makes this process more obvious. This is generally in agreement wit the observation that short fiber additions enhance the "tear strength" of elastomers. "Lo elongation at break" values could be related to the development of crevasses, puncture microscopic cracks surrounding the fillers, and the "emergence of void cavities" induced by localized detachment of the matrix from the fibers [67].

The mean peel strength among the polymer to synthetic adhesive based-fabrics fo leather shavings as fiber in "recycled EVA matrix" was higher and was found to be aroun 1.3125 N/mm, according to Figure 7 "(as the "Adhesion-peel strength" is the 'anti-strippin property from 'polymers' to 'leather'/'Adhesive bond-strength' or 'peel-test' is performe to estimate the 'strength' of a 'bond' by ascertaining the force that the 'force' is necessary t "peel-apart" the 'fused materials')". This is because after milling, a significant amount c leather shavings has been obtained, enhancing the interaction between shaving waste an recycled -EVA. While the 'neat recycled-EVA' was found to have a reduced adhesive pee strength of about 0.675 N/mm.

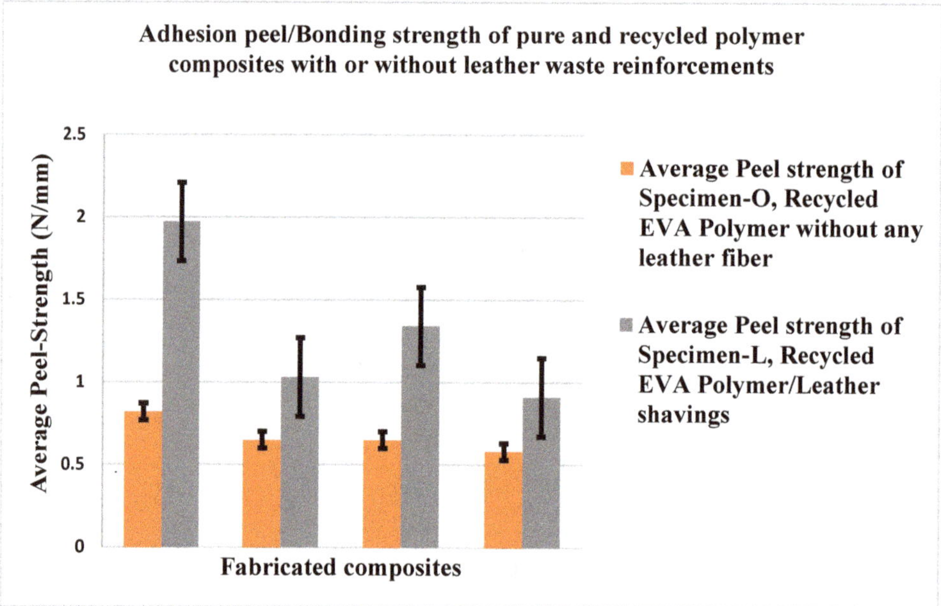

Figure 7. Comparison of "adhesion peel-strength (N/mm)" for "neat-recycled EVA" and "recycle EVA polymer composites".

According to the test results for 'neat recycled-EVA' polymer, the combination of a upper and a sole has weak adherence. In this case, separation or disintegration of th 'adhesive-film' at the surface was noted in relation to textiles made of synthetic adhesiv based materials. This may also be due to 'insufficient roughing', 'inadequate surfac preparation', and 'insufficient drying-time'.

When evaluating the adhesion among "recycled EVA polymer to synthetic adhesiv based-fabrics and solid leather waste reinforced recycled EVA polymer composites t synthetic adhesive based-fabrics", the results clearly show that the upper surface fabric failure occurs in the case of the 'neat recycled-EVA' polymer composites, as shown i Figure 7. While throughout the experiment, the leather fibers in the 'recycled-EVA' polyme composites with synthetic adhesive-based fabrics and leather shaving waste experience adhesion to top surface failure.

"Collagen-hydrolyzates" was produced by chrome-shaving during the early phase, claim Ali Shaikh et al. (2017) [66]. It was a viscid, frequently sticky, slightly cream- or pale-yellowish, odorless gel-like substance with a pH of 8.15. Due to its intense interfacial-bonding strength, this solvent-based substance was extensively used in the leather and garment industries. The drying process takes only around 10 to 15 min because the adhesive is obviously present, which is helpful for future improvement, efficacy, and functioning. As a result, there was a good chance of getting amazing results from the study that was done to create solvent-borne sticking agents. The four samples were used in the studies, which included different amounts of poly-vinyl alcohol (1 to 6 percent) and poly-vinyl acetate (1 to 16 percent). All four of the aforementioned samples, including the one stated above, showed average adhesiveness and bonding ability. After incorporating the adherent into a leather specimen, the "SATRA-TM416 test" was used to measure the bonding strength. The mean "peel strengths" of the specimens were "0.00312, 0.00325, 0.00295 and 0.0025 N/m", according to the results [68].

The thickness of the leather fabric has an impact on "compression resistance", and "hot-press/rolling mill processing", which is intended to improve the finish, is likely to improve the handle. The compressional resilience will decrease as the fiber becomes softer.

According to the findings in Figure 8, "the compression deformation property for leather shaving wastes as fiber in the recycled EVA matrix was greater by about 6.69 percent as compared to the neat recycled EVA matrix, which was found to be about 1".

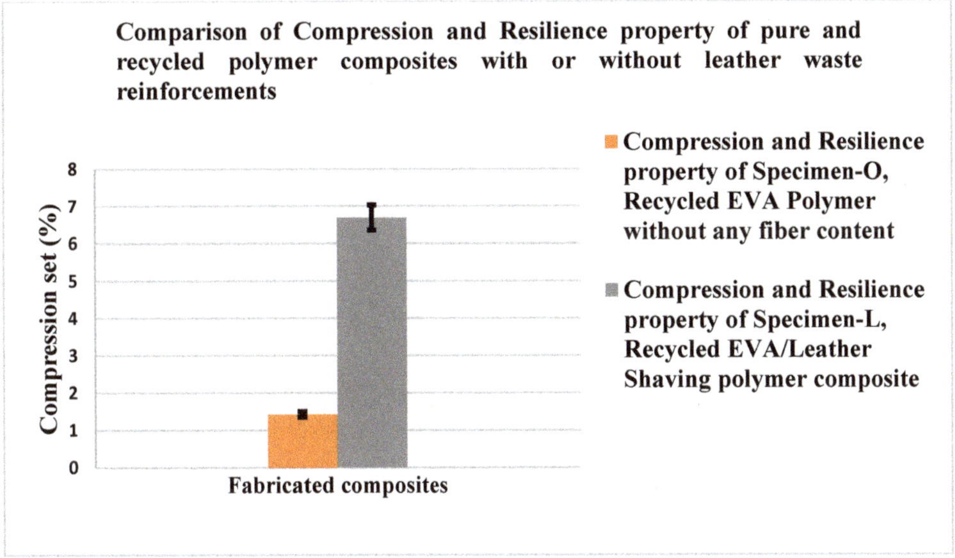

Figure 8. Comparison of "Compression and Resilience (%)" for "neat-recycled EVA" and "recycled-EVA polymer composites".

To determine "the compression deformation properties of "neat recycled-EVA", as well as leather waste/recycled EVA polymer composites, the Compression and Resilience Tester (CRT) of the recycled EVA polymeric composites" was investigated.

This 'CRT test' demonstrates that the 'resilience-energy' and 'load-compression' capabilities of the "leather shavings/recycled EVA composites" are noticeably better than those of the "neat-recycled EVA polymer composites".

Composite materials have superior elastic recovery when the compression set is lower. The fullness and compressibility of shaved leather fibers increase with higher compression energy levels. The values of compression energy rise as the thickness values grow. Or possible indicator of softness is the compression-decompression curve's linearity. A leath fiber with a softer grip has a lower linearity of compression. It follows logically that "th linearity of the compression and compression energy values" should be used to descrit the softness of solid leather fiber wastes.

It has been discovered that as the fiber thickness of solid leather waste increases, bot the "linearity of the compression" and "compression energy" values rise.

To increase compression and toughness, it is essential to prevent chain slides (CRT To lessen chain sliding, both the chemically-bonded polymer and organically modifie nanoclays can be employed (organo-clay). The reduced "molecular-mass tri-methoxy-sily modified poly-butadiene (Silicon Hydride)" has been further used to chemically bond pol mer and organoclay, according to Park et al. (2008)'s analysis [69]. The "poly(ethyleneviny acetate) co-polymer and the ethylene, polymer with 1-butene" can both be silane-grafte using peroxide [70–76]. According to Park et al. (2008) [69], "polymer with 1-butene rac cals" was created as a result of peroxide interactions with "poly (ethylenevinyl-acetate) an ethylene". These produced radicals may combine with "poly-butadiene of silicon hydride This led to the development of ethylene polymer with "1-butene" and "silicon hydrid grafted poly (ethylenevinyl-acetate)" [69]. Organoclay hydroxyl groups and silicon hydrid silanol groups can react. Although CRT is one of the most important qualities for foam appl cations, "poly (ethylene vinyl acetate)/Ethylene, polymer" with "1-butene foams with cla addition" should have better compression-set properties [69]. Park et al. (2008) employe "chemical-bonded polymers and organoclays by Silicon Hydride in his research to restri chain-slip across the clay substrate material since it was stated that the polymeric backbor chains slipping across the clay surface was the reason of the poor elastic recovery" [69 Because of this, "poly(ethylenevinylacetate)"/'Ethylene', 'polymer with 1-butene'/'meth tallow bis(2-hydroxyethyl) quats-nanoclays'/'Silicon Hydride-foams' either with or wit out" 'cis-Butenedioic anhydride-grafting', and 'poly(ethylenevinylacetate)'/'Ethylene polymer containing 1-butene polymeric foams" have significantly lower compressio sets According to CRT results, "methyl tallow bis(2-hydroxyethyl) nanoclays" are sup rior to "di(hydrogenated tallow)di-methyl-ammonium-chloride nanoclays" for enhancin compression-set because they contain hydroxyl-groups within the organo-clay layer an "alkyl-ammonium ion hydroxyl" groups within the active layer [69]. Based on the absenc of the large pinnacle at 3398 cm^{-1} for "poly (ethylenevinyl-acetate)/Ethylene", polyme with "1-butene/methyl tallow bis(2-hydroxyethyl) quats-nanoclays/Silane" and "pol (ethylenevinyl-acetate)/Ethylene", polymer with "1-butene/Di(hydrogenated tallow)d methyl-ammonium-chloride" qu (2008) [69].

Abrasion resistance is a crucial quality requirement in any situation where resilienc is crucial, such as in "automobile" and "upholstery leathers". Abrasion is significantl influenced by the physical-mechanical properties and size of the leather particles. The ma determining factors that affect abrasion are the type of leather, its "fineness or flowabilit and the length of the fiber. Higher elongation, fatigue strength, strain rate, elastic recover and work-of-rupture or breaking characteristics in leather fibers" enable them to withstan frequent, repeating distortions and attain higher levels of abrasion resistance.

Abrasion resistance is significantly predisposed by the "leather fabric thickness" an mass per square meter, which are important structural features of leather materials. Th enhanced values of these parameters determine the relatively greater abrasion resistanc It yields improved "abrasion resistance" due to the high 'surface-area', which maximize 'contact-points' with the recycled-EVA matrix.

Utilizing a rotating drum-type abrasion tester made of leather soles, "solid leathe waste/recycled EVA composites" was tested for abrasion resistance. Results showed th "when compared to neat recycled-EVA polymer composites, the volumetric wear los for the leather shavings used as fibers in the recycled EVA matrix was higher at abou 237.53 mm^3", as shown in Figure 9. "The thread-like collagenous fibrillary-like filamen throughout the composite" have been worn almost in proportion to their content as th percent-wt. of leather has increased, revealing a confirmed "linear increase in wear-rate" as described by Ambrósio et al. (2011) [38].

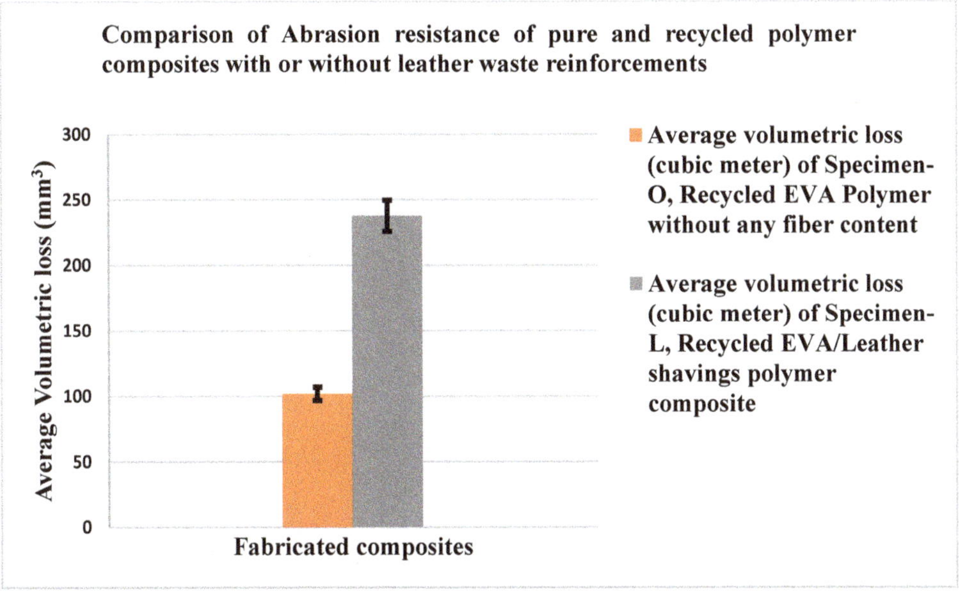

Figure 9. Comparison of "abrasion-resistance (mm^3)" for "neat-recycled EVA" and "recycled-EVA polymer composites".

The 'neat recycled-EVA' composites have exhibited a volumetric wear loss of about 101.75 mm^3 due to their "high elongation, work fracture, and elastic recovery".

"The shape, structure, and type of additive elements, such as zinc octadecanoate and Octadecanoic acid, and the resulting discontinuities in the matrices may serve as evidence of the reduction in wear resistance, tensile strength, percentage elongation, as well as flexing resistance with increasing waste filler inclusion". The void cavities can multiply in size, form, and relationship to one another due to abrasion, strain, and stretch, which results in materials debonding and fracture/rupture failure.

For testing composites' hardness, an indenter driven into the material is used. The modulus of elasticity and visco-elastic properties of materials has an impact on the hardness indent because it is inversely proportional to a penetrating value.

"The Shore A Hardness strength for leather shaving waste used as fiber in the recycled EVA matrix was determined to be approximately 90.5", as shown in Figure 10.

However, when compared to composites made of "recycled EVA and leather shavings, the 'hardness strength' for "neat recycled-EVA" was found to be much lower, at roughly 60. The Beach, The number of leather fibers added significantly increases the hardness of composites made of recycledEVA and solid leather waste. According to Ambrósio et al. (2011) [38], leather filaments have a harder surface than plasticized PVB matrices, which is where the growing trend in hardness originates from.

But the increase in modulus and hardness suggests some degree of interaction between the polymer and the fibers of the leather. These characteristics, together with "the high hardness ratings of these blends, may be due to the uniform distribution of the leather fibers".

"The composite's resistance to elastic deformation on its surface increased as the proportion of chrome-tanned wastes improved from 0 to 40% loading" [66].

This resilience, however, diminished as the proportion of wastes browned in chromium rose.

In general, the "density" and "voids content" of composites have a significant impact on the amount of water absorption. There is a significant increase in "water holding capacity" with increased leather-shavings 'fiber-length'. As a consequence, it is clear that the speed of water absorption increases with leather-fiber loading, and leather shavings have a high percentage of water absorption (about 14.07 percent) compared to "neat recycled EVA polymer composites" (1.56%), as shown in Figure 11.

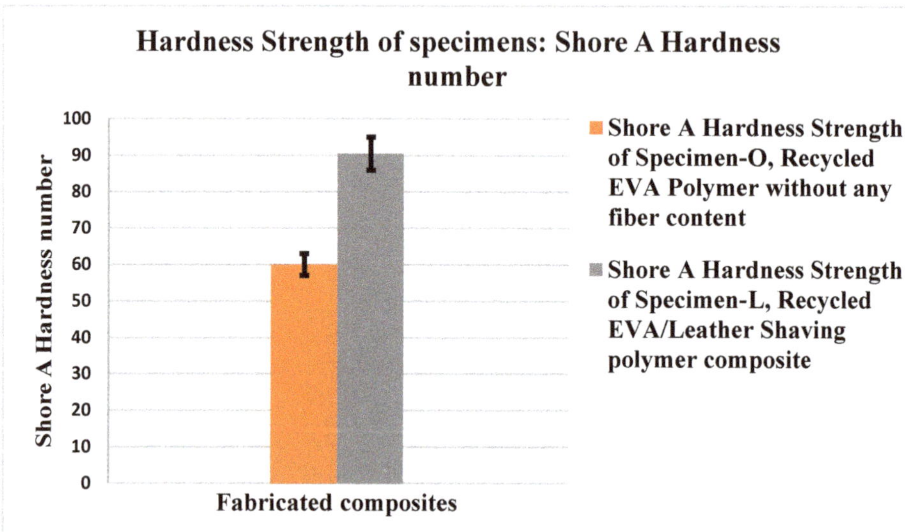

Figure 10. Comparison of "Hardness (Shore A)" for "neat-recycled EVA" and "recycled-EVA polymer composites".

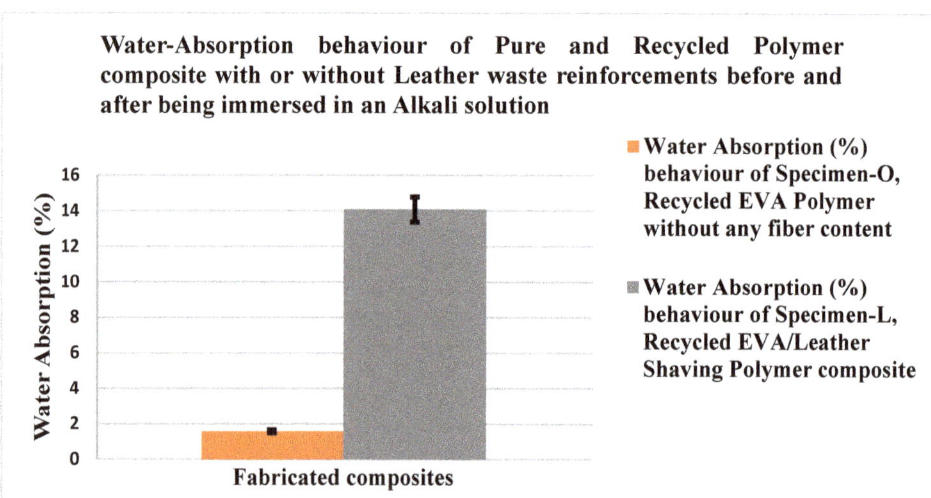

Figure 11. Comparison of "water-absorption (%)" for "neat-recycled EVA", and "recycled-EVA polymer composites".

Given that the leather fibers are hydrophilic by nature, the composite rises in water absorption as the percentage of leather shavings loading has been escalated to 1:1. In the 'footwear' and 'apparel' industries, this property has significant commercial significance.

Although a larger percentage of water uptake is observed in composites developed of chrome-tanned wastes and HDPE, the percentage of water uptake increases when the number of chrome-tanned wastes grows dramatically [66]. This suggested that certain additives, such as zinc octadecanoate and octadecanoic acid, were, in fact, responsible for the development of additional microgaps, abnormalities, irregularities, and other flaws in composite materials. This might also take into account the reduced tensile-modulus that Musa et al. (2017) [66] reported.

10.3. Thermal Studies

10.3.1. Thermo-Gravimetric Analysis (TGA)

The primary elements of leather filaments are "collagenous protein, which is composed of long amino acid chains, chromic oxide, and other biological substances. To maintain chemical stability, the right surface texture, and resistance to deterioration/decomposition by fungus and micro-organisms/microbes during usage, these substances are added to leather". TGA and DSC were used, respectively, to evaluate the thermal properties of the composites. A minimal weight loss of the sample is defined as the temperature in a TGA thermogram that corresponds to a 5% weight loss.

"The lowest weight loss of all the samples was found to be in the 211–289 °C range, indicating that the generated composites are quite thermally stable, at least up to 211 °C".

Figure 12a,b illustrates the TG and DTG curves at distinct heating-rates of the leather-shavings in the N_2 environment, respectively. These graphs exhibit the characteristic form pattern of leather specimens: three-phases could be identified throughout the heating-phase (correlating to areas 1–3 as indicated in Figure 12a). As per the TG-FTIR findings, the very first phase (from 25–150 °C) is predominantly attributable to the emission of water as well as low–molecular–weight volatile chemical byproducts. The second phase ranges from 150–500 °C. This period is characterized by considerable weight-loss (around 50 percent of overall volatile-matter), which correlates to the primary breakdown-phase of collagen-fiber particles [16,21,26,35,37,39,46,47]. There seem to be two shoulders, one of these is noticeable while the other one is feeble, which might well be detected on 'DTG curvatures' in this area, which may have been induced by the collagenous with varying extent magnitudes of cross-linking [16,21,26,35,37,39,46,47]. During the last phase (500–600 °C), the weight-loss of leather-shavings steadily escalated with rising temperature, and the slow yet persistent degradation of the carbon-containing compounds in residuals could be attributed to this phase.

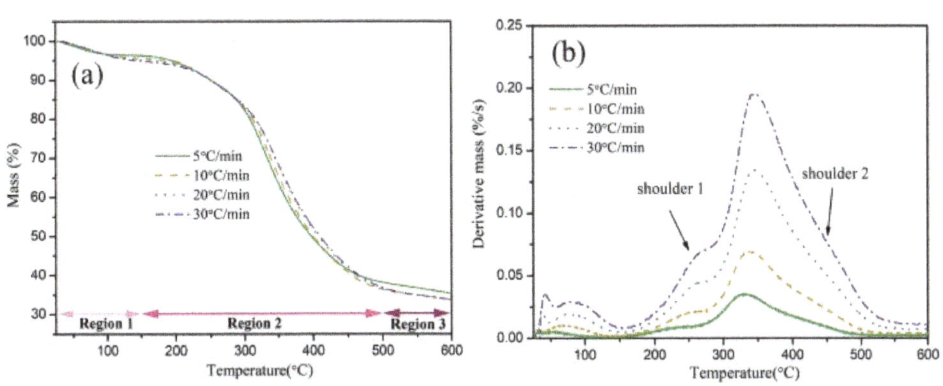

Figure 12. (**a**) TG of leather-shavings and (**b**) DTG curves of leather-shavings Adapted from the reference [77].

According to TGA research, the weight of the specimen remains constant up t 213.47 °C, after which there is a large weight loss (99.09318 percent) for the leather particle for "neat recycled-EVA" polymer composite specimens. In other words, 99.09 percen which also shows that the specimen is 'thermally stable' up to 213.47 °C and begins to los weight (95.651512 percent) after this temperature for the leather shavings/recycled EV. Polymer combination. 95.65 percent, which is also evidence that the specimen is 'thermall stable' up to 213.81 °C, as shown in Figure 13a,b.

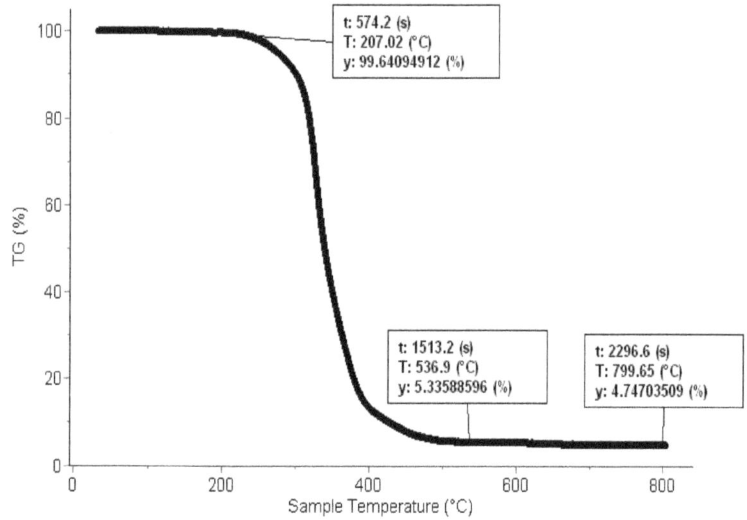

(a)

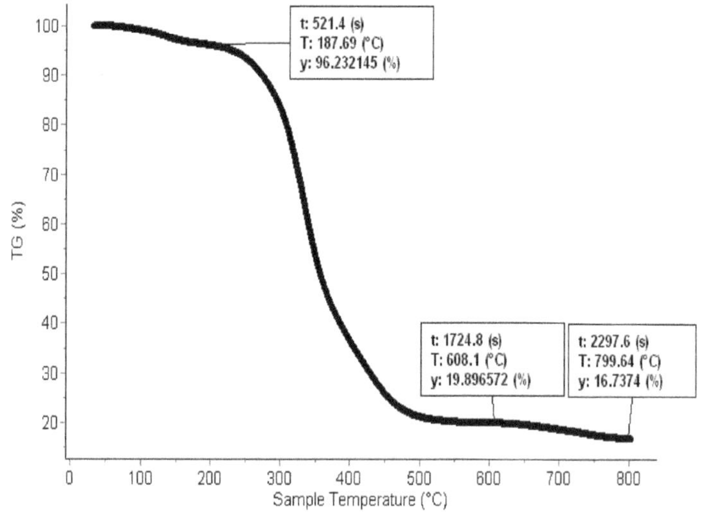

(b)

Figure 13. TGA curve of the (**a**) 'neat recycled-EVA polymer' without any 'leather-fibers' Adapte from the reference [45], and (**b**) 'Leather-shavings'/'recycled-EVA' polymer composites.

Acetic acid is released during the first stage of mass loss. "Acetic acid has alread been completely discharged in the second mass-loss step. The first weight loss (213.47 °C t

300 °C for neat recycled-EVA and 213.81 °C to 300 °C for Leather shavings/recycled EVA polymer composites) is a result of the formation of C=C all throughout the core-polymeric framework structure and has been linked to similar findings" [16,21,26,35,37,39,46,47]. "The second mass loss occurred as a result of the oxidation and volatilization of hydrocarbon compounds as a result of the breakdown of the recycled EVA co-polymer network structure (300 °C to 460 °C for neat recycled EVA and 300 °C to 465 °C for leather shaving/recycled EVA polymer composites)".

The level one degradation relates to the deacylation of the vinyl acetate group with the removal of acetic acid, according to similar findings from in-depth literature. Additionally, level one degradation of vinyl polyethylene chains gives way to level two degradation. Rapid weight loss and a rise in temperature are also signs of the second stage of deterioration. The loss of water molecules is visible in the degradation in the region of 100 °C. According to Tegegn (2018) [48], there are three basic steps to weight loss caused by the heat degradation of collagen hydrolysate. Between 46.05 °C and 220.24 °C, the first stage was detected. There was a weight loss of 9.67%, which may have been caused by the gelatin's moisture evaporating. The degradation of proteins caused the second stage of weight loss, which started at a temperature between 220.2 and 350.98 °C, to lose the most weight (38.84 percent). This finding is marginally higher than that obtained by Camila de Campo et al. in 2017 [40], who reported that the beginning degradation temperature was seen at 200 °C. Collagen hydrolysate degradation began at 220.2 °C. The third stage of weight loss takes place between 350.98 and 431.35 °C in temperature. Using a temperature range of 431.4 to 587.06 °C and a weight reduction of 12.89 percent, the corresponding weight loss was (7.24 percent). At 587.06 °C, it has a residual mass of 30.45%. The greatest weight loss was recorded at 311.75 °C, or 0.766 percent/°C [48].

According to Ravichandran et al. (2005) [78], the structure of leather is bound to alter when leather particles are joined with scrap rubber and solidified at high temperatures under pressure. Additionally, the degradation of the recycled rubber matrix is expected to be accelerated by the leather's breakdown products, particularly trivalent chromium. There has been much discussion over the deterioration of leather-containing scrap rubber vulcanizates [78]. Additionally, the decomposition products were listed in their investigations along with the degradation studies employing thermal gravimetric analysis of the vulcanizates under inert and oxidizing atmospheric conditions.

Additionally, as reported by Ravichandran et al. (2005) [78], "TGA of leather samples were tested between 0 °C and 400 °C in nitrogen environment in order to understand the role of leather on the thermal stability of natural rubber vulcanizates. Two stages of leather degradation below 400 °C were visible in the thermograms of treated and untreated leather. The first stage of weight loss, which takes place below 100 °C, is attributed to released water, while the second stage, which happens between 200 and 400 °C for all samples, is assigned to leather degradation" [78]. Since the temperature at which "all leather samples decomposed was almost the same (between 200 and 300 °C), it was reasonable to assume that they would all have a similar impact on the thermal stability of the parent natural rubber-scrap rubber vulcanisates". The natural rubber vulcanizates' breakdown is influenced by the pieces created during this stage's decomposition [78]. "The TGA of natural rubber vulcanizates was carried out between 50 and 400 °C in a nitrogen environment to examine the effects of untreated, sodium-hydrogen-carbonate processed, azane processed, and carbamide processed leather parts on the 'thermal stability' of latex-reclaimed rubber-vulcanizates [78]". The results of the relevant thermograms better illustrated the thermal stability of latex-reclaimed rubber-vulcanizates up to 350 °C without leather. Although the disintegration patterns for the vulcanizates, including untreated and treated leather, are similar, as shown by the thermograms, a decrease in their 'thermally stable' can be seen when compared to the vulcanizate in the absence of leather. Therefore, it may be inferred that natural rubber vulcanizates without leather are more 'thermally stable' than those that do. The % weight was measured in steps of 50 °C for all the samples in order to compare the weight loss between 50 and 400 °C for the untreated and treated natural leather rubber-scrap rubber vulcanizates. For all of the samples, the TGA result showed how much weight is retained for every 50 °C increase in temperature. The dat comparison showed that the latex-reclaimed rubber-vulcanizates, which contain untreate leather up to 350 °C, retained more than 92 percent of their weight". However, the data fo all the "latex-reclaimed rubber-vulcanizates" that made up the processed leather sample revealed that merely at 300 °C were around 92 percent of the weight retain. The 'thermall stability' of "treated leather natural rubber-scarp rubber vulcanizates" showed a 50 °C temperature decrease, amply proving the perspective [78].

"The weight loss at 400 °C under isothermal conditions was also observed for 46 mi in increments of four minutes in order to compare the rates of deterioration for all th samples. The measured weights for all the samples under isothermal conditions at 400 °C had revealed that there was a slight increase in the rate of degradation in the forme compared to the latter, despite the fact that the rate of degradation for latex-reclaime rubber-vulcanizates containing untreated leather appears to be similar to that for vu canizates containing treated leather". Additionally, it was obvious that "latex-reclaime rubber-vulcanizates" without leather demonstrate a rapid weight loss at a temperatur that is significantly greater than those of "latex-reclaimed rubber-vulcanizates" that con tain 'leather-particulates', as reported by Ravichandran et al. 2005 [78]. At 400 °C, unde 'isothermal conditions', a weight loss of around 50% was seen for all the samples [78].

The results of reclaimed rubber-only (without leather) TGA analyses of natural-rubbe (NR) vulcanizates were conducted between 0 and 600 °C in the air [78]. The investigation i being done to determine how untreated leather affects the thermostability of vulcanizate made from natural rubber and recycled rubber. There was no weight loss below 150 °C Hence the material does not appear to contain any water in the matrix. The in-situ FTIR i applied to the TGA analysis effluents for vulcanizates that do not contain leather at variou time intervals [78]. The four spectra resemble their corresponding untreated leather-loade sample spectra almost perfectly. The thermogravimetric analysis (TGA) of latex-reclaime rubber-vulcanizates, including untreated leather particles, which was conducted betwee 0 °C and 600 °C in air, produced the following results.

Previous research revealed that the thermogram was there even though the presen case's pace of rubber matrix disintegration and the beginning of water release from the ru ber matrix looked to be different [78]. "The rate of degradation is lower for the vulcanizat without any leather than for the loaded sample with untreated leather. Additionally, th thermogram shows the last residue of the leather-loaded sample below 500 °C. Its untreate leather may be to blame for this variance in behaviour. According to Ravichandran et a (2005) [78] the structural water present in leather that takes time to get out of leather due t the likely structural changes in leather may be to blame for the time-dependent water los that was shown to occur up to 200 °C".

Due to the loss of poorly adsorbed water, the TGA trace has shown that weight los might start even below 100 °C [78]. But the material immediately starts to decompos Between 220–350 °C, there is a very slight weight loss (around 18%), followed by tw further stages between 350 and 500 °C. In this temperature range, the overall weight los was equivalent to about 42%. Starting at 500 °C, the final residue begins to lose weigh extremely quickly. Given that the TGA trace does not completely disappear, there mus be a thermally stable residue, most likely inorganic chromium oxide, as chromium is frequent element found in leather [78].

According to the TGA data for leather vulcanizates treated with sodium bicarbonat "water is continuously released below 200 °C, and the vulcanizates begin to decompose a 240 °C. From this, it can be inferred that the effect of sodium bicarbonate-treated leathe vulcanizates on the thermal properties of "latex-reclaimed rubber matrix" would be com parable to that of untreated leather vulcanisates" [78].

The TGA results of azane-processed leather vulcanizates were comparable to thos of urea-treated leather vulcanizates [78]. The discharged water remained unchanged, an the rubber matrix's breakdown had already begun. At 500 °C, the residue underwen two stages of disintegration. This finding demonstrated that the degradation of "latex-reclaimed rubber-vulcanizates" including leather that had been azane-processed, would behave almost identically to leather that had been urea-treated [78].

The TGA results of the composites made from urea-treated leather showed similarities between them and the composites made from raw leather, but there were also clear discrepancies [78]. The effect of leather treated with urea on the thermal durability of a natural rubber-scrap rubber matrix was amply proven by this material's two-stage disintegration of the residue at temperatures of 500 and 540 °C. "The release of water was evident below 200 °C, and the rubber matrix started to break down at 240 °C, as with composites consisting of raw leather. The second stage decomposition of the vulcanised containing carbamide processed leather at 540 °C may be caused by the strong mixing and interaction between the matrix and leather that leads to a co-operative deterioration that can only occur at a high temperature" [78].

10.3.2. Differential Scanning Calorimetry (DSC)

According to research, a glass transition is, in fact, a second-order endothermic transformation that manifests as a gradual changeover shift in consecutive DSC thermogram heating graphs. At this moment, changes in chain mobility have caused a polymeric transition's physical and mechanical properties to change from elastic to brittle. It was found that the T_g values for the "neat recycled-EVA" were around –20 °C and were unaffected by the cross-link density or vinyl acetate concentration. As shown in Figure 14a,b, the T_g of leather shavings/recycled EVA polymer composites was discovered to be approximately −16 °C.

Exothermic crystallization causes a significant exothermic peak to form during "the quick cooling process at around 215 °C (for neat recycled EVA polymer composites) and 218 °C (for leather shavings/recycled EVA polymer composites)".

An 'endothermic peak maximal' on 'DSC heating graphs' could be viewed as a melting-temperature (Tm), a "first-order transformation". The Tm of 'neat recycled-EVA' was determined to be approximately 88.5 °C. At the same time, the Tm of the blend of recycled EVA polymer and leather shaving composites was discovered to be about 118 °C.

The specimens with higher water content and more volatile chemicals in the gradually progressing heating than the control-board could be the cause of the lower endothermic transitioning result. Comparing "neat recycled-EVA" polymer composites to leather shavings, the neat composites had greater endothermic values (Tm1) than the latter. This might be a result of the fact that all the fibers were spread very equally throughout the leather matrix, which caused their melting phase to shift toward a higher temperature. Due to the structural heterogeneity of leather, which includes remnants of tanned components that may interact with fibrous filaments, the second and third 'endothermic peak values' of such a composite are also converted to exceptionally high-temperatures.

Regardless of the presence or absence of leather, the polymeric chain's arrangement loses its stability and homogeneity. The addition of a plasticizer inside of polymers increases the chain motility, which in turn affects the T_g of the polymer.

The DSC analysis also revealed that "solid leather fibers lose water at an endothermic transition around 100 °C, are thermostable up to 211 °C, and start to break down collagen around 332.56 °C for 'neat recycled-EVA' samples and 318.47 °C for leather shavings/recycled EVA polymer composite samples, respectively. The glass transition temperature (T_g), which was measured for the developed composites and was found to be between −16 and 30 °C, is a valuable feature for understanding the processing parameters as indicated by DSC". Between 325 and 500 °C, during the leather preparation process, a significant mass loss developed as a result of protein loss and the calcination of a material. "This mass loss persisted between 130 and 150 °C and was attributed to unstable volatile compounds such as lubricants (oils) and low molecular-weight greases observed throughout the leather fibrils".

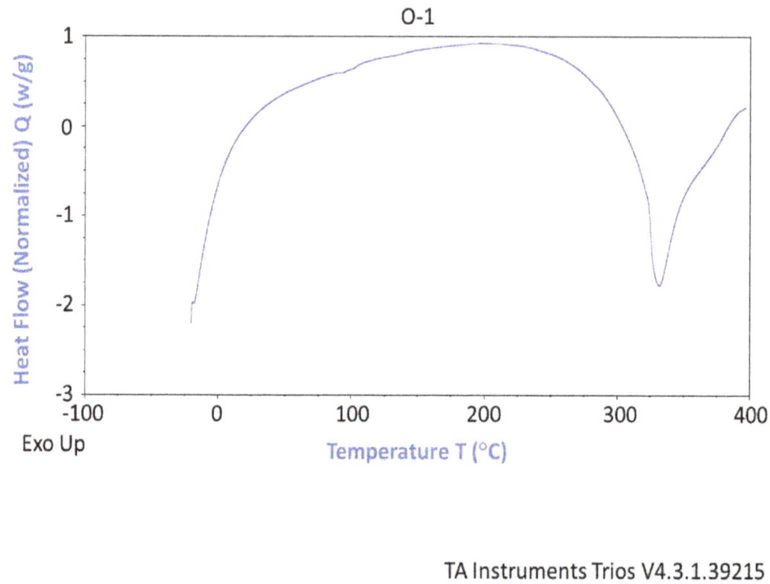

(**a**)

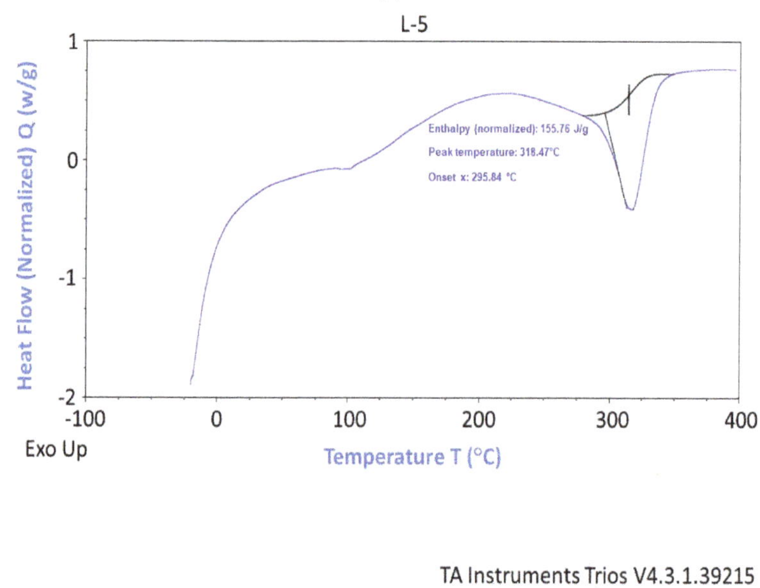

(**b**)

Figure 14. "DSC thermograms for (**a**) Neat recycled EVA polymer without any leather fibers Adapte from the reference [45], and (**b**) Leather shavings/recycled EVA polymer composites".

According to Joseph et al. (2017) [41], the leather waste was fully amorphous in natu because "there was no melting peak for the pure leather sample according to comparab results for the DSC melting thermograms". A significant melting peak for the pure pol caprolactone (PCL) sample was noted at 57.14 °C. Other compositions that contained PC and leather only showed one peak, which is indicative of PCL melting, and there wa no change in the peak's position. This outcome also demonstrated that PCL and leather wastes did not interact. With an increase in leather content, the peak's intensity drops proportionally. Leather has a very small impact on the 'melting-temperature' of composites. The 'melting-point' is between 56.6 °C and 58.2 °C.

The influence of 'leather content' on the temperature of composite crystallization was demonstrated using crystallization thermograms. The crystallization temperature was raised to 34 °C by adding leather waste to the PCL matrix by 2 °C. Additionally, because it will shorten the chilling process and consequently the entire cycle time, this trait can be advantageous for processing this mixture by 'extrusion' or 'injection molding' in final products [41].

"DSC analysis of latex-reclaimed rubber-vulcanizates without leather, untreated leather, and urea-treated leather samples is carried out between 0–400 °C, at a heat-rate of 20 °C/min, according to Ravichandran et al. (2005) [78]. DSC trace of latex-reclaimed rubber-vulcanizates without leather, untreated leather, and urea-treated leather samples". For natural rubber-reclaimed-rubber and vulcanizates without leather as well as for all the samples treated with leather, the modest endothermic elevations up to 200 °C are due to the release of retained water. An endothermic hump has been observed between 200 and 250 °C for vulcanizates that contain treated leather such as urea, sodium hydrogen carbonate, and azane, as well as for vulcanizates devoid of leather [78]. The endothermic chain dis-entanglement of the polymer chains is what causes this hump. In practically all of the samples, the subsequent endothermic hump between 200 and 300 °C can be attributed to little decomposition. The thermograms for 'leather' and "latex-reclaimed rubber-vulcanizates" that contained leather could not adequately resolve this decomposition [78].

"The DSC traces of all the samples showed a significant endothermic elevation over 300 °C, which appeared to perfectly correlate with the 'decomposition-stage' of "latex-reclaimed rubber-vulcanizates" contained with and without 'leather samples'. A thorough examination of the principal endotherm's origin for each sample has revealed that "natural rubber-scrap rubber vulcanizates" exhibit more stability in the absence of leather than 'latex-reclaimed rubber vulcanizates' do, as the former's endotherm origin is located above the latter [78].

11. Fourier-Transform Infrared-Spectroscopy (FT-IR)

As it is apparently observed in this monograph that surface-layer changes in 'recycled-EVA polymer' may be correlated with absorption regions C-H (3100–2914.88 cm^{-1}), which, under humid conditions, increase in intensity. As a result of hydrogen bonding between vinyl acetate and water, this increment occurs. The 'stretching vibration' of the 'C=O' is caused by the terminal "trans-vinylene double bond" (1701.87–1736.58 cm^{-1}), the 'C=C' and the "methylene stretch" (1432.85–1466.6 cm^{-1}), 'CH_3' (1371.14 cm^{-1}), and the 'C-O' (1296.89 cm^{-1}). In comparison to the "deformation bands" of the '$CH_{2'}$ group (1464 cm^{-1}), the band corresponds to a stretch mode in which 'C-O-C' is ascribed stretch mode. The EVA with VA content in it can be monitored using this band as an integral standard. As demonstrated in Figure 15a, there is a predominant absorption band at 1739 cm^{-1} for the acetate function.

These peaks are attributed to the EVA groups of alcohols, phenols, and carboxylic acids, which are primarily in a change of the material's strength behavior. "The spectra of the 'neat recycled-EVA' co-polymers show absorption peaks at approximately 2917.77 cm^{-1} and 2869.56 cm^{-1}, which are connected to the asymmetric and symmetric stretching of the co-polymers' methylene group (-CH_2). The VA groups' distinctive absorption peaks are as follows: The stretching vibration of the -C=O band is attributed to 1732.73 cm^{-1}; the asymmetric stretching vibration of the C-O band is attributed to 1130.08–1104.05 cm^{-1}; the symmetric stretching vibration of the C-O-C band is attributed to 996.053 cm^{-1}; and the inner rocking vibration of the methylene group is attributed to 810.92 cm^{-1}. The contributions from both VA and ethylene (CH_2) units are substantially responsible for the observed absorption maxima at 1379.82 cm^{-1}. As shown in Table S5, it is expected that the intensities of the absorption peaks at 810.92, 996.053, 1130.08–1104.05, 1379.82, 2869.56, an 2917.77 cm^{-1} seem to rise as the VA concentration rises".

"The solid leather wastes and recycled EVA composites treated using a combinatio two-roll mill and hot press yielded the ATR-FTIR results". The recycled EVA was create when polyethylene and vinyl acetate reacted. The curve compares the spectra of recycle EVA composites that were treated using both a hot press and a two-roll mill.

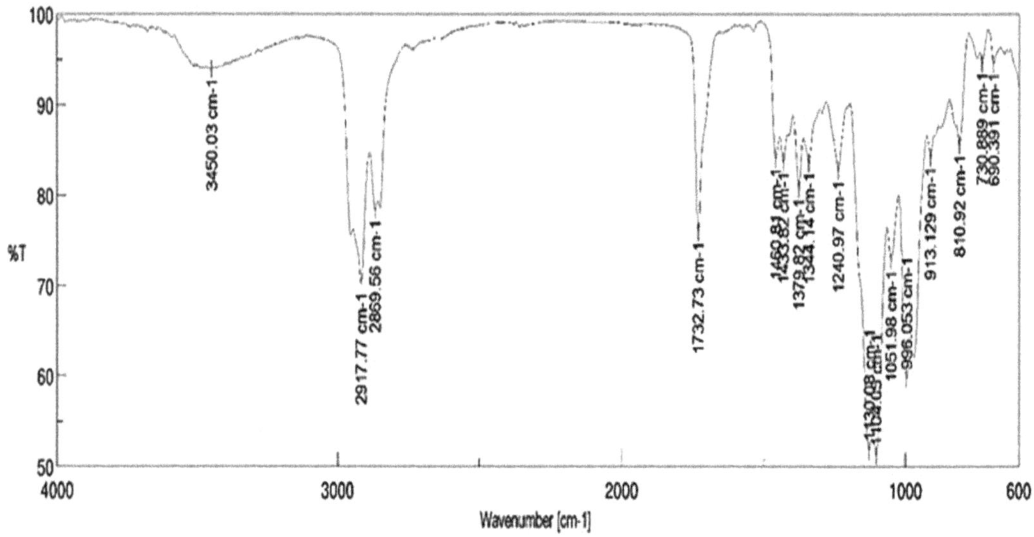

(**a**)

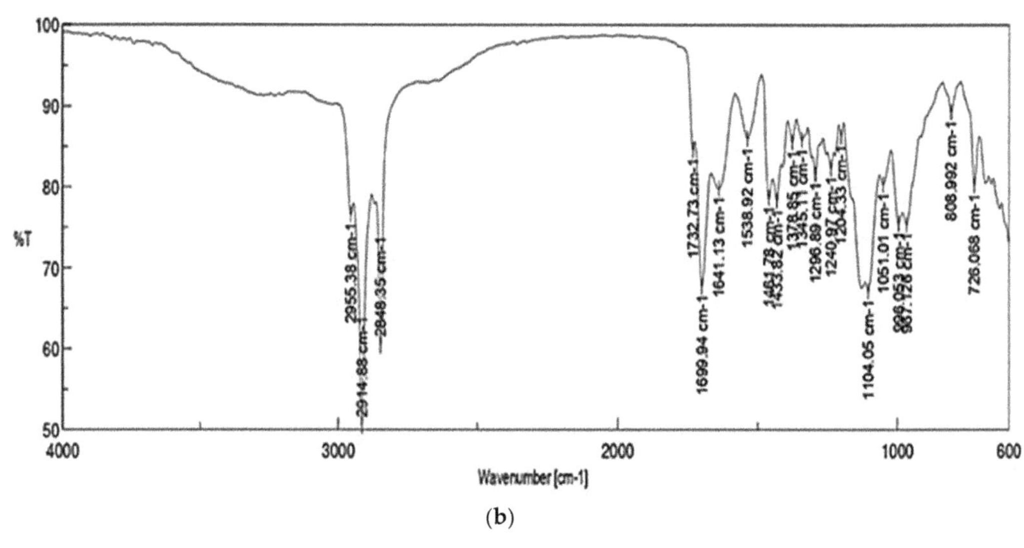

(**b**)

Figure 15. FT-IR spectra of (**a**) Neat recycled EVA polymer without any leather fibers Adapted fror the reference [45], and (**b**) Leather shavings/recycled EVA polymer composites.

"While in ATR-FTIR monograph, molecular structure of Leather shavings/recycled EVA polymer composite are studied relative to the absorption regions O-H in side chains and terminal groups (3200.15–3500.40 cm^{-1}); (2914.88–2848.35 cm^{-1}) indicating the absence of fatty substances; A band of deformation vibration of first and second order amide (-NH) appears at 1641.13–1732.73 cm^{-1}; 1433.82–1461.78 cm^{-1} is due to -COO^{-} groups; 1345.11–1378.85 cm^{-1} is due to -C-O- group; 1204.33–1296.89 cm^{-1} is due to deformation vibration of -C=O group; 967.126–1051.01 cm^{-1} is due to -C-O-C- ether group and below 726.068 to 600 cm^{-1} is due to Cr-O bonds as demonstrated in the Figure 15b and Table S6".

As the comparable outcomes previously documented by Ambrósio et al. (2011) [38] have reported that "the co-polymer chains of the recycled EVA and of the plasticizers/additives do not lead to improvements in their molecular interaction geometries after being manufactured by the combined processing effect of two-roll mill and hot press". "The spectra of recycled-EVA and solid leather waste processed in a two-roll mill appear to be extremely similar".

According to previous research by 'Joseph et al. (2017)', the intensity of these peaks rises in leather-PCL composites as the wt.-composition of leather-fibers escalates. There are no other peaks in the leather-PCL composite other than the distinctive peaks for leather and PCL, confirming that there was no 'chemical interaction' between 'leather' and 'PCL', as had previously been validated by the DSC data examined by Joseph et al. (2017) [41].

Tegegn (2018) reported an analogue result for the FT-IR spectra of the "gelatin-starch cross-linked film" with 'KBr pellets' using the "Shimazu IR affinity-IS spectrometer" [75], which were used to analyze chemical properties. "FTIR analysis of a cross-linked composite film made of gelatin, starch, polyvinyl alcohol (PVA), and glucose from leather wastes revealed the following amide: A (3254.05 cm^{-1}) No discernible shift in the 'absorbance' of 'band amide A', it was (3252.12 cm^{-1}) associated with "NH stretching", and 'amide B' (2922.28 cm^{-1}) band inextricably correlated with "stretching vibration" of 'CH_2 bonds', it was (2920.35 cm^{-1}, 1645.35 cm^{-1}, 1543.12 cm^{-1}, and 1238.35 cm^{-1} inextricably correlated with 'amide: I, II, and III', as well as 'amide I' associated with 'C=O stretching'. As a result of "CN stretching" and "NH bonding", a slight shift is seen in 'amide: I' (1649.21 cm^{-1} to 1645.35 cm^{-1}) and in 'amide: II' and 'III' (1649.21 cm^{-1} to 1645.35 cm^{-1}). A minor change occurred in 'amide III' (1238.35 cm^{-1} to 1242.21 cm^{-1}), and a comparable alteration occurred as a result of the interaction between cross-linked leather-waste reinforced composite polymers" [79]. With 'CH_2 bending', and 'C-O-C stretching', respectively, there are bands at 1411.95 cm^{-1} and 1097.54 cm^{-1} [40].

According to Ravichandran et al. (2005) [78], "in-situ FTIR analysis of the breakdown products for 'vulcanizates' including "untreated leather" was performed between 500 and 4000 cm^{-1} at varied time intervals in order to understand the "nature of decomposition". The IR spectra of the breakdown products recorded between 0 and 23 min were illustrated using the spectra acquired for various time intervals. The thermal-property -based investigation has revealed the emission of "hydrocarbon-waste fragments" by a very lesser-intense peak slightly below 3000 cm^{-1}.

Water was partially released at 3500 cm^{-1} due to -OH stretch, and CO_2 was obviously released at 2354.5 cm^{-1} due to its 'stretching vibration'. At 666.7 cm^{-1}, its 'bending mode' was also visible. These findings have unequivocally proven that the materials degraded through combustion throughout the specified time period. The IR spectra showed more intense peaks for the products released between 24 and 46 min, further demonstrating the release of more fragments and oxidized products, including CO_2 and water [78]. It could be explained by increasing activation as the temperature escalates.

The IR spectra of the released-products between 46 and 52 min have shown that the peaks caused by the aforementioned fragments were significantly more intense than those between 24 and 46 min [78]. Additionally, the maxima of the characteristic curves for water and CO_2 are right above 3500 cm^{-1} and 2354. 3 cm^{-1}, respectively, were very clear indicators of the discharge of these gases. Once more, the high temperature may be to inflict such dramatic peaks since more oxidation may occur [78].

The IR spectra caused by the released products between 52 and 60 min revealed decrease in the intensity of CO_2 vibration at 2354.3 cm^{-1} due to the number of release hydrocarbon particles and water [78]. With this spectrum, it is important to note that les water was released, as evidenced by the decline in peak intensity above 3500 cm^{-1}. Thi observation has demonstrated that because the majority of the decomposition was alread complete, the residue that has undergone decomposition at this specific stage may be c lower quantity.

According to Ravichandran et al. (2005) [78], "the in-situ FTIR spectra of the combu tion products generated for vulcanizates containing sodium bicarbonate treated leathe at different time intervals showed more release of CO_2 and water than organic particles Between 46 and 52 min, there was an immense release of CO_2 and water; as a result, th temperature range associated with this decomposition may be more "combustion" an more rubber matrix. Between 52 and 60 min of combustion, there is also a decrease release of CO_2 and water, which is demonstrated by the lower level of cross-linked matri residue [78].

The higher-volume of CO_2, hydrocarbons, and water was released when in-situ FTI measurement of the combustion products for vulcanizates containing ammonia-treate leather was performed at various time intervals. The possible causes include a rise i temperature between periods of 30 and 50 min. The spectra also show that the residu produced when these combustion products were released in smaller quantities had greate thermal stability. However, the results have shown that more CO_2 and water have bee released in some spectra and less volume of organic-fragments [78]. As a result, the residu needed additional thermal energy as the combustion process came to an end.

According to Ravichandran et al. (2005) [78], the FTIR analysis of the combustio products for vulcanizates comprising urea-treated leather over various time periods wa demonstrated, showing the emission of CO_2, hydrocarbon fragments, and water. Thi was confirmed by inflections in the spectrum caused by "-OH stretch of water" just abov 3500 cm^{-1}, "C-H stretch of organic-fragments" to just below 3000 cm^{-1}, and CO_2's presenc at 2354.3 cm^{-1}. The portion of the spectrum shows the well-dissolved sharper, more intens peaks for CO_2 at 2363.6 cm^{-1}, hydrocarbon at 2927.6 cm^{-1}, and water at 3600 cm^{-1}. Th release of CO_2 and water is higher, while the release of organic-fragments is less, as see by the other spectra. This could be a result of the cross-linked rubber matrix structure an leather residue coming together well [78].

12. Scanning Electron Microscopy with Energy-Dispersive Analysis of X-ray (SEM-EDAX)

The elemental composition and molecular dispersion of the entrapped solid leathe fibers were revealed by EDX tests. "The SEM images have indicated that the particles ar in the uniformly-dispersed and of spherical structure morphology, while E-DAX explore the stoichiometric-compounds and percent-chemical purity of the specimens to establis the occurrence/existence of chemical-elemental compositions".

As depicted in the micro-images, "recycled-EVA polymer" with leather-shaving fiber exhibits a structure that is characterized by the major elements ('Carbon', 'Sulphur', an 'Oxygen') as determined by EDAX analysis. According to Figure 16a, the surface of the 'nea recycled-EVA' matrix comprises C, S, Cl, O, and P. Saikia et al. (2017) [41] revealed simila results as well. "The incorporation of leather fibers and additives during the processin of combined-effect of rolling as well as hot-press compression molding has led to th removal of some chemical elements such as S, P, and Cl, according to the EDX of leathe shavings/recycled EVA polymer composite surfaces" (Figure 16b) [35,37,39,41,46,47].

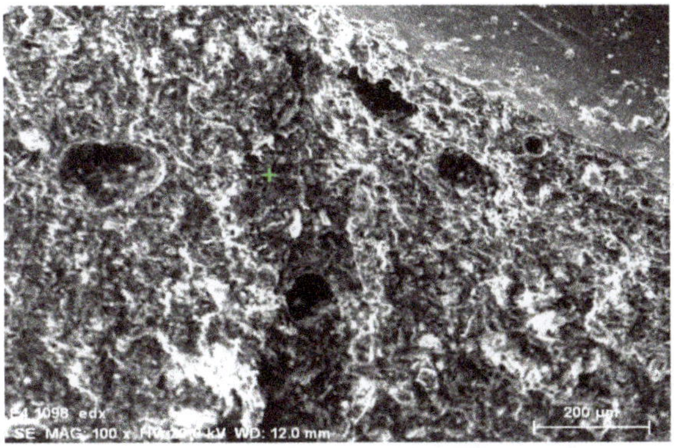

E4 1098Date:9/15/2020 3:55:00 PM Image size:600 x 450Mag:100xHV:20.0kV

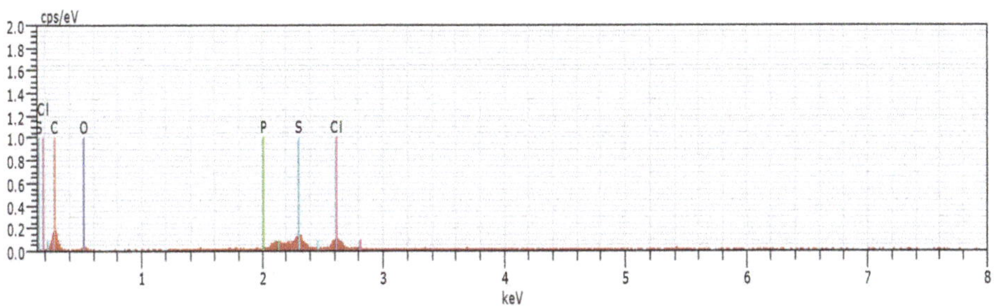

Spectrum: 02 1235 Date:9/15/2020 3:55:23 PM HV:20.0Kv Puls th.:0.23kcps

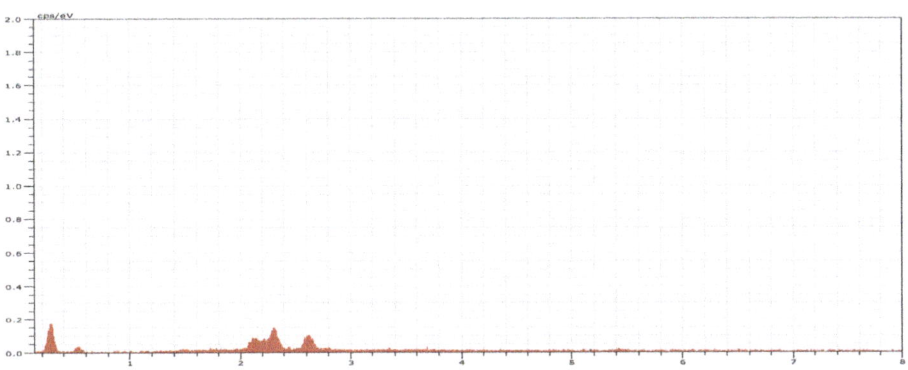

(a)

Figure 16. *Cont.*

PL5 1054Date:9/3/2020 4:38:39 AM Image size:600 x 450Mag:200xHV:15.0kV

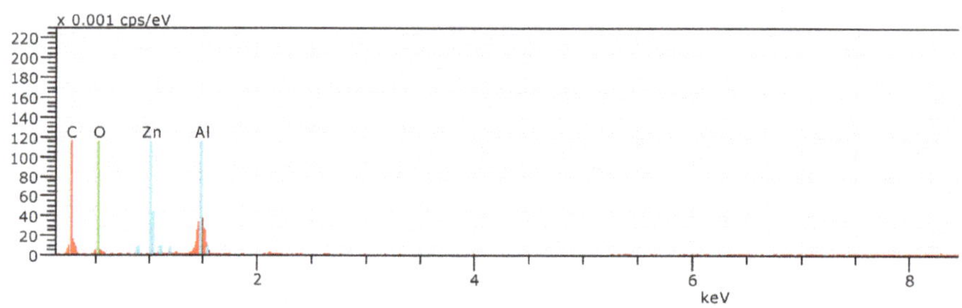

Spectrum: L-3 1181 Date:9/3/2020 4:39:01 AM HV:15.0kV Puls th.:0.11kcps

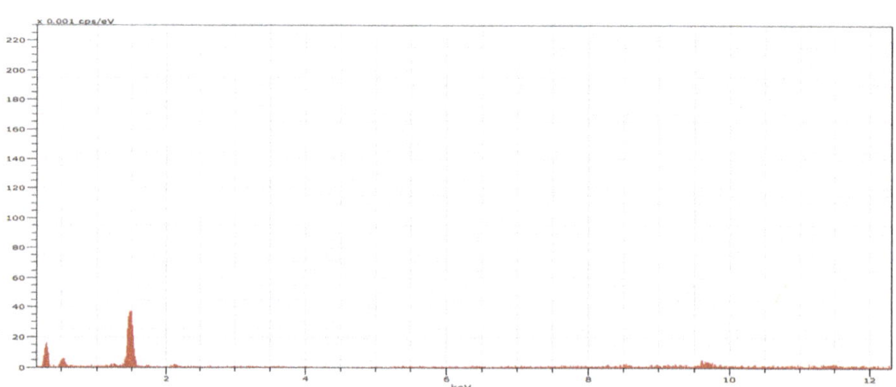

(**b**)

Figure 16. (**a**) SEM-EDAX analysis of the surface of neat recycled EVA polymer without any leather fibers Adapted from the reference [45]. (**b**) SEM-EDAX analysis of the surface of Leather shavings/recycled EVA polymer composites.

In the SEM-EDAX analysis, the occurrence of "trivalent chromium (0.1%)" in the leather shaving fibers, which have a "proteinaceous collagenous fibril-type threaded-form rugged structure", was identified. This is believed to result from the "tanning-process". The existence of "oxygen" can be attributed to the "collagenous in the leather-shaving fibers", which can serve as an "actively binding-site" for "trivalent chromium". These "binding-sites" typically include "carboxylate groups", which form "organic-salts" with "trivalent chromium". The appearance of "sulfur" and "chloride" is thought to be owing to the existence of "inorganic-salts" of "trivalent chromium". According to the EDX analysis of the "leather shavings/recycled EVA polymeric composites", the processing, which involved the inclusion of "leather fibers" and "additives", along with a combination of "rolling", and "hot-press compression molding", resulted in the eradication of certain "chemical elements" such as, "sulfur (S)", "phosphorus (P)", and "chloride (Cl)". SEM, as exhibited in Figure 16a, has illustrated that the recycled-EVA matrix typically reported the 'uniformity' or 'compatibility' of the "recycled-EVA matrix," revealing an excellent "interfacial bonding strength". The SEM, as exhibited in the Figure 16b, has illustrated the 'trace fragments' of recycled EVA on a layer of leather fabric aggregates, demonstrating that the recycled EVA is binding the agglomeration along with the fact that the recycled EVA and the leather fibers have a strong, reasonable interfacial bonding and adhesion strength. The potential for the interface to transmit matrices stress to the coalescence of leather fibers is shown because, in this case, the composites' cryogenically breaking or deformation/rupture would have occurred inside the coalescing amalgamates rather than through the interfacial-framework.

13. SEM Morphological Analysis

The micrographs of the 'neat recycled-EVA composites' are shown in Figure 17a,b, while the cross-sectional areas and microstructures of the leather shavings/recycled EVA composites are shown in Figures 18a,b and 19a,b, respectively. The recycled-EVA matrix typically contains an even distribution of the leather fiber 'agglomeration'. However, the 'uniformity' or 'compatibility' of the 'leather fibers' aggregates within the "recycled-EVA matrix" reveals an excellent "interfacial bonding strength". Some 'crevasses', 'openings', and 'voids' in the "surface morphology" are present and appear to be related to the 'de-fibrillation' and 'extraction' of the 'leather fiber aggregates'. Figure 17a,b indicates the locations of "interface-layer adhesion" or 'bonding' with 'interface consistency' or 'cohesion'. Due to the physicochemical differences between 'hydrophilic fibers' (made of "collagenous macro-molecules") and 'hydrophobic thermoplastic matrix', it is typically challenging to achieve strong 'compatibility', better 'bonding', excellent 'stability', and strong 'adherence binding' between "thermoplastic matrices" and "leather fibers". By adding stabilizers, additives, binding agents, or compatibilization agents to 'polypropylene blends' for finely ground wood, it is possible to increase the "interfacial bonding-strength", "biocompatibility", "miscibility", and produce "uniform", "contiguous surface areas" in 'thermoplastic composites' made of 'natural fabrics'. One such compatibilization agent has been utilized, named "cis-butenedioic anhydride" [38,78,80].

Figures 18a,b and 19a,b shows the micrographs of the recycled EVA/leather shavings composite with 50% leather fiber. In Figures 18a and 19a, an 'agglomeration' of leather fibers is 'encapsulated' within a recycled EVA skin. Such recycled EVA skin implies strong compatibility, interface adhesion, or close contact between the recycled EVA matrix and the leather fiber, including the possibility that the interphase may actually be firmer than the aggregation of leather fibers. The structural layout of leather fibers, including microfibrils of size less than 1 m, has been illustrated by the higher-magnification representation of a leather fabric thread-like, coalesces or 'agglomerate' in Figures 18b and 19b (symbolized via agglomerations throughout Figures 18a and 19a) [38].

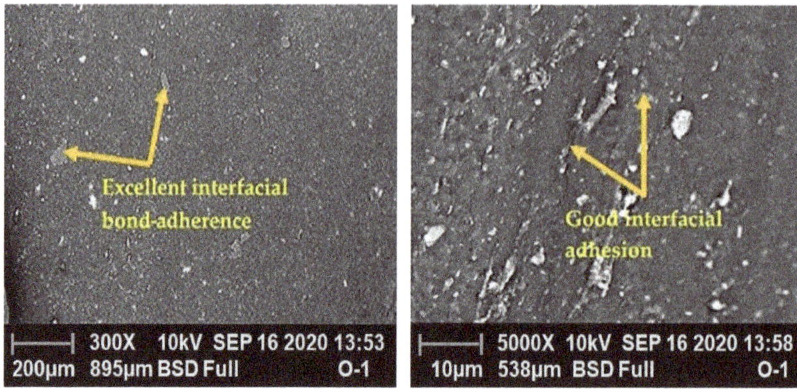

(**a**) (**b**)

Figure 17. Micrographs of Neat recycled EVA polymer without leather fibers at a magnification of (**a**) 300× and; (**b**) 5000× Adapted from the reference [45].

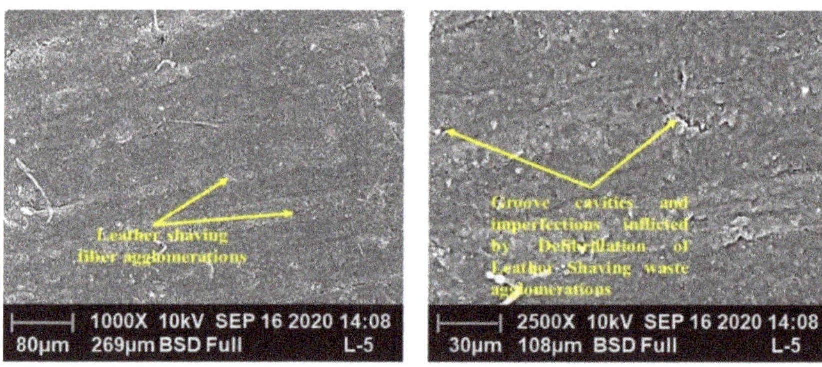

(**a**) (**b**)

Figure 18. Micrographs of the surface of Leather shavings/recycled EVA polymer composites at magnification of (**a**) 1000× and; (**b**) 2500×.

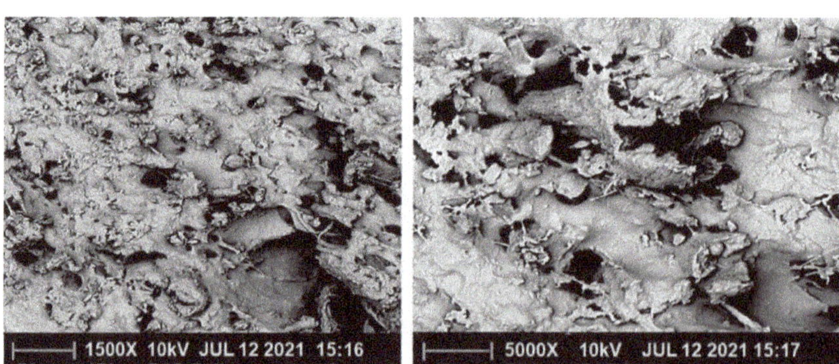

(**a**) (**b**)

Figure 19. Micrographs of a cross-section of Leather shavings/recycled EVA polymer composites at a magnification of (**a**) 1500× and; (**b**) 5000×.

Figures 18b and 19b show 'trace fragments' of recycled EVA on a layer of leather fabric aggregates, demonstrating that the recycled EVA is binding the agglomeration along with the fact that the recycled EVA and the leather fibers have a strong, reasonable interfacial bonding and adhesion strength. The potential for the interface to transmit matrices stress to the coalescence of leather fibers is shown because, in this case, the composites' cryogenically breaking or deformation/rupture would have occurred inside the coalescing amalgamates rather than through the interfacial-framework.

As the proportion of leather-particles within recycled EVA matrix increases, as illustrated in Figures 18a and 19a, the leather fibrous aggregates seem to be reasonably close to and in contact with one another. If the leather particles were properly filled with recycled EVA matrix, the grooves, discontinuities, and voids in the microstructure seen in Figures 18a,b and 19a,b, which were reportedly generated by de-fibrillation and isolated/detached, would be eased. A twin-screw extruder can improve the blend formulation even though the blending employing the combined effects of two roll-milling and hot-press compression molding operations was successful. For dissipating and diffusing leather-fibrous agglomerations and then uniformly/homogeneously/spreading the leather shavings textiles throughout the recycled EVA matrix, a twin-screw extrusion or injector with conveying/kneading blocks are preferred.

The degree of fiber alignment, homogeneity of fiber dispersion, and degree of fiber elastomer adherence have all been studied using scanning electron microscopy [81,82]. Excellent fiber orientation in cellulose fiber rubber composites has been found through SEM analysis. Fibers are really close to being parallel [83]. The degree of fiber alignment and fiber-elastomer adhesion is found to be poor with 'nylon' and 'aramid fibers' among the 'short fibers' such as 'glass', 'carbon', 'aramid', and 'nylon', and 'cellulose' [84]. Both good dispersion and the good adherence of cellulose fibers to the elastomer matrix have been demonstrated. Specific interactions between the ingredients may be partially responsible for the increased adherence of cellulose fibers to elastomers. When used with polar polymers such as nitrile rubber and its blends with PVC, leather, a highly polar material, may likewise display comparable interactions. Additionally, it should be noted that the leather has a very high cohesive strength relative to its adhesive strength with the matrix.

The cut-section of the neat recycled EVA matrix is smooth and even on the surface, while the fibers of the leather shavings are evenly dispersed over the surface. However, in some areas, the binding, bonding, and compaction of the leather shavings to the recycled EVA matrix are insufficient. Contiguous split leather fibers are visible in smaller quantities; this could be the result of shear stresses being applied during rolling and hot-pressing of the composition, or it could be caused by twisting forces. More amalgamation, fragmentation, and trapping of leather fibers are observed as the volume of fibers rises. The penetration of recycled EVA matrix through entanglement, which is primarily attributed to the 'physical adherence' between matrix and fibers, is shown. As a result, the tensile strength has been enhanced with a maximum content of 1:1 in leather-fiber.

Incorporating leather shavings into recycled EVA composite has increased the amount of fibrillar of recycled EVA matrix that was collected, and it has also caused the volume of fibrillar-interaction structures to tend toward being higher in volume (1:1), as observed in micrographs taken at a magnification of 5000×. Additionally, the lower magnification at 300× demonstrates the uniform distribution and intricate structure of the leather shavings, which appear to be completely covered in fibrillar structures made from recycled EVA matrix. This shows that the fibers made from recycled EVA matrix and leather shavings have good binding and adherence.

Collagen strands are tightly woven together to form leather. Researchers have examined the degree of adhesion between leather shaving fibers and recycled EVA blends using scanning electron microscopy (SEM). The interface between the fibers and the matrix in recycled EVA polymer that has been blended with leather shaving fibers at high temperatures is greatly dispersed, according to the material's morphology. As the fibers are incorporated into the matrix at high temperatures, a substantial fusion of the fibers with the

matrix has been observed, revealing that the fibers are entirely covered by the matrix [85
Even though leather shavings have been employed in this study in the form of particles, i
is anticipated that they will form a significant portion of the recycled-EVA polymer.

The physical characteristics of leather shavings as they are formed in the fibrou
form are shown in Figures 18a,b and 19a,b. The closely knitted fibrous leather is easil
disseminated into an EVA matrix, especially when there is a significant amount of recyclec
EVA polymer present.

According to Ravichandran et al. (2005) [67,78], the same result has been observed fo
the effect of untreated as well as neutralized leather particle morphology of NR vulcanizate
included 500 parts of a scrap of the same particle size (200–300 μm). The leather appear
to have existed as a co-continuous stage across the matrix with well-defined boundarie
between the two phases in the morphology of scrap rubber vulcanizates, which containe
untreated leather particles. This was a direct effect of the 'untreated leather particle
'rigid' and 'closely-knitted structure'. Although leather is typically a firmly knit fibrou
material, treated leather particles were discovered to have a loose-bounded structure [67,78
Depending on the type of treatment, the treated leather particles appear to occur as a cc
continuous stage in the rubber matrices with various degrees of interfacial adhesion [67,78

According to Joseph et al. (2017) [41], the comparable same SEM micrograph finding
showed that the leather fibers were evenly dispersed over the surface as well as a wea
adhesion, bond strength, and permeability of shavings to the Poly-caprolactone-matrix. A
confined, segregated, and isolated fiber may be seen at lower concentrations; this could be du
to shear tension loads applied during the different composition formulations in the extrusio
process. The above-mentioned 'leather-fibers' with enhanced composition have led to furthe
fiber accumulation, amalgamation, and entrapment. The improvement in tensile strength wa
achieved by the PCL-matrix penetrating through entanglement, which was attributable t
forming of a physical adhesion between the matrix and fibers [41,77,79–84].

Huang et al. have examined the comparable findings about relevance with exis
ing materials' morphologies, thermostability, and multi-objective optimization of prc
cess/operating parameters. With the use of a melt mixing technique and a variety of HD
compounds, ref. [86] has developed PLA and PBS mixture-blends. Analysis of the impact o
HDI composition on the morphology and distinctive properties of the mixture was the ain
of the previous research. The investigators have noted from the SEM results that when th
HDI contents have increased, the size of PLA and PBS has decreased. In addition, the PB
has contributed to accelerating the rate at which PLA crystallizes. "Additionally, becaus
of improved interface bonding between PLA and PBS, the 'toughness' of newly produce
composites' has greatly increased. A comprehensive overview on the use of various catalys
substances during the catalytic pyrolysis of waste materials including plastics (such as P
PE, PS, and PVC) along with the waste from the petroleum industries" was published b
Pan et al. in [87]. (petroleum mud). The authors investigated how to handle petroleun
waste and 'plastic-wastes' by using molecular sieves, carbons, metal oxides, and M-serie
catalysts. In the review article, the mechanism, difficulties, and potential advancement
of CP were also discussed. The authors came to the conclusion that more emphasis need
to be paid to systematic planning for the reuse of waste products from the petroleum an
plastics industries. Metal microcapsules (MMCs) and polymer microcapsules (PMCs) use
in epoxy-based composites were investigated by Sun et al. in terms of their tribologica
behavior. The investigators noted that under typical conditions, MMCs have more con
pressive strength, wear loss, and modulus than PMCs. In comparison to MMCs, PMC
had a lower friction coefficient. Additionally, the friction value was decreased in PMC
with an increase in microcapsule concentration. However, the MMCs exhibit the contrar
findings. The impact of nano clay in the glass fiber-reinforced PMCs was assessed b
Prabhu et al. [88]. Using the resin infusion technique under vacuum-conditions, the hybri
polyester-based PMCs were fabricated. The composition of the clay was changed fron
1 to 55 by weight in increments of 1. Various common testing methods were used t
investigate mechanical and machining behavior. The newly fabricated composites wer

characterized using TEM, tensile testing, impact testing, SEM testing, and fracture testing. Drilling was used to carry-out the machining behavior. The investigators have found that as clay content was increased, tensile strength also escalated. A composite with a 3% of "clay-composition", in comparison with the others, possessed better mechanical properties. According to a microstructure study, this composite exhibits less delamination at a high feed rate. Additionally, hybrid composites had higher fracture toughness and strain energy than traditional composites. Using the hand lay-up technique, Gopinath et al. [89] have developed glass and jute fiber-reinforced vinyl ester polymer matrix composites. The authors have analyzed various 'deck-configurations', and the investigators have used ANSYS and FEM to validate the findings. The authors noticed that the "stiffness" performed better in the alteration of 'responses' when they were placed in the proper location and shape. The 'V' form stiffener deck layout and the 'V and U' shape combination produced better results than other configurations. Additionally, the web joint—where there was the least deflection and strain—benefitted from the addition of stiffener, producing more remarkable outcomes than other 'locations'. Guo et al. [90] fabricated HDPE-based polymer matric composites that had glass and wood fibers as reinforcements. When developing PMCs, the authors used the injection molding process. The "mechanical behavior", and "water-absorption capability" of newly formed composites' were tested. The same conditions were used to compare hybrid composites to wood-fiber reinforced composites, and the authors found that there has been an enhancement of 40% and 253% for the tensile strength and modulus, respectively. The limitations of greater water absorption by WPCs are fiber swelling and loss of interface bonding. Additionally, the mechanical characteristics of the PMCs, as compared to WPC, were greatly enhanced by the addition of glass fibers. The nHA-reinforced PEEK polymer matrix composites were developed by Ma et al. [91] using the 3D braiding-self retention-hot pressing method. The weight percentage of the nHA varied between 6.5 and 14.5 percent. The modulus, hardness, and strength attained in the PMC with 6.5 percent nHA were 8.3 GPa, 3.34 GPA, and 155.32 MPa, respectively. These values were significantly higher than those of the 'base-matrix', according to the researchers. A reduction of 23.6 percent was observed with the addition of 14.5 percent; however, the toughness was raised up to 54.9 percent with the addition of 6.5 percent HA. This decrease was owing to the poor interfacial bonding that the rise in concentrations of nHA has attained. Glass-fiber (GF), Kevlar-fiber (KF), and carbon-fiber (CF) reinforced nylon polymeric matrix composites were developed by Mei et al. [92], and their effects on the properties of the basic matrix were assessed. As a part of the fabrication process, the 'rings' and 'layers' of fiber were used. SEM and tensile testing were used to investigate the newly developed composite materials. The investigators unveiled that, in comparison with the others that the CF-reinforced composites had superior tensile strength (110 MPa) and modulus (3941 MPa). The concentration of fiber rings and layers was increased, which improved the mechanical behavior. Pan et al. [93] studied various methods for recycling 'plastic-wastes' in order to get rid of plastic waste. The researchers have also discussed the benefits and drawbacks of the techniques used to recycle and get rid of 'plastic-wastes'. The authors argued that in order to reduce air pollution and protect human life, it is necessary to develop innovative techniques for disposing of or recycling 'plastic-wastes'. The related outcomes have been revealed by the prior studies [59–61,65,94–128].

14. X-ray Diffraction Analysis (XRD)

An X-ray diffraction study was conducted to identify the chemical-constituents or phases involved throughout the crystal-structures of recycled EVA polymer composites.

Figure 20 exhibits the X-ray analysis findings for the peak position of the recycled polymer composite samples, respectively. The positional-angles, peaked-heights, peaked-widths at half-maximal (FWHM), atomic d-spacing, and relative-intensity of each specimen peaking have been presented. The results from Tables 1 and 2 have illustrated that the main noteworthy variation between the spectrum seems to be the crystallinity value of the neat-recycled EVA polymeric composite sample has been reported to be 7%, and the amorphous value has been unveiled to 93%. In contrast the crystallinity value of th leather-shavings/recycled EVA Polymeric-composite sample has been reported to be 19.3% and the amorphous value has been unveiled to 80.7%, which corresponds to orthorhomb crystalline-forms. This novel outcome furthermore shows that the crystallinity value o leather-loaded recycled EVA composites is maximum and the neat-recycled EVA composite are minimum, which eventually confirms that leather-filled recycled EVA composite are having more hardness value as compared to neat recycled EVA composite sample Figure 20 displays the locations (2θ) of the recycled polymeric composite's spectra, with it inter-planar spacing and Miller-Indices. Among all samples, there was also a match fo stearic-acid, zinc-stearate, and other additives.

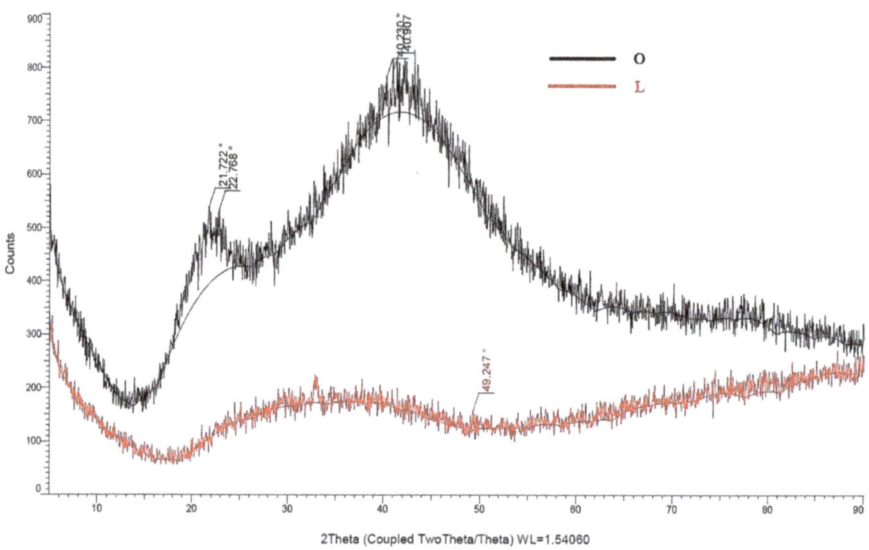

Where 'O' stands, "neat-recycled EVA polymeric composites", and 'L' refers "leather-shavings/recycled EVA Polymeric-composites".

Figure 20. XRD spectra-pattern of neat-recycled EVA polymeric composites and leathe shavings/recycled EVA Polymeric-composites.

The impact of leather wastes on the crystalline structure of recycled EVA polyme composites was explored utilizing XRD-Diffraction. The intense reflections observe in recycled EVA polymeric were seen to be in leather-wastes/recycled EVA Polymeri composites. Chen and Wu reported comparable outcomes utilizing PCL-multiwalle carbon-nanotubes. Investigators discovered that incorporating MWCNT into PCL ha no influence on the crystalline formation of PCL [83,84]. Jiang et al. found that the siz of crystallites orthogonal to the (hkl) plane in PCL/silica composites rises progressivel as the PCL wt.% rises. This phenomenon has been ascribed to insufficiently crystallize macromolecules or tiny and metastable-crystalline that are firmly bound by rigid silic network-frameworks [83,84].

This novel outcome furthermore shows that the crystallinity value of leather-loade recycled EVA composites is maximum and the neat-recycled EVA composites are minimum which eventually confirms that leather-filled recycled EVA composites have more hardnes value as compared to neat recycled EVA composite samples. The relative-intensity of peak on another peak was determined and recorded versus the leather-waste-loading a depicted in Figure 20.

To recapitulate, the noisy lines observed from XRD Diffraction peaks are due to th presence of the extracellular fibrous protein collagenous in the leather fibers. The XRI analysis of leather composite materials revealed distinct considerable diffraction noisy pattern-peaks caused by the prolonged-range ordered sequence of collagenous fibrillar molecules, with repetitive gap/overlap locations [118–124].

Table 1. XRD findings for chosen peak-regions with Net-intensity values, 2-theta values, relative-intensities, full-width half-maximum values (FWHM), and lattice-parameters (d).

Neat-Recycled EVA Polymeric Composites					Leather-Shavings/Recycled EVA Polymeric-Composites				
2θ (°)	FWHM	Net Intensity	Relative Intensity	d Line-Spacing	2θ (°)	FWHM	Net Intensity	Relative Intensity	d Line-Spacing
21.722	1.032	145	100%	4.08798 Å	49.247	0.509	34.6	100%	1.84877 Å
22.768	0.903	122	84.2%	3.90256 Å					
40.230	0.774	71.3	49.2%	2.23984 Å					
40.907	1.011	77.5	53.5%	2.20434 Å					

Table 2. %age Crystallinity test for the fabricated polymeric composites.

Samples	Crystallinity % Age	Amorphous % Age
Neat-recycled EVA polymeric composites	7.0	93.0
Leather-shavings/recycled EVA Polymeric-composites	19.3	80.7

Comparable findings have been unveiled in numerous studies. For instance, the XRD patterns of composites containing leather shavings and hybrid leather powder reveal that the crystalline structure collapsed due to the presence of SiO_2 particles in hybrid leather power-reinforced composites, which block the role between leather fibers in the ordered region. However, both composites possessed the triple helical intrinsic fibrous structure, and it is an indication of the use of ultrafine leather powder as a functional reinforcement [125]. The leather fibrous powder can also be effectively used as an adsorbent to remove dyes from the wastewater, and the X-ray diffraction analysis reveals that cow leather powders retain their protein structure [126]. A thin film X-ray diffraction analysis is also performed by researchers to examine the collagen fiber spacing and to identify the phases of isolated and commercial collagens [127]. The results of this study suggested the elimination of acetic acid residues from the isolated collagen for the purpose of a stable environment for the cells and their applications. X-ray diffraction patterns of Bacterial Cellulose/collagen hydrogel confirm the introduction of collagen in it and suggest a more amorphous structure than the Bacterial Cellulose alone [128]. When the XRD spectra of leather fibers with chromium-tanned, the intensity of Bragg peaks continuously increases. There seem to be two probable reasons for this: an enhancement in the overall long-order structural-arrangement of the collagenous fibrils and/or the inclusion of Cr(III) ions in the collagenous framework, which further strengthens the electron-density contrasts [119,121–124]. Thus accordingly, Maxwell et al., if the improvement in absorption diffraction peak-intensities in the XRD-spectrum of chromium-tanned leather composites was primarily exacerbated by the increased electronic-density contrasts induced by Cr(III), therefore the adhesion-binding of Cr(III) all along the length of the collagenous-network molecule must be considered to even affect the same absorption spectra intensity dispersal of the peaks similar to that of the untanned leather fibers. If collagen binding is restricted to specific amino acids, then these amino-acids. If collagen-binding interaction is restricted to specific amino-acid residues, therefore these organic-molecules should always be dispersed reasonably uniformly throughout the collagenous network molecule [119]. The amplitude of the maxima peak attributable to intermolecular lateral packing was seen to reduce as chromium content was increased. After the inclusion of chrome-content, the inter-molecular laterally pack distances dropped somewhat, and the full-width at half-maximum (FWHM) of such a peak-point increased, implying a significantly wider dispersion. The drop in diffraction absorption peak seemed to be associated with the decline in inter-molecular pack-distanc ordered arrangement [119].

Earlier studies have revealed the outcomes from a comparative assessment of previou XRD-outcomes. Similarly, Ammasi et al. [129] proposed a new method to reuse solid was materials obtained from leather industries. The XRD patterns of untreated and polypeptid treated leather samples illustrated the high intensity of the diffraction peaks. This type c pattern was obtained due to their highly e crystallinity. XRD patterns also indicated tha the crystallinity of treated samples was better than the untreated ones. The noisiness i the patterns is also attributed to the alteration in the inter/intra molecular packing of th leather molecules. Cardona et al. [43] utilized leather waste materials in the natural rubbe and characterized. The authors observed similar spectra of leather waste material an urea-treated leather waste materials. This type of formation is due to the addition of sodiur carbide and sodium sulfate during leather tanning, which makes sure the interaction c chrome into collagen networks. Wang and Jin [130] synthesized waterborne polyurethan in the form of fillers, which were used in the development of synthetic leather. XRI patterns revealed that the micro-phase was separated (increasing intensity in XRD peaks This is due to the enhancement in steric hindrance and hydrogen bonding of the ingredien molecules. Ma et al. [131] prepared chromium-tanned leather porous carbon sample using different mass ratios of alkali and carbon. The authors observed that as-prepare carbon exhibits broad diffraction peaks from 15° to 30° of theta angle due to amorphou crystalline structures. However, the pattern of BWLPC-2 indicated the presence of CrO a 36° and 64° theta angles. Also, with the introduction of KOH, diffraction peaks of activate carbon weakened continuously until they disappeared. Kanagaraj et al. [132] prepare protein-based material using trimming wastes; this material was then used in the chrom tanning process to increase the exhaustion level of the Cr. The increasing intensity in XRI patterns revealed enhanced exhaustion of the Cr contents. This improvement in exhaustio was due to the increasing interaction among the collagen sites and Cr contents. Ribeir et al. [133] explored the utilization of tannery waste materials to resolve environmenta issues. Authors treated leather saving on immobilizing chromium ions from the Portlan cement matrix. XRD patterns revealed the loss of a large amount of organic material durin ignition, and Cr_2O_3 is the inorganic phase that existed in the material, which is formed du to the oxidation of chromium. Dey et al. [134] introduced ZnO nano particulates in buffin dust to develop anti-bacterial characteristic footwear sols. The broad diffraction peak a $2\theta = 26°$ is observed in the XRD pattern of the ZnO-reinforced composite. This broadenin nature was due to the elongation of cross-linked chains and the interaction of collagen wit amorphous crystals that ensures the growth of ZnO particulates on the buffing dust. L et al. [44] developed polyvinyl alcohol-based composite films, which were reinforced wit leather fibers. Polyvinyl alcohol indicates strong peaks at $2\theta = 19.5°$ due to better crystallin performance and low-intensity peaks at 11.8° because of closer packing arrangement. Saiki et al. [135] used industrial leather waste to prepare composite products and evaluated thei properties. Dye-trimmed waste materials, together with natural fibers (jute and cotton were used during fabrication. The authors observed a composite containing a 1:1 blen ratio of leather waste and natural fiber waste possessed the best characteristic propertie among others. XRD pattern of fabricated composites indicated weaker and broader pea intensities than base reinforcing materials due to structural-modifications in it and also du to interactions among the collagen and natural fibers during formulation.

15. Atomic Force Microscopy (AFM) Analysis

AFM has been utilized to determine the size (thickness, width, etc.) of the interphas and its stiffness relative to the bulk phase of neat and leather-shavings/recycled EVA polyme composites. AFM has been utilized to examine the AFM Roughness Analysis and AFM Grai Analysis of the fiber-reinforced with the leather-shavings/recycled EVA polymer composite Phase imaging was focused on individual fibers to map the leather fibers/recycled EV polymer composites interface and to see the fracture behavior (cracks etc.).

AFM has been utilized to explore the interfacial adhesion properties of Mechanical tested specimens of neat and leather-shavings/recycled EVA polymer composite surface, their surface topography mapping, phase-imaging analysis, and lateral forces of polymer composites that have leather-shavings.

AFM has been utilized to analyze the surface topography mapping, phase-image analysis, and lateral forces of neat and recycled EVA polymer composites that have leather-shavings as fibers.

AFM has been utilized to analyze the worn surface of the leather-shavings/recycled EVA polymer composites and to determine the elastic modulus characterization of neat polymer and leather-shavings/recycled EVA polymer composites. AFM phasing image analyzes variability in compositions/constituents, adherence, frictional, rheological characteristics, and certain other characteristics in addition to fundamental topographical map imaging. AFM has been employed to determine the surface structure, roughness of neat, and leather-shavings/recycled EVA polymer composites, leather fiber geometry, leather fiber diameter, filler distribution, leather fiber-dimensions and leather fiber-recycled EVA matrix adhesion in leather-shavings/recycled EVA polymer composites.

AFM analysis reveals a surface-layer of a neat recycled EVA and leather-shavings loaded recycled EVA polymeric-composites with several void-spaces, cavities, and significantly larger-depths, as well as peak-position ranges, whereas the inclusion of additives and lubricants, such as zinc octadecanoate and Octadecanoic acid into the recycled EVA polymeric matrix reveals a relatively uniformly smooth-surface with fewer void-spaces in material, but much more prominent peak was observed as showed in Figure 21.

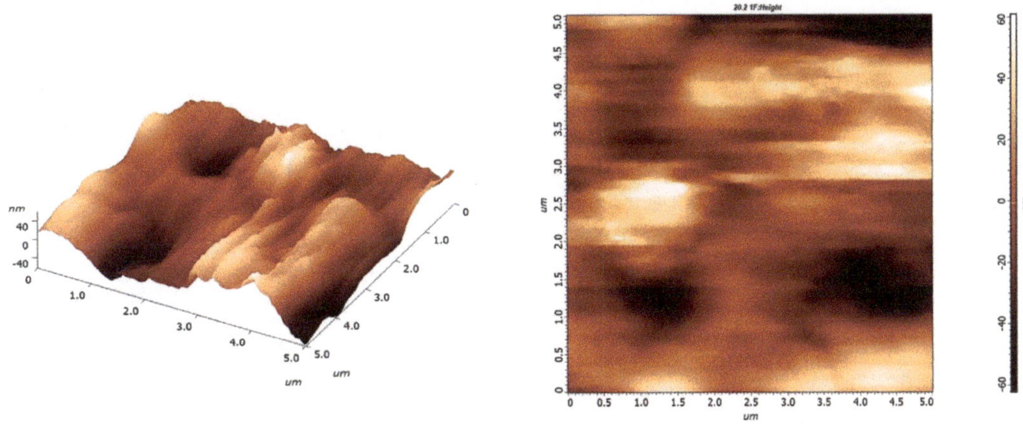

Figure 21. AFM topographic pictograph illustrating the surface and micro-hole profilometry for neat recycled EVA polymer composites.

Figure 22 shows that the surface-roughness values rise with the incorporation of Leather-shavings into the blend, indicating that the interfacial-contact among the leather-wastes, as well as the recycled EVA polymeric-matrix, is weaker, resulting from inadequate distribution and compliance. This revealed that leather-shavings, as well as the recycled EVA polymeric-matrix were compatible. It is claimed that the inclusion of additives, such as plasticizers and fillers, has certainly enhanced the interaction of leather-shaving wastes with recycled EVA polymeric-matrix and high interfacial surface-smoothness.

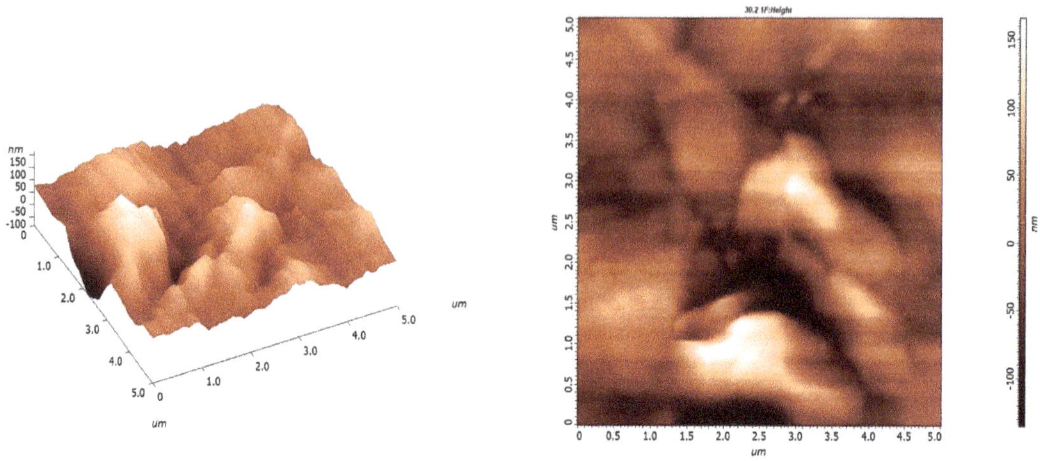

Figure 22. AFM topographic pictograph illustrating the surface and micro-hole profilometry fc Leather shavings wastes/recycled EVA polymer composites.

16. Applications of the Developed Leather Waste/Recycled EVA Polymer Composites

As a result of developing sustainable functional or viable products from waste materia there will be a synergistic influence on the surrounding environment, in addition to produ ing 'value-added products' derived from "waste materials" accumulated in industries suc as the 'Leather' and 'Polymer' sectors. The utilization of 'leather wastes', and 'recycle polymers' are likewise explored, which implies the widespread deployment of "leathe wastes" technology for the manufacture of composites with enhanced characteristics fc a diverse spectrum of application sectors, particularly the 'footwear industry', 'appare accessories', 'automobile', 'construction', and 'industrial usages'.

Numerous multipurpose applications for the unique class of manufactured composite can be found in the 'footwear' and 'apparel sectors', 'composite sheets', and 'Leather-lik personal-goods'. The 'floorings', 'personal safety products', applications for 'leather i decorative furniture', 'leather boards', and 'inexpensive adsorbents', 'Activated Carbon 'energy generation' and 'energy recovery', 'adhesives in the woodworking industry', 'ce ramic glaze pigments', 'production of biogas', 'Agriculture uses', 'biodegradable materials and 'polymer films', 'Poultry feeds', 'cosmetics', 'biomedical uses', 'biocomposites', an the ability to 'microencapsulate drugs', as well as 'construction' and 'building materials 'automotive interior trim molding components', 'thermal' and 'acoustical insulation-panels 'cushioning boards', 'shoe soles', 'floor covering', and 'moldings' with remarkably excep tional 'physical properties', 'air permeance', and 'desirable aesthetics'.

17. Concluding Remarks

In this study, recycled EVA polymer matrices and the high-weight content of leathe shavings with 1:1 were mixed with fabricating the flexible composite sheets through th combined effects of two-roll milling and hot-press compression molding with uniforr blending. Observations based on thorough, detailed characterization investigations includ the following:

i. The physico-mechanical properties of "leather shavings/recycled EVA" compos ites exhibited to be significantly influenced by the leather-fibrous loading with composition of 1:1. The tensile strength tends to rise slightly when the proportio of leather shavings in composites has been increased to 1:1. As the volume c leather fibers within the composites has increased, the modulus of elasticity of th composites has significantly improved. Leather shavings were discovered to have stronger compressional deformation property in the recycled EVA matrix than th 'neat-recycled EVA' matrix, which was reported to be higher by about 7.7%. The average peel strength between the polymer and leather (for leather shavings as fiber) in the recycled EVA matrix was determined to be approximately 0.9575 N/mm, according to the results obtained.

ii. The TGA investigation showed that 'recycled-EVA' polymer was thermostable up to 213.47 °C, whereas leather fibers showed no discernible major weight loss nearly comparable to 211 °C. According to the DSC results, the 'release of moisture' from the 'leather shavings' through an endothermic transition that occurs at about 100 °C is thermostable up to 211 °C and begins to decompose collagen at 332.56 °C for 'neat-recycled EVA' samples and 327.23 °C for "leather shavings/recycled EVA" polymer composite samples, respectively.

iii. In reference to the absorbance bands O-H (3100–2914.88 cm^{-1}) that increased in intensity under humid conditions due to hydrogen bonding interactions between the carbonyl group in vinyl acetate and water, the ATR-FTIR spectra have explored surface-interface-layer alterations and abnormalities in recycled EVA co-polymer. While the molecular structure of "leather shavings/recycled EVA" composites can be seen in the ATR-FTIR monograph as the band at 3314.07 cm^{-1} corresponding to the vinyl alcohol -OH group; 2849.31–2916.81 cm^{-1} is due to the -CH_2, -CH_3 groups inside chains and terminal groups; and below 723.175 to 600 cm^{-1} is due to Cr-O bonds.

iv. Leather fibril particles are widely present as conglomerate aggregate clusters and are effectively distributed or interspersed throughout the recycled EVA matrices, according to the SEM surface morphology examination of the "leather shavings/recycled EVA" composites. The micrograph results show numerous interfaces with remarkable bonding strength and interfacial contact between the recycled EVA matrix and the leather shavings' residual particles.

v. According to "XRD-analysis", the "crystallinity value" of the 'neat-recycled EVA' polymeric composite sample is 7%, and the amorphous content is revealed to be 93%. While the crystallinity of the leather-shavings/recycled EVA polymeric composite sample was revealed to be 19.3% and the amorphous content to be 80.7 percent. This unexpected result has also demonstrated that the crystallinity value of the leather-filled recycled EVA composites is higher than that of the neat recycled EVA composite samples, which eventually confirms that the leather-filled recycled EVA composites have higher values for strength, modulus, and hardness.

vi. The inclusion of additives and lubricants into the recycled EVA polymeric matrix has revealed a relatively uniformly smooth-surface with fewer void-spaces in the material, but a much more prominent peak has been observed.

To recapitulate, it is necessary to conclude that these leather shavings waste/recycled EVA polymer composites with lower cost can be employed for multipurpose applications as well as in the reduction of environmental pollution.

18. Suggestions for Future Work

The subsequent suggestions were proposed premised on aforesaid experimental outcomes.

The leather shavings are merely employed in the current study to reuse, reprocess, and fabricate flexible composites. However, this study also strongly recommends that future research can be done to assess the various other leather residues which have been produced during the reutilization and reprocessing techniques.

By using various cross-linkers, modifications can be made to the "leather fibers reinforced recycled EVA" polymer composites in order to broaden their application areas and improve their mechanical properties.

The precise degradation mechanisms would be made clearer by a combined TGA-FTIR investigation of such model composition (1:1). In order to explore the energy changes associated with the degradation features of these model compounds, advanced calorimetric techniques may also be used. It is equally important to combine recycled EVA polymer with leather wastes in suitable reactors on a large scale and, furthermore, to study the various processing parameters for their pyrolysis in controlled environments, even though suc microscopic analysis could increase understanding of the fundamental degradation proces

While an increase in contact points between the constituents could significantly d grade the recycled EVA matrix, a reduction in particle size and the resulting increase i surface area would improve reinforcing. Therefore, it may be possible to study the influenc of surface area on the potential degradation of the recycled EVA matrix. Additionally, th processes that would eliminate chromium from the leather shavings might theoreticall avert recycled EVA matrix deterioration. It is worthwhile to look at the characteristics o leather composites with recycled EVA polymer if the leather shavings are to be utilize directly without any processing. Thus, it might be conceivable to incorporate the recyclin of 'recycled-EVA polymer' and the 'tannery wastes' in this process.

Supplementary Materials: The following supporting information can be downloaded at: http //www.mdpi.com/article/10.3390/su15054333/s1, Figure S1. (a) Leather-shavings employed a reinforcing-particulates, and (b) Recycled EVA granules Adapted from the reference [45]. Figure S Process flow diagram for recycling thermoplastic elastomer and polymer composites from soli leather wastes. Figure S3. (a) "Neat recycled-EVA polymeric-composites Adapted from the refe ence [45]", and (b) "Leather-shavings/recycled EVA-polymeric composites". Figure S4. "Dumb-be shaped' Tensile Test Sample Adapted from the reference [45]. Figure S5. (a,b) "Variation of te sile load (N) against extension (mm) of recycled EVA polymer composites with or without leathe shavings". Figure S6. Comparison of "Compressive strength (MPa)" for "neat-recycled EVA", an "recycled-EVA polymer composites". Table S1. Compositions of polymer composite samples i wt.-ratios. Table S2. Standard test methods, equipment details with model/machine numbers, an standards were followed in this investigation. Table S3. Absorbance values were obtained wit the phosphate buffer testing procedure. Table S4. Absorbance values were obtained with a wate testing procedure. Table S5. Presence of functional group in neat recycled EVA without incorporatin leather fiber reinforcements in FTIR analysis. Table S6. Presence of functional group in Leathe shavings/recycled EVA polymer composites in FTIR analysis.

Author Contributions: Conceptualization, S.S. (Shubham Sharma); methodology, S.S. (Shubhar Sharma); formal analysis, S.S. (Shubham Sharma), P.S., J.S., S.M.R., S.S. (S. Siengchin); investigatio S.S. (Shubham Sharma); resources, S.S. (Shubham Sharma), P.S., J.S., S.M.R., S.S. (S. Siengchin writing—original draft preparation, S.S. (Shubham Sharma); writing—review and editing, S.S. (Shu ham Sharma), P.S., J.S., S.M.R., S.S. (S. Siengchin); supervision, S.S. (Shubham Sharma), P.S., J.S S.M.R., S.S. (S. Siengchin); project administration, S.S. (Shubham Sharma), P.S., J.S. All authors hav read and agreed to the published version of the manuscript.

Funding: This article received no external fundings from any of the sources.

Institutional Review Board Statement: Not applicable.

Informed Consent Statement: Not applicable.

Data Availability Statement: No data were used to support this study.

Acknowledgments: The author Shubham Sharma wishes to acknowledge the Department of RI IK Gujral Punjab Technical University, Kapurthala, Punjab, India, for providing an opportunity t conduct this research task. The author P. Sudhakara gratefully acknowledges the support from Scienc and Engineering Research Board (SERB, YSS/2015/001294), New Delhi, India and CLRI-MLP-2 Furthermore, the authors are thankful to the department of CATERS, Chennai, for providing th characterization facilities (CSIR-CLRI Communication no. 1671).

Conflicts of Interest: The authors declare no conflict of interest.

References

1. Sundar, S.; Sinha, P.; Rai, M.; Verma, D.; Nawin, K.; Alam, S. Comparison of short-course multidrug treatment with standar therapy for visceral leishmaniasis in India: An open-label, non-inferiority. *Lancet* **2011**, *377*, 477–486. [CrossRef] [PubMed]
2. Ramasami, T. Approach towards a unified theory for tanning: Wilson's dream. *J. Am. Leather Chem. Assoc.* **2001**, *96*, 290–304.
3. Kanagaraj, J.; Senthilvelan, T.; Panda, R.C.; Kavitha, S. Eco-friendly waste management strategies for greener environme towards sustainable development in leather industry: A comprehensive review. *J. Clean. Prod.* **2015**, *89*, 1–17. [CrossRef]

Jiang, H.; Liu, J.; Han, W. The status and developments of leather solid waste treatment: A mini-review. *Waste Manag. Res.* **2016**, *34*, 399–408. [CrossRef]

Sundar, V.J.; Raghavarao, J.; Muralidharan, C.; Mandal, A.B. Recovery and Utilization of Chromium- Tanned Proteinous Wastes of Leather Making: A Review. *Crit. Rev. Environ. Sci. Technol.* **2011**, *41*, 2048–2075. [CrossRef]

Newhall, H. *South Essex Sewerage District. Personal Communication with J. Margolis*; JRB Associates: Framingham, CA, USA, 1982.

Cabeza, L.F.; Taylor, M.M.; Dimaio, G.L.; Brown, E.; Marmer, W.N.; Carrió, R.; Celma, P.J.; Cot, J. Processing of leather waste: Pilot scale studies on chrome shavings. Isolation of potentially valuable protein products and chromium. *Waste Manag.* **1998**, *18*, 211–218. [CrossRef]

Krecke, E. Process for the Production of Leather Base Material. U.S. Patent US4536430, 25 May 1982.

Waite, E.; Franklin, F.B. Improvement in Fabrics from Waste Leather. U.S. Patent US114373A, 2 May 1871.

Ostberg, A.J. Method of Manufacturing Substitutes for Leather-Board. U.S. Patents US976827A, 22 November 1910.

Mathieu, A.N. Improvement in Manufacture of Leather Pasteboard and Paper. U.S. Patents US20020A, 13 April 1855.

Suzuki, M.; Sasaki, M. Synthetic Resin Leather. U.S. Patent US20030190438A1, 5 October 2004.

Adams, A., Jr. Flexible-Leather-Board Composition. U.S. Patent US1188600A, 27 June 1916.

Albert, L. Fibrous Material and Method of Making the Same. U.S. Patent US1269905A, 12 June 1918.

Wyler, A. Process for Bonding Leather to Leather. U.S. Patent US5028285A, 2 July 1991.

Madera-Santana, T.J.; Aguilar-Vega, M.; Marquez, A.; Moreno, F.; Richardson, M.; Machin, J. Production of leather-like composites using short leather fibers. II. Mechanical characterization. *Polym. Compos.* **2004**, *23*, 991–1002. [CrossRef]

Andreopoulos, A.; Tarantili, P. Waste leather particles as a filler for poly (vinyl chloride) plastisols. *JMS-Pure Appl. Chem.* **2007**, *37*, 1353–1362. [CrossRef]

Garcia, N.G.; Reis, E.A.P.; Budemberg, E.R.; Agostini, D.L.D.S.; Salmazo, L.O.; Cabrera, F.C.; Job, A.E. Natural rubber/leather waste composite foam: A new eco-friendly material and recycling approach. *J. Appl. Polym. Sci.* **2014**, *132*. [CrossRef]

Jimenez, J.P.; Mari, E.; Villena, E.M.; Cabangon, R.J. Utilization of Spent Tea Leaves and Waste Plastics for Composite Boards. *Philipp. For. Prod. J.* **2013**, *4*, 29–36.

Senthil, R.; Hemalatha, T.; Kumar, B.S.; Uma, T.S.; Das, B.N.; Sastry, T.P. Recycling of finished leather wastes: A novel approach. *Clean Technol. Environ. Policy* **2014**, *17*, 187–197. [CrossRef]

Parisi, M.; Nanni, A.; Colonna, M. Recycling of Chrome-Tanned Leather and Its Utilization as Polymeric Materials and in Polymer-Based Composites: A Review. *Polymers* **2021**, *13*, 429. [CrossRef] [PubMed]

Ponsubbiah, S.; Suryanarayana, S.; Gupta, S. Composite from Leather Waste. *Int. J. Latest Technol. Eng. VII* **2018**, *7*, 77–80.

Zăinescu, G.; Deselnicu, V.; Constantinescu, R.; Georgescu, D. Biocomposites from tanned leather fibres with applications in construction. *Leather Footwear J.* **2018**, *18*, 203–206. [CrossRef]

Dwivedi, S.P.; Petru, M.; Saxena, A.; Sharma, S.; Mishra, M.; Pramanik, A.; Singh, S.; Li, C.; Ilyas, R.A. Recovery of Cr from chrome-containing leather wastes to develop aluminum-based composite material along with Al_2O_3 ceramic particles: An ingenious approach. *Nanotechnol. Rev.* **2022**, *11*, 3218–3234. [CrossRef]

Shanmugam, V.; Mensah, R.A.; Försth, M.; Sas, G.; Restás, Á.; Addy, C.; Xu, Q.; Jiang, L.; Neisiany, R.E.; Singha, S.; et al. Circular economy in biocomposite development: State-of-the-art, challenges and emerging trends. *Compos. Part C Open Access* **2021**, *5*, 100138. [CrossRef]

Pan, S.; Zhao, M.; Andrawes, B.; Zhao, H.; Li, L. Compressive behavior of cylindrical rubber buffer confined with fiber reinforced polymer. *J. Low Freq. Noise Vib. Act. Control* **2020**, *39*, 470–484. [CrossRef]

Dwivedi, S.P.; Agrawal, R.; Sharma, S. Effect of Friction Stir Process Parameters on Mechanical Properties of Chrome Containing Leather Waste Reinforced Aluminium Based Composites. *Int. J. Precis. Eng. Manuf. -Green Technol.* **2021**, *8*, 935–943. [CrossRef]

SLTC. *Society of Leather Technologist and Chemists*; Official Methods of Analysis: Northampton, UK, 1996.

IULTCS Chem. Test Method: Determ. *Chromium VI Content* **1997**, *81*, 109.

Compounds, E.B. Our range of EVA-based Compounds were developed to meet market demands for applications where lightness is paramount. Learn more about our APIZERO™ and APIFIVE™ range of products. Available online: https://www-v2.trinseo.com/Solutions/Thermoplastic-Elastomers/EVA-based-Compounds (accessed on 19 November 2022).

Kanagaraj, J.; Velappan, K.C.; Babu, N.K.C.; Sadulla, S. Solid Wastes Generation in the Leather Industry and Its Utilization for Cleaner Environment. *ChemInform* **2006**, *37*, 541–548. [CrossRef]

Xie, J.; Wang, S.; Cui, Z.; Wu, J. Process Optimization for Compression Molding of Carbon Fiber- Reinforced Thermosetting Polymer. *Materials* **2019**, *12*, 2430. [CrossRef] [PubMed]

Singh, J.K.; Ching, Y.; Abdullah, L.; Ching, K.; Razali, S.; Gan, S. Optimization of Mechanical Properties for Polyoxymethylene/Glass Fiber/Polytetrafluoroethylene Composites Using Response Surface Methodology. *Polymers* **2018**, *10*, 338. [CrossRef] [PubMed]

Govindaraju, R.; Jagannathan, S. Optimization of mechanical properties of silk fiber-reinforced polypropylene composite using Box-Behnken experimental design. *J. Ind. Text.* **2018**, *47*, 602–621. [CrossRef]

Kale, R.D.; Jadhav, N.C. Utilization of waste leather for the fabrication of composites and to study its mechanical and thermal properties. *SN Appl. Sci.* **2019**, *1*, 1–9. [CrossRef]

Ibarra, L.; Jorda, C. Effect of a Diazide as Adhesive Agent in Elastomeric Matrix-Short Polyamide Fibers Composite. *J. Appl. Polym. Sci* **1993**, *48*, 375–381. [CrossRef]

37. Ambrósio, J.D.; Lucas, A.A.; Otaguro, H.; Costa, L.C. Preparation and characterization of poly (vinyl butyral)-leather fib composites. *Polym. Compos.* **2011**, *32*, 776–785. [CrossRef]
38. Nanni, A.; Parisi, M.; Colonna, M.; Messori, M. Thermo-Mechanical and Morphological Properties of Polymer Composit Reinforced by Natural Fibers Derived from Wet Blue Leather Wastes: A Comparative Study. *Polymers* **2021**, *13*, 1837. [CrossRe
39. Ravichandran, K.; Natchimuthu, N. Vulcanization characteristics and mechanical properties of natural rubber-scrap rubb compositions filled with leather particles. *Polym. Int.* **2005**, *54*, 553–559. [CrossRef]
40. Joseph, S.; Ambone, T.S.; Salvekar, A.V.; Jaisankar, S.N.; Saravanan, P.; Deenadayalan, E. Processing and characterization of was leather based polycaprolactone biocomposites. *Polym. Compos.* **2017**, *38*, 2889–2897. [CrossRef]
41. Dwivedi, S.P.; Sahu, R.; Saxena, A.; Dwivedi, V.K.; Srinivas, K.; Sharma, S. Recovery of Cr from chrome-containing leather was and its utilization as reinforcement along with waste spent alumina catalyst and grinding sludge in AA 5052-based metal matr composites. *Proc. Inst. Mech. Eng. Part E J. Process Mech. Eng.* **2022**, *236*, 160–170. [CrossRef]
42. Sun, D.; Yan, J.; Ma, X.; Lan, M.; Wang, Z.; Cui, S.; Yang, J. Tribological Investigation of Self-Healing Composites Containin Metal/Polymer Microcapsules. *ES Mater. Manuf.* **2021**, *14*, 59–72. [CrossRef]
43. Cardona, N.; Velásquez, S.; Giraldo, D. Characterization of leather wastes from chrome tanning and its effect as filler on th rheometric properties of natural rubber compounds. *J. Polym. Environ.* **2017**, *25*, 1190–1197. [CrossRef]
44. Li, C.; Feng, X.; Ding, E. Preparation, properties, and characterization of novel fine leather fibers/polyvinyl alcohol composite *Polym. Compos.* **2015**, *36*, 1186–1194. [CrossRef]
45. Sharma, S.; Sudhakara, P.; Petru, M.; Singh, J.; Rajkumar, S. Effect of nanoadditives on the novel leather fiber/recycle poly(ethylene-vinyl-acetate) polymer composites for multifunctional applications: Fabrication, characterizations, and mu tiobjective optimization using central composite design. *Nanotechnol. Rev.* **2022**, *11*, 2366–2432. [CrossRef]
46. Natchimuthu, N.; Radhakrishnan, G.; Palanivel, K.; Ramamurthy, K.; Anand, J.S. Vulcanization characteristics and mechanic properties of nitrile rubber filled with short leather fibres. *Polym. Int.* **1994**, *33*, 329–333. [CrossRef]
47. Musa, E.T.; Hamza, A.; Ahmed, A.S. Investigation of the Mechanical and Morphological Properties of High-Density Polyethyler (HDPE)/Leather Waste Composites. *IOSR J. Appl. Chem.* **2017**, *10*, 48–58.
48. Ravichandran, K.; Natchimuthu, N. Natural Rubber: Leather Composites. *Polímeros* **2005**, *15*, 102–108. [CrossRef]
49. Montgomery, D.C. *Design and Analysis of Experiments*; John Wiley: New York, NY, USA, 2012.
50. Draper, N.R. On the Experimental Attainment of Optimum Conditions. In *Breakthroughs in Statistics*; Springer Series in Statistic Springer: Berlin/Heidelberg, Germany, 1951.
51. Myers, R.H.; Khuri, A.I.; Carter, W.H. Response surface methodology. *Technometrics* **1966**, *31*, 137–157.
52. SATRA Tm137, Tensile Properties of Plastics and Rubbers. June 1995. Available online: https://www.satra.com/test_methods detail.php?id=101 (accessed on 19 November 2022).
53. *ASTM D3410/D3410M-16e1*; Standard Test Method for Compressive Properties of Polymer Matrix Composite Materials wit Un-Supported Gage Section by Shear Loading. ASTM: West Conshohocken, PA, USA, 2016.
54. SATRA Tm-218, Tear Strength of Rubbers and Plastics—Trouser Method. June 1999. Available online: https://www.satra.com test_methods/detail.php?id=174 (accessed on 19 November 2022).
55. SATRA Tm 205, A. Shore, Hardness of Others-Soling Materials, Rubber Materials, PU Sole Materials, and Leather Sole. 201 Available online: https://www.satra.com/test_methods/detail.php?id=161 (accessed on 19 November 2022).
56. SATRA Tm 401, Peel Strength of Adhesive Bonds. 2000. Available online: https://www.satra.com/test_methods/detail.php?id 280 (accessed on 19 November 2022).
57. SATRA Tm 64, Compression Set Apparatus Test Machine/Equipment (GW-053). 1996. Available online: https://www.satr com/test_methods/detail.php?id=54 (accessed on 19 November 2022).
58. Kandar, M.M.; Akil, H.M. Application of Design of Experiment (DoE) for Parameters Optimization in Compression Moulding fc Flax Reinforced Biocomposites. *Procedia Chem.* **2016**, *19*, 433–440. [CrossRef]
59. Ramaraj, B. Mechanical and thermal properties of ABS and leather waste composites. *J. Appl. Polym. Sci.* **2006**, *101*, 3062–306 [CrossRef]
60. Ismail, H.; Nizam, J.M.; Khalil, H.P.S. The effect of a compatibilizer on the mechanical properties and mass swell of white ric husk ash filled natural rubber/linear low density polyethylene blends. *Polym. Test.* **2001**, *20*, 13–14. [CrossRef]
61. Ismail, H.; Edyham, M.R.; Wirjosentono, B. Bamboo fibre filled natural rubber composites: The effects of filler loading an bonding agent. *Polym. Test.* **2002**, *21*, 60–65. [CrossRef]
62. Dodwell, G. *Components—Guidelines and Characteristics Opportunities*; Leao, A.L., Carvalho, F.X., Frollini, E., Eds.; SATRA Bu Dongguan, China, 1989; pp. 208–209.
63. Covington, A.D.; Chemis, T. *The Science of Leather*; Royal Society of Chemistry: Cambridge, UK, 2009.
64. Stael, G.; Rocha, M.; Menezes, S.; Almeida, J.R.; Ruiz, N.M.D.S. Analysis of the mechanical properties and characterization b solid state 13C NMR of recycled EVA copolymer/silica composites. *Mater. Res.-IBERO-Am. J. Mater.* **2005**, *8*. [CrossRef]
65. Hang, L.T.; Viet, D.Q.; Linh, N.P.D.; Doan, V.A.; Dang, H.L.T.; Dao, V.D.; Tuan, P.A. Utilization of leather waste fibers in polyme matrix composites based on Acrylonitrile-Butadiene rubber. *Polymers* **2021**, *13*, 117. [CrossRef] [PubMed]
66. Aftab, M.; Shaikh, A.; Deb, A.K.; Akter, E.; Ferdous, T.; Abu, M.; Mia, S. Resource addition to leather industry: Adhesive fror chrome shaving. *J. Sci. Innov. Res.* **2017**, *6*, 138–141.

Park, K.W.; Kim, G.H.; Chowdhury, S.R. Improvement of compression set property of ethylene vinyl acetate copolymer/ethylene-1-butene copolymer/organoclay nanocomposite foams. *Polym. Eng. Sci.* **2008**, *48*, 1183–1190. [CrossRef]

Shieh, Y.T.; Tsai, T.H. Silane grafting reactions of low-density polyethylene. *J. Appl. Polym. Sci.* **1998**, *69*, 255. [CrossRef]

Shieh, Y.T.; Liu, C.M. Silane grafting reactions of LDPE, HDPE, and LLDPE. *J. Appl. Polym. Sci.* **1999**, *74*, 3404. [CrossRef]

Parent, J.S.; Geramita, K.; Ranganathan, S.; Whitney, R.A. Silane-modified poly (ethylene-co-vinyl acetate): Influence of comonomers on peroxide-initiated vinylsilane grafting. *J. Appl. Polym. Sci.* **2000**, *76*, 2–7. [CrossRef]

Sirisinha, K.; Meksawat, D. Comparison in processability and mechanical and thermal properties of ethylene-octene copolymer crosslinked by different techniques. *J. Appl. Polym. Sci.* **2004**, *93*, 1179–1185. [CrossRef]

Razavi-Nouri, M.; Sabet, A.; Tayefi, M. Effect of dynamic curing time on thermal, mechanical and rheological behavior of organoclay-containing nanocomposite based on ethylene-octene copolymer. *J. Polym. Res.* **2017**, *24*, 101. [CrossRef]

Kuan, H.C.; Kuan, J.F.; Ma, C.C.M.; Huang, J.M. Thermal and mechanical properties of silane-grafted water crosslinked polyethylene. *J. Appl. Polym. Sci.* **2005**, *96*, 2383–2391. [CrossRef]

Sirisinha, K.; Meksawat, D. Preparation and properties of metallocene ethylene copolymer crosslinked by vinyl trimethoxysilane. *Polym. Int.* **2005**, *54*, 1014–1020. [CrossRef]

Tegegn, W.B. Preparation of Biodegradable Film Using Gelatin from Wet Blue Shaving and Starch from Tamarind Seed. Master's Thesis, Bahir Dar University, Bahir Dar, Ethiopia, 2018.

De Campo, C.; Pagno, C.H.; Costa, T.M.H.; Rios, A.D.O.; Flôres, S.H.; Costa, T.M.H.; Rios, A.D.O.; Flôres, S.H. Gelatin capsule waste: New source of protein to develop a biodegradable film. *Polímeros* **2017**, *27*, 100–107. [CrossRef]

Liu, J.; Luo, L.; Zhang, Z.; Hu, Y.; Wang, F.; Li, X.; Tang, K. A combined kinetic study on the pyrolysis of chrome shavings by thermogravimetry. *Carbon Resour. Convers.* **2020**, *3*, 156–163. [CrossRef]

Masilamani, D.; Srinivasan, V.; Ramachandran, R.K.; Gopinath, A.; Madhan, B.; Saravanan, P. Sustainable packaging materials from tannery trimming solid waste: A new paradigm in wealth from waste approaches. *J. Clean. Prod.* **2017**, *164*, 885–891. [CrossRef]

Oksman, K.; Clemons, C. Mechanical properties and morphology of impact modified polypropylene-wood flour composites. *J. Appl. Polym. Sci.* **1998**, *67*, 1503–1513. [CrossRef]

Setua, K.; De, S.K. Short silk fiber reinforced natural rubber composites. *Rubber Chem. Technol.* **1983**, *56*, 808–826. [CrossRef]

Coran, A.Y.; Boustany, K.; Hamed, P. Short-Fiber-Rubber Composites: The Properties of Oriented Cellulose-Fiber-Elastomer Composites. *Rubber Chem. Technol.* **1974**, *47*, 396–410. [CrossRef]

Connor, J.E. Short-Fiber-Reinforced Elastomer Composites. *Rubber Chem. Technol.* **1977**, *50*, 945–958. [CrossRef]

Chen, E.C.; Wu, T.M. Isothermal crystallization kinetics and thermal behavior of poly(ε-caprolactone)/multi-walled carbon nanotube composites. *Polym. Degrad. Stab.* **2007**, *92*, 1009–1015. [CrossRef]

Jiang, S.; Ji, X.; Lijia, A.; Jiang, B. Crystallization behavior of PCL in hybrid confined environment. *Polymer* **2001**, *42*, 3901–3907. [CrossRef]

Mamoune, A.; Saouab, A.; Ouahbi, T.; Park, C.H. Simple models and optimization of compression resin transfer molding process. *J. Reinf. Plast. Compos.* **2011**, *30*, 1629–1648. [CrossRef]

Huang, J.; Zou, W.; Luo, Y.; Wu, Q.B.; Lu, X.; Qu, J. Phase Morphology, Rheological Behavior, and Mechanical Properties of Poly(lactic acid)/Poly (butylene succinate)/Hexamethylene Diisocyanate Reactive Blends. *ES Energy Environ.* **2021**, *12*, 86–94. [CrossRef]

Islam, A.; Dwivedi, S.P.; Dwivedi, V.K.; Sharma, S.; Kozak, D. Development of Marble Dust/Waste PET Based Polymer Composite Material for Environmental Sustainability: Fabrication and Characterizations. *J. Mater. Perform. Charact.* **2021**, *10*, 538–552. [CrossRef]

Prabhu, P.; Iqbal, S.M.; Balaji, A.; Karthikeyan, B. Experimental investigation of mechanical and machining parameters of hybrid nanoclay glass fiber-reinforced polyester composites. *Adv. Compos. Hybrid Mater.* **2019**, *2*, 93–101. [CrossRef]

Gopinath, R.; Poopathi, R.; Saravanakumar, S.S. Characterization and structural performance of hybrid fiber-reinforced composite deck panels. *Adv. Compos. Hybrid Mater.* **2019**, *2*, 115–124. [CrossRef]

Guo, G.; Finkenstadt, V.L.; Nimmagadda, Y. Mechanical properties and water absorption behavior of injection-molded wood fiber/carbon fiber high-density polyethylene hybrid composites. *Adv. Compos. Hybrid Mater.* **2019**, *2*, 690–700. [CrossRef]

Ma, J.; Li, Z.J.; Xue, Y.Z.B.; Liang, X.Y.; Tan, Z.J.; Tang, B. Novel PEEK/nHA composites fabricated by hot-pressing of 3D braided PEEK matrix. *Adv. Compos. Hybrid Mater.* **2020**, *3*, 156–166. [CrossRef]

Mei, H.; Ali, Z.; Ali, I.; Cheng, L. Tailoring strength and modulus by 3D printing different continuous fibers and filled structures into composites. *Adv. Compos. Hybrid Mater.* **2019**, *2*, 312–319. [CrossRef]

Sharma, T.; Singh, S.; Sharma, S.; Sharma, A.; Shukla, A.K.; Li, C.; Zhang, Y.; Eldin, E.M.T. Studies on the Utilization of Marble Dust, Bagasse Ash, and Paddy Straw Wastes to Improve the Mechanical Characteristics of Unfired Soil Blocks. *Sustainability* **2022**, *14*, 14522. [CrossRef]

Qin, Y.; Yu, Q.; Yin, X.; Zhou, Y.; Xu, J.; Wang, L.; Wang, H.; Chen, Z. Photo-polymerized trifunctional acrylate resin/magnesium hydroxide fluids/cotton fabric composites with enhancing mechanical and moisture barrier properties. *Adv. Compos. Hybrid Mater.* **2019**, *2*, 320–329. [CrossRef]

Rudresh, B.M.; Kumar, B.R.; Madhu, D. Combined effect of micro-and nano-fillers on mechanical, thermal, and morphological behavior of glass-carbon PA66/PTFE hybrid nano-composites. *Adv. Compos. Hybrid Mater.* **2019**, *2*, 176–188. [CrossRef]

96. Sengwa, R.J.; Choudhary, S.; Dhatarwal, P. Investigation of alumina nanofiller impact on the structural and dielectric propertie of PEO/PMMA blend matrix-based polymer nanocomposites. *Adv. Compos. Hybrid Mater.* **2019**, *2*, 162–175. [CrossRef]
97. Sun, H.; Fang, Z.; Li, T.; Lei, F.; Jiang, F.; Li, D.; Zhou, Y.; Sun, D. Enhanced mechanical and tribological performance of PA6 nanocomposites containing 2D layered α-zirconium phosphate nanoplatelets with different sizes. *Adv. Compos. Hybrid Mate* **2019**, *2*, 407–422. [CrossRef]
98. Ashok, R.B.; Srinivasa, C.V.; Basavaraju, B. Dynamic mechanical properties of natural fiber composites—A review. *Adv. Compo Hybrid Mater.* **2019**, *2*, 586–607. [CrossRef]
99. Das, T.K.; Ghosh, P.; Das, N.C. Preparation, development, outcomes, and application versatility of carbon fiber-based polyme composites: A review. *Adv. Compos. Hybrid Mater.* **2019**, *2*, 214–233. [CrossRef]
100. Dwivedi, S.; Yadav, R.; Islam, A.; Dwivedi, V.; Sharma, S. Synthesis, Physical and Mechanical Behavior of Agro-Waste RHA an Eggshell-Reinforced Composite Material. *J. Inst. Eng. (India) Ser. C.* **2022**, *103*. [CrossRef]
101. Liu, C.; Yin, Q.; Li, X. A waterborne polyurethane-based leather finishing agent with excellent room temperature self-healin properties and wear-resistance. *Adv. Compos. Hybrid Mater.* **2021**, *4*, 138–149. [CrossRef]
102. Etaati, A.; Pather, S.; Cardona, F.; Wang, H. Injection molded noil hemp fiber composites: Interfacial shear strength, fiber strengt and aspect ratio. *Polym. Compos.* **2014**, *37*, 213–220. [CrossRef]
103. Kabir, M.M.; Wang, H.; Lau, K.T.; Cardona, F. Chemical treatments on plant-based natural fibre reinforced polymer composite An overview. *Compos. Part B Eng.* **2012**, *43*, 2883–2892. [CrossRef]
104. Ku, H.; Wang, H.; Pattarachaiyakoop, N.; Trada, M. A review on the tensile properties of natural fiber reinforced polyme composites. *Compos. Part B Eng.* **2011**, *42*, 856–873. [CrossRef]
105. Liu, C.; Zhang, T.; Luo, Y.; Wang, Y.; Li, J.; Ye, T.; Guo, R.; Song, P.; Zhou, J.; Wang, H. Multifunctional polyurethane spong coatings with excellent flame retardant, antibacterial, compressible, and recyclable properties. *Compos. Part B Eng.* **2021**, *21* 108785. [CrossRef]
106. Lu, T.; Liu, S.; Jiang, M.; Xu, X.; Wang, Y.; Wang, Z.; Gou, J.; Hui, D.; Zhou, Z. Effects of modifications of bamboo cellulose fiber on the improved mechanical properties of cellulose reinforced poly(lactic acid) composites. *Compos. Part B Eng.* **2014**, *62*, 191–19 [CrossRef]
107. Battig, A.; Sanchez-Olivares, G.; Rockel, D.; Maldonado-Santoyo, M.; Schartel, B. Waste not, want not: The use of leather waste i flame retarded EVA. *Mater. Des.* **2021**, *210*, 110100. [CrossRef]
108. Siakeng, R.; Jawaid, M.; Asim, M.; Fouad, H.; Awad, S.; Saba, N.; Siengchin, S. Flexural and Dynamic Mechanical Properties c Alkali-Treated Coir/Pineapple Leaf Fibers Reinforced Polylactic Acid Hybrid Biocomposites. *J. Bionic Eng.* **2021**, *18*, 1430–143 [CrossRef]
109. Kumari, M.; Chaudhary, G.R.; Chaudhary, S.; Umar, A. Rapid Analysis of Trace Sulphite Ion Using Fluorescent Carbon Dot Produced from Single Use Plastic Cups. *Eng. Sci.* **2021**, *17*, 101–112. [CrossRef]
110. Singh, M.V. Conversions of waste tube-tyres (WTT) and waste polypropylene (WPP) into diesel fuel through catalytic pyrolys using base SrCO3. *Eng. Sci.* **2020**, *13*, 87–97. [CrossRef]
111. Pan, D.; Su, F.; Liu, H.; Liu, C.; Umar, A.; Castañeda, L.; Algadi, H.; Wang, C.; Guo, Z. Research progress on catalytic pyrolysi and reuse of waste plastics and petroleum sludge. *ES Mater. Manuf.* **2021**, *11*, 3–15. [CrossRef]
112. Mu, L.; Dong, Y.; Li, L.; Gu, X.; Shi, Y. Achieving high value utilization of bio-oil from lignin targeting for advanced lubricatio *ES Mater. Manuf.* **2021**, *11*, 72–80. [CrossRef]
113. Pan, D.; Su, F.; Liu, C.; Guo, Z. Research progress for plastic waste management and manufacture of value-added products. *Ad Compos. Hybrid Mater.* **2020**, *3*, 443–461. [CrossRef]
114. Jawaid, M.; Awad, S.A.; Asim, M.; Fouad, H.; Alothman, O.Y.; Santulli, C. A comparative evaluation of chemical, mechanical, an thermal properties of oil palm fiber/pineapple fiber reinforced phenolic hybrid composites. *Polym. Compos.* **2021**, *42*, 6383–639 [CrossRef]
115. Sarmin, S.N.; Jawaid, M.; Awad, S.A.; Saba, N.; Fouad, H.; Alothman, O.Y.; Sain, M. Olive fiber reinforced epoxy composite Dimensional Stability, and mechanical properties. *Polym. Compos.* **2021**, *43*, 358–365. [CrossRef]
116. Jawaid, M.; Awad, S.; Fouad, H.; Asim, M.; Saba, N.; Dhakal, H.N. Improvements in the thermal behaviour of date palm/bambo fibres reinforced epoxy hybrid composites. *Compos. Struct.* **2021**, *277*, 114644. [CrossRef]
117. Xu, D.; Huang, G.; Guo, L.; Chen, Y.; Ding, C.; Liu, C. Enhancement of catalytic combustion and thermolysis for treatin polyethylene plastic waste. *Adv. Compos. Hybrid Mater.* **2021**, *5*, 113–129. [CrossRef]
118. Maxwell, C.A.; Bell, N.; Kennedy, C.J.; Wess, T.J. X-ray diffraction and FT-IR study of caprine and ovine hide. *Pap. Conserv.* **200** *29*, 55–62. [CrossRef]
119. Maxwell, C.A.; Smiechowski, K.; Zarlok, J.; Sionkowska, A.; Wess, T.J. X-ray studies of a collagen material for leather productio treated with chromium salt. *J. Am. Leather Chem. Assoc.* **2006**, *101*, 9–17.
120. Maxwell, C.A.; Wess, T.J.; Kennedy, C.J. X-ray diffraction study into the effects of liming on the structure of collagen. *Biomacr molecules* **2006**, *7*, 2321–2326. [CrossRef]
121. Zhang, Y.; Ingham, B.; Cheong, S.; Ariotti, N.; Tilley, R.D.; Naffa, R.; Holmes, G.; Clarke, D.J.; Prabakar, S. Real-Time Synchrotro Small-Angle X-ray Scattering Studies of Collagen Structure during Leather Processing. *Ind. Eng. Chem. Res.* **2018**, *57*, 63–6 [CrossRef]

22. Zhang, Y.; Ingham, B.; Leveneur, J.; Cheong, S.; Yao, Y.; Clarke, D.J.; Holmes, G.; Kennedy, J.; Prabakar, S. Can sodium silicates affect collagen structure during tanning? Insights from small angle X-ray scattering (SAXS) studies. *RSC Adv.* **2017**, *7*, 11665–11671. [CrossRef]
23. Zhang, Y.; Mansel, B.W.; Naffa, R.; Cheong, S.; Yao, Y.; Holmes, G.; Chen, H.L.; Prabakar, S. Revealing Molecular Level Indicators of Collagen Stability: Minimizing Chrome Usage in Leather Processing. *ACS Sustain. Chem. Eng.* **2018**, *6*, 7096–7104. [CrossRef]
24. Zhang, Y.; Snow, T.; Smith, A.J.; Holmes, G.; Prabakar, S. A guide to high-efficiency chromium (III)-collagen cross-linking: Synchrotron SAXS and DSC study. *Int. J. Biol. Macromol.* **2019**, *126*, 123–129. [CrossRef] [PubMed]
25. Li, C.; Xue, F.; Ding, E. Fabrication and characterization of ultrafine leather powder: A functional reinforcement containing SiO_2 particles. *Micro Nano Lett.* **2014**, *9*, 308–311. [CrossRef]
26. Xia, L.; Li, C.; Zhou, S.; Fu, Z.; Wang, Y.; Lyu, P.; Zhang, J.; Liu, X.; Zhang, C.; Xu, W. Utilization of waste leather powders for highly effective removal of dyes from water. *Polymers* **2019**, *11*, 1786. [CrossRef]
27. Bak, S.Y.; Lee, S.W.; Choi, C.H.; Kim, H.W. Assessment of the influence of acetic acid residue on type I collagen during isolation and characterization. *Materials* **2018**, *11*, 2518. [CrossRef]
28. Moraes, P.R.F.D.S.; Saska, S.; Barud, H.; Lima, L.R.D.; Martins, V.D.C.A.; Plepis, A.M.D.G.; Ribeiro, S.J.L.; Gaspar, A.M.M. Bacterial cellulose/collagen hydrogel for wound healing. *Mater. Res.* **2016**, *19*, 106–116. [CrossRef]
29. Ammasi, R.; Victor, J.S.; Chellan, R.; Chellappa, M. Amino acid enriched proteinous wastes: Recovery and reuse in leather making. *Waste Biomass Valorization* **2020**, *11*, 5793–5807. [CrossRef]
30. Wang, Y.; Jin, L. Preparation and characterization of self-colored waterborne polyurethane and its application in eco-friendly manufacturing of microfiber synthetic leather base. *Polymers* **2018**, *10*, 289. [CrossRef]
31. Ma, F.; Ding, S.; Ren, H.; Peng, P. Preparation of chrome-tanned leather shaving-based hierarchical porous carbon and its capacitance properties. *RSC Adv.* **2019**, *9*, 18333–18343. [CrossRef]
32. Kanagaraj, J.; Panda, R.C.; Prasanna, R. Sustainable chrome tanning system using protein-based product developed from leather waste: Wealth from waste. *Polym. Bull.* **2022**, *79*, 10201–10228. [CrossRef]
33. Ribeiro, D.V.; Yuan, S.Y.; Morelli, M.R. Effect of chemically treated leather shaving addition on characteristics and microstructure of OPC mortars. *Mater. Res.* **2012**, *15*, 136–143. [CrossRef]
34. Dey, T.K.; Hossain, A.; Jamal, M.; Layek, R.K.; Uddin, M. Zinc Oxide Nanoparticle Reinforced Waste Buffing Dust Based Composite Insole and Its Antimicrobial Activity. *Adv. Polym. Technol.* **2022**. [CrossRef]
35. Saikia, P.; Goswami, T.; Dutta, D.; Dutta, N.K.; Sengupta, P.; Neog, D. Development of a flexible composite from leather industry waste and evaluation of their physico-chemical properties. *Clean Technol. Environ. Policy* **2017**, *19*, 2171–2178. [CrossRef]

sustainability

Article

Preparation and Characterization of Black Seed/Cassava Bagasse Fiber-Reinforced Cornstarch-Based Hybrid Composite

Walid Abotbina [1], S. M. Sapuan [1,*], R. A. Ilyas [2,3,4], M. T. H. Sultan [5] and M. F. M. Alkbir [6,7]

1 Advanced Engineering Materials and Composites Research Centre, Department of Mechanical and Manufacturing Engineering, Universiti Putra Malaysia, Serdang 43400, Selangor, Malaysia
2 Sustainable Waste Management Research Group (SWAM), School of Chemical and Energy Engineering, Faculty of Engineering, Universiti Teknologi Malaysia, Johor Bahru 81310, Johor, Malaysia
3 Centre for Advanced Composite Materials (CACM), Universiti Teknologi Malaysia, Johor Bahru 81310, Johor, Malaysia
4 Laboratory of Biocomposite Technology, Institute of Tropical Forest and Forest Products, Universiti Putra Malaysia, Serdang 43400, Selangor, Malaysia
5 Department of Aerospace Engineering, Universiti Putra Malaysia, Serdang 43400, Selangor, Malaysia
6 Advanced Facilities Engineering Technology Research Cluster, Malaysian Institute of Industrial Technology (MITEC), University Kuala Lumpur, Persiaran Sinaran Ilmu, Bandar Seri Alam, Masai 81750, Johor, Malaysia
7 Facilities Maintenance Engineering Section, Malaysian Institute of Industrial Technology, Universiti Kuala Lumpur, Persiaran Sinaran Ilmu, Bandar Seri Alam, Johor Bahru 81750, Johor, Malaysia
* Correspondence: sapuan@upm.edu.my; Tel.: +60-19-3863191

Citation: Abotbina, W.; Sapuan, S.M.; Ilyas, R.A.; Sultan, M.T.H.; Alkbir, M.F.M. Preparation and Characterization of Black Seed/Cassava Bagasse Fiber-Reinforced Cornstarch-Based Hybrid Composites. *Sustainability* **2022**, *14*, 12042. https://doi.org/10.3390/su141912042

Academic Editor: Abu Naser Md Ahsanul Haque

Received: 19 August 2022
Accepted: 20 September 2022
Published: 23 September 2022

Abstract: Great advances have been made in the preparation of bioplastics and crude oil replacement to create a better and more sustainable and eco-friendly future for all. Here, we used cassava bagasse fibers at different ratios as reinforcement material to enhance the properties of black seed w-cornstarc films using the facile solution casting technique. The reinforced films showed compact and relatively smoother structures without porosity. The crystallinity values increased from 34.6 ± 1.6% of the control to 38.8 ± 2.1% in sample CS-BS/CB 9%, which reflects the mechanical properties of the composite. A gradual increase in tensile strength and elastic modulus was observed, with an increase in loading amounts of 14.07 to 18.22 MPa and 83.65 to 118.32 MPa for the tensile strength and elastic modulus, respectively. The composite film also exhibited faster biodegradation in the soil burial test, in addition to lower water absorption capacity. Using bio-based reinforcement material could significantly enhance the properties of bio-based packaging materials. The prepared hybrid composite could have a promising potential in food packaging applications as a safe alternative fo conventional packaging.

Keywords: cassava bagasse fiber; black seed; cornstarch; hybrid composite film; cellulose

1. Introduction

The past few years witnessed greatly advanced in sustainable and ecologically friendl materials due to the huge environmental problems associated with conventional non biodegradable plastics [1–3]. The conversion of biomass into valuable materials has attracted the attention of scientists, which has both substantial economic and environmental relevance [4–8]. Throughout the last two decades, several developments have been mad on biocomposites, making them functioning and interesting alternatives to conventional materials [9–12]. Cornstarch (CS), among the biomaterials, is a highly preferable polysaccharide polymer in bioplastics and biomaterials fabrication due to its sustainability, high availability, and ability to form a continuous matrix [13–17]. Nevertheless, CS exhibits several drawbacks, including its relatively strong hydrophilic character, in addition to its poor mechanical properties, which limited its applications for packaging purposes [18,19

Several attempts have used numerous bioresources and natural fibers to enhance the mechanical properties of biocomposites but still mimic that of synthetic and crude oil plastic [20–23]. Black seed (*Nigella sativa*) is a famous medicinal plant that has been used for centuries for therapeutic purposes due to its anti-microbial, anti-diabetic, anti-inflammatory, anti-cancer, and anti-hypertensive activities [24]. The large-scale black seed oil industry produces a significant amount of waste every year (about 70% of the raw material) [25,26]. Most of these wastes are landfilled and not being furtherly utilized. Thymoquinone is one of the main active compounds in black seed that possess most of the therapeutic activities of the seeds and is characterized by hydrophobic properties, making it poorly soluble in water [27]. It has been reported that most black seed oil extraction methods allow a significant loss of waste, making this waste highly valuable for further utilization [25,28]. Black seed fiber was used in this study to utilize as a protective material to resist soil acidity and microbial attack.

Cassava bagasse is another cheap and broadly available fiber of cassava plant (*Manihot esculenta*) in tropical countries, which is considered a byproduct of cassava starch production [29]. Cassava bagasse is mainly composed of cellulose fibers along with residual starch, which ranges 15 and 50 wt%, making it highly attractive in several industrial applications. The industrial exploitation of cassava mainly includes the elimination of soluble sugars in addition to fiber separation. This result in the formation of two materials; purified cassava starch and cassava bagasse fibers [30]. Apart from containing cellulose and hemicelluloses, cassava bagasse also contains a high amount of starch, a natural polymer that has a high polarity due to the presence of large amounts of hydroxyls in its macromolecules, which interact with lignocellulosic fibers, resulting in improved mechanical properties [31].

Hence, the present study aims to use cassava bagasse fibers (CB) as reinforcement material to improve the mechanical and biodegradability properties of black seed fiber/cornstarch films by using a facile solution casting technique. The morphological and mechanical properties were analyzed and related to water absorption and biodegradation profile. Surface morphology and crystallinity were also analyzed.

2. Materials and Methods

2.1. Materials

The Cornstarch was obtained from Thye Huat Chan Sdn. Bhd. located at Sungai Buloh, Selangor, Malaysia, while the black seed (*Nigella sativa*) was obtained from Berkat Madinah Sdn. Bhd. (Selangor, Malaysia), after extracting the oil from the seeds. Cassava bagasse was purchased from NSK Trade City Kuchai Lama (Kuala Lumpur, Malaysia).

2.2. Preparation of the Film

The composite films were prepared using the conventional solution casting technique; 10 g of pure cornstarch (CS) was dissolved in 180 mL distilled water and heated for 20 min at 85 °C with continuous stirring (using a thermal-magnetic mixture) to allow starch gelation. Two types of plasticizer (fructose, and glycerol) were added to the solution in a similar ratio of 30% (*w*/*w* powder starch) along with black seed (BS) 9% (*w*/*w* powder starch) [5]. Different loading of cassava bagasse fibers (0, 3, 6, 9%) of dry starch-based was used as a hybridized agent. The final mixture was heated for an additional 20 min with continuous stirring until a gelatinized solution formed, which was then discharged evenly in a thermal petri dish. The dish with a casted solution was desiccated in an air circulation oven for 15 h at 65 °C. The formed films were removed gently from the dishes and kept at ambient conditions for 7 days prior to characterization. The obtained films were labeled according to their compositions and concentration of cassava bagasse fibers (CB), as shown in Table 1.

Table 1. Mixing proportion of different black seed/cassava bagasse fiber-reinforced cornstarch hybri composite film.

Film	Fructose and Glycerol g/g Dry Starch	Starch g/180 mL Distilled Water	BS g/100 g of Dry Starch	CB g/100 g of Dry Starch
Control(CS/BS)	0.3	10	9	0
CS-BS/CB3%	0.3	10	9	3
CS-BS/CB6%	0.3	10	9	6
CS-BS/CB9%	0.3	10	9	9

2.3. *Characterization of Prepared Films*

2.3.1. Physical and Morphological Analysis

The thickness of each film was calculated by an electronic caliper (Mitutoyo-Co Kanagawa, Japan), using an inch accuracy of ±0.001. The density was directly measure from the weight and volume. Moisture content (MC) was calculated by weighing a constan dimension of each film (20 × 15 mm) and placed for 24 h in the dehydration oven at 105 °C Then, the samples were weighed again and the difference in weights was determined as th MC of the sample. The film solubility was calculated using the same dimensions and dryin step mentioned earlier and following the method described in Shojaee-Aliabadi et al. [32 SEM (Hitachi S-3400 N, Nara, Japan) was used to observe the surface morphology of all th prepared samples by coating them with a thin golden layer to conduct electricity.

2.3.2. Surface Functional Groups and X-ray Diffraction Analysis

Fourier Transform Infrared Spectroscopy (FTIR) type (Bruker vector 22, Lancashir UK) was used to investigate the surface functional group of prepared films, using 16 scan per sample and over a frequency range of 400 to 4000 cm^{-1}. In order to determine th crystallinity index of each sample, a 2500 X-ray diffractometer (Rigaku, Tokyo, Japan) wa used to analyze the XRD diffraction and calculate the crystallinity index in the same metho described by [33].

2.3.3. Mechanical Properties

The mechanical properties of the films were investigated by tensile test according t D882 (ASTM, 2002) standards. The test was performed at room temperature (30 °C) usin 5KN INSTRON tensile machine. Constant strips (10 × 70 mm) were prepared from eac film and then firmly mounted between tensile clamps. Five replicates for each sample wer done, using 2 mm/min crosshead speed, tensile strength, elongation at break, and elasti modulus was finally calculated.

2.3.4. Thermal Properties

The thermal gravimetric analyzer (TGA) instrument type (Q500 V20.13 Build 3 Bellingham, WA, USA) was used to study the thermal stability of the prepared film sample Each sample was subjected to a temperature ranging from room temperature to 600 °C a an ascending rate of 10 °C/min.

2.3.5. Water Absorption and Soil Burial Test

Water absorption was calculated at room temperature by preparing constant dimen sions (20 × 15 mm) of all the films and oven dried them for 3 h at 105 °C to eliminate all th moisture. The samples then were immediately weighted and steeped in distilled water fo 3 h and the excess was removed from the surface by using a soft cloth. Water absorptio was then determined using the following equation:

$$\text{Water absorption } (\%) = \frac{\text{water loadedfilm} - \text{ initial weight}}{\text{initial weight}} \times 100$$

A soil burial test was conducted in polyethylene vials at room temperature in normal soil with the same method in [34].

2.4. Statistical Analyses

The statistical analyses of the findings were performed using Microsoft Excel 365, and the obtained data were plotted using Origin® 8.5 software (OriginLab Corporation, Northampton, MA, USA) for the graphical presentation of the results.

3. Results and Discussion

3.1. Physical and Morphological Analysis

The results of physical properties including moisture content (MC), density, thickness, and water solubility of the prepared film samples are presented in Table 2. The control had the highest MC, density, and water solubility, but it exhibited the lowest thickness. The addition of 3% CB increased the MC of the film from 7.54 ± 0.6 to only 6.81 ± 0.4%, but it reduced over again to 6.40 ± 0.2 and 6.02 ± 0.3% for CS-BS/CB6% and CS-BS/CB6% respectively. This could be attributed to the interaction that occurred between the BS and the CB; at a small amount, the interaction was limited, which led to an insufficient amount of interaction that was enhanced with loading more CB. The thickness clearly increased with the increase in loading amount from 0.272 ± 0.01 to 0.35 ± 0.04 µm, which also result in the reduction of the density value of the samples. The solubility of the film decreased from 34.23 to 31.47 ± 1.1% as the CB concentration increased, which could be due to the interaction that occurred between the materials that led to resistance of the water diffusion and enhance the film integrity [35].

Table 2. Physical characteristics of black seed/cassava bagasse fiber reinforced cornstarch hybrid composite film.

Sample Name	MC (%)	Density (g/cm^3)	Thickness (µm)	Solubility (%)
Control	7.546 ± 0.6	1.34 ± 0.02	0.272 ± 0.01	34.23 ± 2.0
CS-BS/CB3%	6.813 ± 0.4	1.31 ± 0.07	0.274 ± 0.08	32.93 ± 1.4
CS-BS/CB6%	6.406 ± 0.2	1.29 ± 0.01	0.332 ± 0.05	32.45 ± 0.9
CS-BS/CB9%	6.026 ± 0.3	1.25 ± 0.06	0.35 ± 0.04	31.47 ± 1.1

Figure 1 presents the scanning electron microscope (SEM) of the prepared film samples at same magnification. The difference in surface morphology between the control, which appeared more rough and porous compared with the reinforced ones, can be seen. Reinforced films showed compact and relatively smoother structures without porosity; the interaction between CB, BS and the CS could be the reason for the smooth surfaces and absence of porosity. The highest homogenous surface can be seen in the sample with the highest loading amount (CS-BS/CB9%), which supports our hypothesis. Although the CS-BS/CB3% sample was found to be more homogenous and smoother than the control, weak interfacial interaction, which is mainly related to the formation of hydrogen bonds through hydroxyl groups, could be the reason for this, compared with the higher loading samples [36,37]. The correspond results of mechanical and FTIR analysis to the increase of homogeneity will be remarked on in their own subsection.

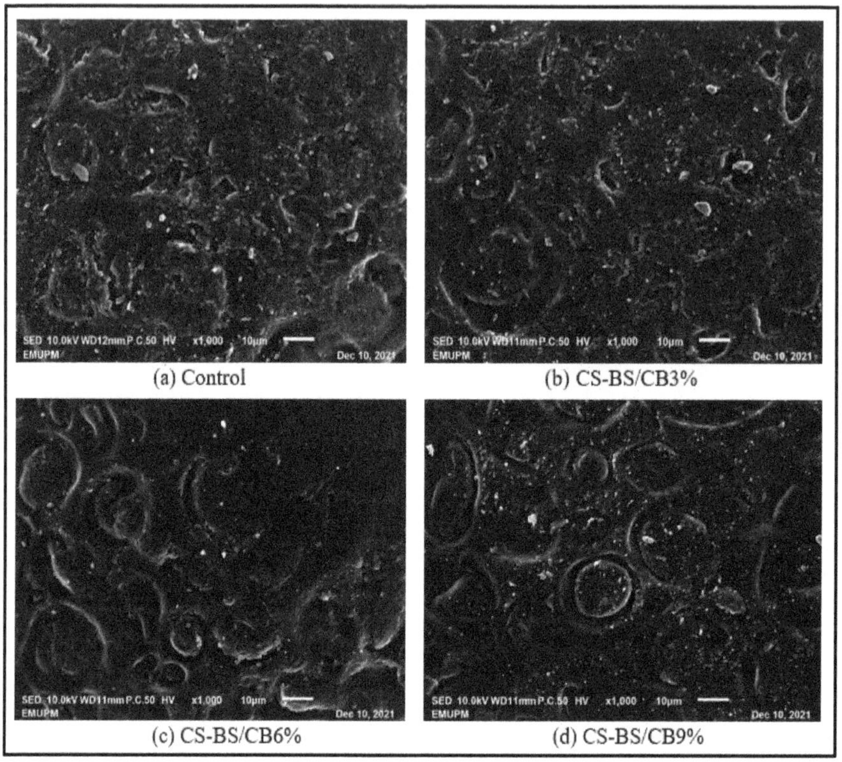

Figure 1. Surface morphology black seed/cassava bagasse fiber reinforced cornstarch hybrid com posite film.

3.2. *Surface Functional Groups*

Since the chemical composition and the portions of all the films are well know the FT-IR spectra (Figure 2) exhibited characteristic absorption peaks associated with th three materials of the hybrid components. The well-dispersed fillers caused the formatio of similar spectra patterns, indicating the formation of alike chemical bonds within al samples. All the films showed the typical characteristic bands of starch O-H stretching C-H, and C-O stretching at 3298, 2998, and 1151 cm^{-1}, with slight shifting among them The peak at 989.48 cm^{-1} appeared in all the samples corresponding to alkenes, which are the characteristics of black seed active compounds in addition to the aliphatic amine at 1033 and 1089 cm^{-1} [38]. Although the reinforced films did not show a significan difference between the three loading amounts, the slight shifting of peaks between th control and CS-BS/CB9% can be observed, which could be due to the effect of interactio between the three materials in the film. Higher CB loading formed greater hydrogen bond quantitatively generated higher peak intensity. The band 1423 cm^{-1} appeared in all th samples, which corresponds to glycerol, while the ones between 1944 and 1649 cm^{-1} refe to bound water [39].

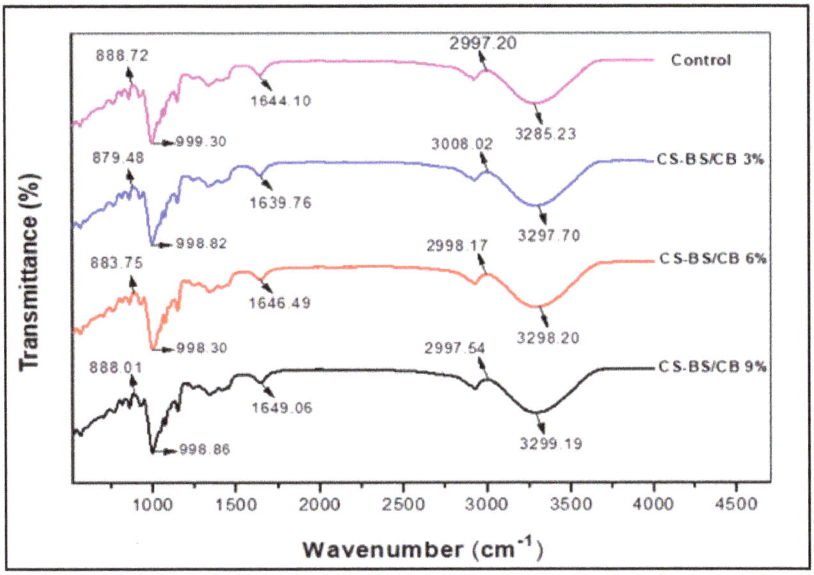

Figure 2. FT−IR curves of black seed/cassava bagasse fiber reinforced cornstarch hybrid composite film.

3.3. *X-ray Diffraction (XRD)*

The XRD curves of the black seed/cassava bagasse fiber reinforced cornstarch film composites are presented in Figure 3. The difference between the control and the three reinforced samples can be observed; the control increased the intensity of the two main peaks at 17 and 20°. However, among the reinforced fibers, CS-BS/CB 9% had significantly higher intensity compared with the other two samples. Owing to the gelatinization of starch molecules, sharper peaks can be observed in CS-BS/CB 9% at angles 17.82° and 20.47° in addition to 22.72°, which is the typical pattern of A-type plant starches [40]. The crystallinity values of the control were only 34.6 ± 1.6%, which increased with the increase of reinforcement percentage to become 38.8 ± 2.1% in sample CS-BS/CB 9%. The difference in crystallinity values could be attributed to the effect of starch reinforcement on the structure, which was also explained earlier by the mechanical properties. Natural fibers of black seed are oriented materials; they are able to combine with the starch molecules and thus improve the crystallinity of the film composite [41].

3.4. *Mechanical Properties*

The mechanical properties of the film are very important in order to be applicable for most practical applications. In this study, tensile testing of the film composites was done to measure each tensile strength, elastic modulus, and elongation at the break. As illustrated in Figure 4, it can be observed that the lowest tensile strength was reported for the control with only 14.07 MPa, which increased to 15.04, 16.65, and 18.22 MPa for the CS-BS/CB3%, CS-BS/CB6% and CS-BS/CB9%, respectively. Similarly, the elastic modulus increased from 83.65 MPa to 118.32 MPa. The gradual increase in tensile strength and elastic modulus is due to the homogeneous mixture and the interaction that occurred between the three materials, leading to an increase in rigidness and stiffness of the composite film with less flexibility, which was confirmed by the reduction in elongation at the break values [42]. The enhancement in the mechanical properties could have occurred due to the relative crystallinity, as we reported earlier and described by Salaberria et al.; the increase in the crystallinity of the film promotes the rigidity and stiffness of the film, leading to a reduction in elongation at the peak and thus better performance [43]. Black seed fibers also play a significant role by bonding the hydroxyl group with both CS and CB, leading to bette transfer of stress from the film matrix to the fibers and thus higher tensile strength [33 Furthermore, the significant reduction in film elongation can be explained by rebuilding th intermolecular bonding of the film, which improves its stiffness and rigidity, as describe by da Rosa Zavareze et al., and also decreases the composite film flexibility by eliminatin chain mobility [44–48]. Table 3 expresses the tensile strength and elongation at the break c several starch-based composites.

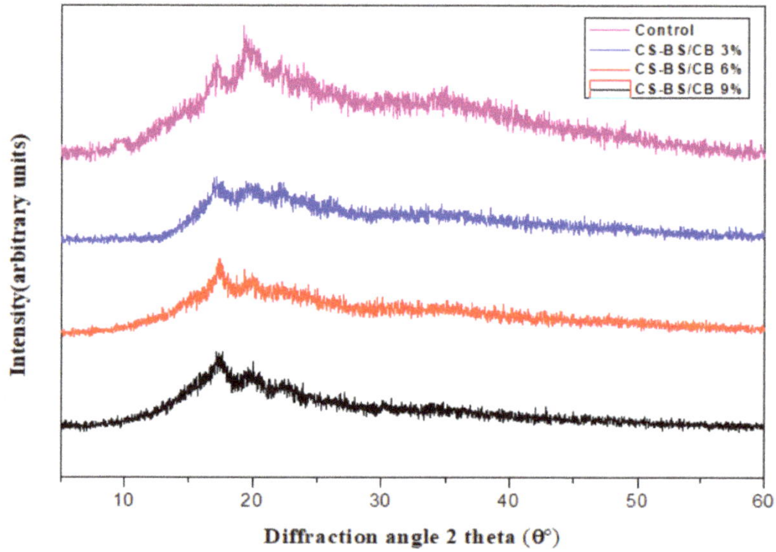

Figure 3. X-ray diffraction curves of black seed/cassava bagasse fiber reinforced cornstarch hybri composite film.

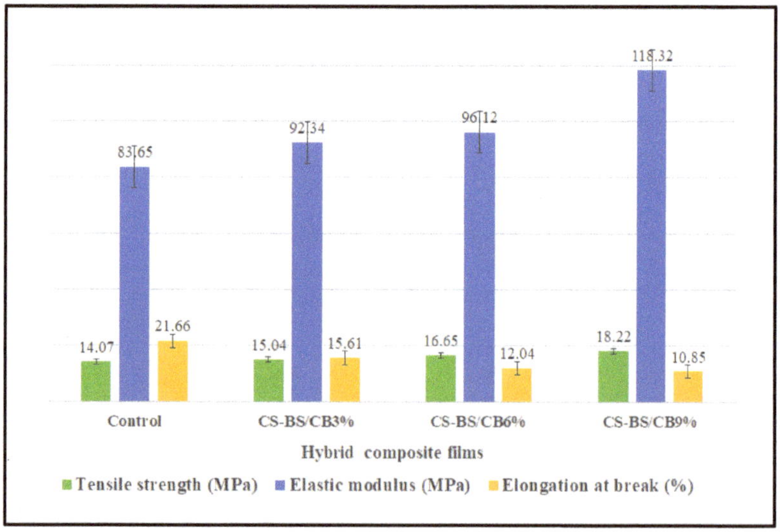

Figure 4. The mechanical properties of black seed/cassava bagasse fiber-reinforced cornstarch hybri composite film.

Table 3. The tensile strength (TS) and elongation at break (E) of various starch-based composites.

Polymer	Fiber	Plasticizer	TS	E	Reference
Cassava starch	Cassava bagasse (6 wt.%)	Fructose	10.78 MPa	3.19 mm	[49]
Corn starch	Corn Husk (8 wt.%)	Fructose	12.84 MPa	3.7%	[50]
Corn starch	Kenaf fiber (6 wt.%)	Sorbitol	17.74 MPa	48.79%	[51]
Cassava starch	Banana pseudostem powder (10 wt.%)	-	16 MPa	113.5%	[52]
Dioscorea hispida starch	*Dioscorea hispida* Fiber (6 wt.%)	Sorbitol	9.29 MPa	25.44%	[53]
Sugar palm starch	Sugar palm cellulose fiber (10 wt.%)	Glycerol and sorbitol	19.68 MPa	32.8%	[54]

3.5. Thermal Properties

Thermal degradation of the prepared film composites mainly occurs in three different stages (Figure 5). The first stage corresponds to the moisture removal and sample dehydration, which started from 52 ± 2.6 °C until 163.5 ± 3.2 °C. At this stage, a clear difference can be observed in weight loss between the control, which had the highest weight loss, and CS-BS/CB9%, which showed the lowest. The second stage corresponds to the decomposition of starch molecules within the films, which occurred between 282 and 340 °C. The initial decomposition at 282 °C is attributed to the amylopectin and amylose differential degradation rate [55,56]. A similar effect for the reinforcement can be observed at this stage; the addition of CB into the film enhanced its thermal stability and reduced the weight loss due to the interactions in the composite matrix between the three materials. The greatest differences between the four investigated samples were observed in the third phase of degradation, which correlated to several oxygen-based reactions of carbonaceous residues—known as "glowing combustion" [57]. The control had the most significant weight loss at this stage; however, the reinforcement enhanced the stability and reduce the decomposition, in a similar way to what was reported by Florencia et al. [1]. Higher reinforcement samples CS-BS/CB6% and CS-BS/CB9% were clearly more stable by 12% than the control, owing to their higher residual carbohydrate content.

3.6. Water Absorption and Soil Burial Test

Water absorption is an important characteristic in packaging materials and is considered a major drawback in many bio-based films. Figure 6 presents the results of the water absorption investigation in this study, in which can be seen the effect of reinforcement on the water sensitivity of the films. At the initial assay (20 min of immersion), CS-BS/CB3% exhibited more water absorption than the control, which then reduced after 40 min of immersion to 63% compared to the control, which reported 67%. However, after the films reached the maximum degree of water saturation (180 min of immersion), a huge difference can be observed: the control showed the highest water absorption, at 64.7%, followed by CS-BS/CB3%, at 62.8%. The 6 and 9% CB reinforcement significantly reduced the water absorption to 51 and 42%, respectively, which could be due to the hydrogen bonding that occurred, resulting in the slight hydrophobic nature of the films [58]. Furthermore, the reduction of water absorption rate caused by CB reinforcement could be also attributed to the creation of interfacial bonding in the hybrid composite, which was confirmed earlier by the results of mechanical analysis. Such interfacial bonding hinders the water penetration through the film matrix and thus reduces its water absorption [59].

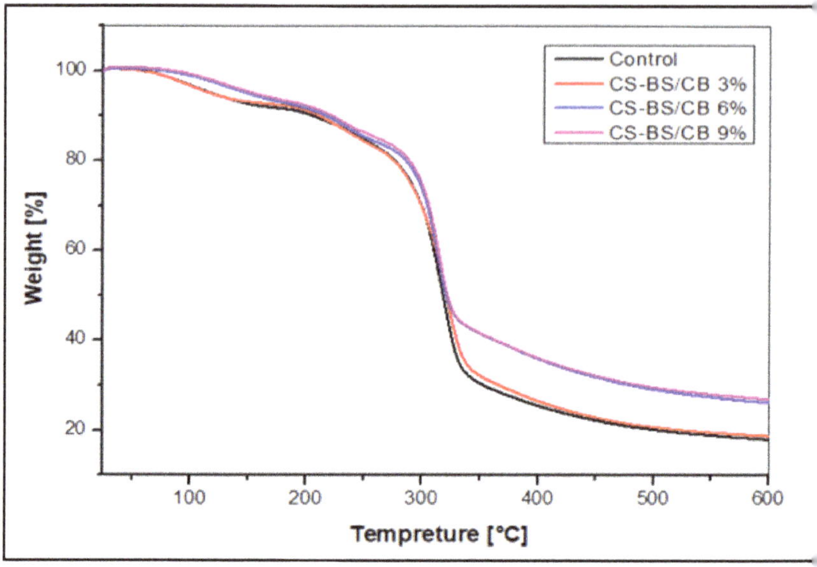

Figure 5. TGA curves of black seed/cassava bagasse fiber reinforced cornstarch hybrid composite film.

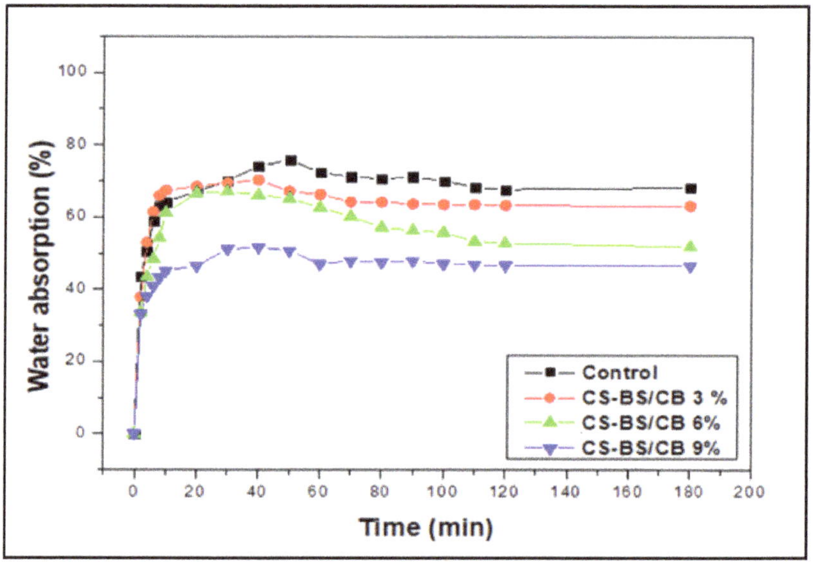

Figure 6. Water absorption of black seed/cassava bagasse fiber reinforced cornstarch hybrid composite film.

A soil burial test was conducted in normal soil at room temperature over a period of 12 days. Figure 7 presents the results of films' biodegradability, in which can be seen the difference between them in the degradation rate. The CB reinforcement enhanced the biodegradation rate of the films; the highest rate was reported for CS-BS/CB9% with 99% degradation after 12 days. A previous study reported that the degradation of cassava film is greater than the corn-based one [60]; in our study, the biodegradation enhancement that occurred by CB reinforcement could be due to Cola cordifolia, which was found in both cassava and corn. Furthermore, the consumption of hydrogen bonding makes the film highly sensitive to microbial attacks and thus biodegrades easily [61]. In the case of control and CS-BS/CB3%, which absorbed a higher amount of water, the microorganisms are known for CO_2 and water production as a result of bio-based material consumption. However, the Solvation of CO_2 in the soil leads to the formation of H_2CO_3, forming an acidic environment around the film, which eventually reduces the microbial activity and thus limits the film's biodegradation [62]. Although the presence of black seeds fiber helps in resisting soil acidity and microbial attack [63], the significant biodegradability of CS-BS/CB9% could be due to their limited water absorption, which maintains microbial activity, and also to the crystalline structure of the CB reinforcement.

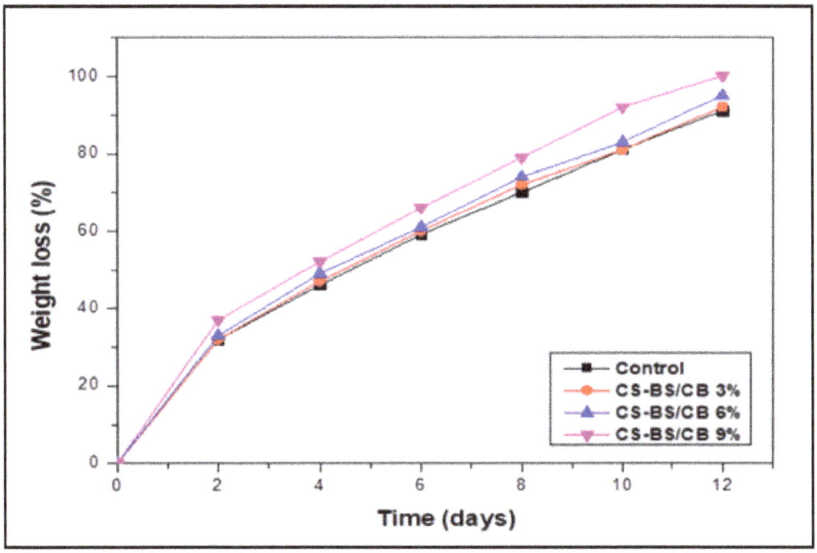

Figure 7. Soil burial test of black seed/cassava bagasse fiber reinforced cornstarch hybrid composite film.

4. Conclusions

Intensive studies have been conducted on waste utilization and the development of enhanced materials able to mimic synthetic plastic to overcome its associated environmental issue. In this study, cassava bagasse fibers at different ratios were used as a reinforcement to enhance the properties of black seed fiber-cornstarch films. The addition of fibers significantly enhanced the mechanical properties of the film and a gradual increase in tensile strength and elastic modulus was observed with an increase in loading amount. The composite film also exhibited faster biodegradation in the soil burial test in addition to lower water absorption capacity. Using fibers as reinforcement could significantly enhance the properties of bioplastics in terms of mechanical, water absorption, and biodegradation properties. The prepared hybrid composite could have promising potential in food packaging applications as a safe alternative to conventional packaging.

Author Contributions: Conceptualization, W.A. and S.M.S.; methodology, W.A.; formal analysis, W.A. and S.M.S.; writing original draft preparation, W.A.; writing—review and editing, W.A., S.M.S., M.T.H.S., M.F.M.A. and R.A.I.; visualization, W.A.; project administration, W.A., S.M.S. and R.A.I.; funding acquisition, S.M.S., W.A. and R.A.I. All authors have read and agreed to the published version of the manuscript.

Funding: This research was funded by University Putra Malaysia for financial support throug Geran Putra Berimpak (UPM/800-3/3/1/GPB/2019/9679800) and the grant Inisiatif Putra Siswaza (GP-IPS/2021/9697100). The authors would like express gratitude for the financial support r ceived from the Universiti Teknologi Malaysia, the project "The impact of Malaysian bamboo chemical and fibre characteristics on their pulp and paper properties, grant number PY/2022/02318-Q.J130000.3851.21H99". The research has been carried out under the program Research Excellenc Consortium (JPT (BPKI) 1000/016/018/25 (57)) provided by the Ministry of Higher Educatio Malaysia (MOHE).

Institutional Review Board Statement: Not applicable.

Informed Consent Statement: Not applicable.

Data Availability Statement: Not applicable.

Acknowledgments: The authors would like to thank University Putra Malaysia for financial suppo through Geran Putra Berimpak (UPM/800-3/3/1/GPB/2019/9679800) and the grant Inisiatif Putr Siswazah (GP-IPS/2021/9697100). The authors would like express gratitude for the financial suppo received from the Universiti Teknologi Malaysia, the project "The impact of Malaysian bamboo chemical and fibre characteristics on their pulp and paper properties, grant number PY/2022/02318-Q.J130000.3851.21H99". The research has been carried out under the program Research Excellenc Consortium (JPT (BPKI) 1000/016/018/25 (57)) provided by the Ministry of Higher Educatio Malaysia (MOHE).

Conflicts of Interest: The authors declare no conflict of interest.

References

1. Florencia, V.; López, O.V.; García, M.A. Exploitation of by-products from cassava and ahipa starch extraction as filler c thermoplastic corn starch. *Compos. Part B Eng.* **2020**, *182*, 107653. [CrossRef]
2. Tarique, J.; Sapuan, S.; Khalina, A.; Ilyas, R.; Zainudin, E. Thermal, flammability, and antimicrobial properties of arrowroc (*Maranta arundinacea*) fiber reinforced arrowroot starch biopolymer composites for food packaging applications. *Int. J. Bic Macromol.* **2022**, *213*, 1–10. [CrossRef] [PubMed]
3. Zentou, H.; Abidin, Z.Z.; Yunus, R.; Biak, D.R.A.; Issa, M.A. Optimization and modeling of the performance of polydimethylsilo ane for pervaporation of ethanol—water mixture. *J. Appl. Polym. Sci.* **2020**, *138*, 50408. [CrossRef]
4. Rizal, S.; Olaiya, F.; Saharudin, N.; Abdullah, C.; Olaiya, M.K.; Haafiz, M.M.; Yahya, E.; Sabaruddin, F.; Ikramullah; Khali H.P.S.A. Isolation of Textile Waste Cellulose Nanofibrillated Fibre Reinforced in Polylactic Acid-Chitin Biodegradable Composit for Green Packaging Application. *Polymers* **2021**, *13*, 325. [CrossRef]
5. Abotbina, W.; Sapuan, S.M.; Ilyas, M.T.H.; Alkbir, M.F.M.; Ilyas, R.A. Development and Characterization of Cornstarch-Base Bioplastics Packaging Film Using a Combination of Different Plasticizers. *Polymers* **2021**, *13*, 3487. [CrossRef]
6. Ilyas, R.A.; Zuhri, M.Y.M.; Aisyah, H.A.; Asyraf, M.R.M.; Hassan, S.A.; Zainudin, E.S.; Sapuan, S.M.; Sharma, S.; Banga S.P.; Jumaidin, R.; et al. Natural Fiber-Reinforced Polylactic Acid, Polylactic Acid Blends and Their Composites for Advance Applications. *Polymers* **2022**, *14*, 202. [CrossRef]
7. Ilyas, R.A.; Zuhri, M.Y.M.; Norrahim, M.N.F.; Misenan, M.S.M.; Jenol, M.A.; Samsudin, S.A.; Nurazzi, N.M.; Asyraf, M.R.M Supian, A.B.M.; Bangar, S.P.; et al. Natural Fiber-Reinforced Polycaprolactone Green and Hybrid Biocomposites for Variou Advanced Applications. *Polymers* **2022**, *14*, 182. [CrossRef]
8. Ilyas, R.A.; Aisyah, H.A.; Nordin, A.H.; Ngadi, N.; Zuhri, M.Y.M.; Asyraf, M.R.M.; Sapuan, S.M.; Zainudin, E.S.; Sharma, S.; Abra H.; et al. Natural-Fiber-Reinforced Chitosan, Chitosan Blends and Their Nanocomposites for Various Advanced Application *Polymers* **2022**, *14*, 874. [CrossRef]
9. Abotbina, W.; Sapuan, S.; Sultan, M.; Alkbir, M.; Ilyas, R. Extraction, Characterization, and Comparison of Properties of Cassav Bagasse and Black Seed Fibers. *J. Nat. Fibers* **2022**, 1–14. [CrossRef]
10. Nazrin, A.; Sapuan, S.M.; Zuhri, M.Y.M.; Tawakkal, I.S.M.A.; Ilyas, R.A. Water barrier and mechanical properties of sugar paln crystalline nanocellulose reinforced thermoplastic sugar palm starch (TPS)/poly(lactic acid) (PLA) blend bionanocomposite *Nanotechnol. Rev.* **2021**, *10*, 431–442. [CrossRef]
11. Syafiq, R.M.O.; Sapuan, S.M.; Zuhri, M.Y.M.; Othman, S.H.; Ilyas, R.A. Effect of plasticizers on the properties of sugar paln nanocellulose/cinnamon essential oil reinforced starch bionanocomposite films. *Nanotechnol. Rev.* **2022**, *11*, 423–437. [CrossRef
12. Nazrin, A.; Sapuan, S.M.; Zuhri, M.Y.M.; Tawakkal, I.S.M.A.; Ilyas, R.A. Flammability and physical stability of sugar palr crystalline nanocellulose reinforced thermoplastic sugar palm starch/poly(lactic acid) blend bionanocomposites. *Nanotechno Rev.* **2021**, *11*, 86–95. [CrossRef]
13. Surya, I.; Olaiya, N.; Rizal, S.; Zein, I.; Sri Aprilia, N.A.; Hasan, M.; Yahya, E.B.; Sadasivuni, K.K.; Abdul Khalil, H.P.S. Plasticize enhancement on the miscibility and thermomechanical properties of polylactic acid-chitin-starch composites. *Polymers* **202** *12*, 115. [CrossRef] [PubMed]

4. Ganguly, S.; Maity, T.; Mondal, S.; Das, P.; Das, N.C. Starch functionalized biodegradable semi-IPN as a pH-tunable controlled release platform for memantine. *Int. J. Biol. Macromol.* **2017**, *95*, 185–198. [CrossRef] [PubMed]
5. Goldshtein, J.; Margel, S. Synthesis and characterization of polystyrene/2-(5-chloro-2H-benzotriazole-2-yl)-6-(1,1- dimethylethyl)-4-methyl-phenol composite microspheres of narrow size distribution for UV irradiation protection. *Colloid Polym. Sci.* **2011**, *289*, 1863–1874. [CrossRef]
6. Ganguly, S.; Mondal, S.; Das, P.; Bhawal, P.; Maity, P.P.; Ghosh, S.; Dhara, S.; Das, N.C. Design of psyllium-g-poly(acrylic acid-co-sodium acrylate)/cloisite 10A semi-IPN nanocomposite hydrogel and its mechanical, rheological and controlled drug release behaviour. *Int. J. Biol. Macromol.* **2018**, *111*, 983–998. [CrossRef]
7. Medina, D.D.; Goldshtein, J.; Margel, S.; Mastai, Y. Enantioselective Crystallization on Chiral Polymeric Microspheres. *Adv. Funct. Mater.* **2007**, *17*, 944–950. [CrossRef]
8. Ghanbarzadeh, B.; Almasi, H.; Entezami, A.A. Improving the barrier and mechanical properties of corn starch-based edible films: Effect of citric acid and carboxymethyl cellulose. *Ind. Crops Prod.* **2011**, *33*, 229–235. [CrossRef]
9. Abotbina, W.; Sapuan, S.M.; Sulaiman, S.; Ilyas, R.A. Review of corn starch biopolymer. In *Proceedings of the 7th Postgraduate Seminar on Natural Fibre Reinforced Polymer Composites*; Institute of Tropical Forest and Forest Products (INTROP), Universiti Putra Malaysia: Serdang, Malaysia, 2020; Volume 2020, pp. 37–40.
0. Norfarhana, A.; Ilyas, R.; Ngadi, N. A review of nanocellulose adsorptive membrane as multifunctional wastewater treatment. *Carbohydr. Polym.* **2022**, *291*, 119563. [CrossRef]
1. Bangar, S.P.; Harussani, M.; Ilyas, R.; Ashogbon, A.O.; Singh, A.; Trif, M.; Jafari, S.M. Surface modifications of cellulose nanocrystals: Processes, properties, and applications. *Food Hydrocoll.* **2022**, *130*, 107689. [CrossRef]
2. Ilyas, R.; Sapuan, S.; Harussani, M.; Hakimi, M.; Haziq, M.; Atikah, M.; Asyraf, M.; Ishak, M.; Razman, M.; Nurazzi, N.; et al. Polylactic Acid (PLA) Biocomposite: Processing, Additive Manufacturing and Advanced Applications. *Polymers* **2021**, *13*, 1326. [CrossRef] [PubMed]
3. Haris, N.I.N.; Hassan, M.Z.; Ilyas, R.; Suhot, M.A.; Sapuan, S.; Dolah, R.; Mohammad, R.; Asyraf, M. Dynamic mechanical properties of natural fiber reinforced hybrid polymer composites: A review. *J. Mater. Res. Technol.* **2022**, *19*, 167–182. [CrossRef]
4. Ahmad, M.F.; Ahmad, F.A.; Ashraf, S.A.; Saad, H.H.; Wahab, S.; Khan, M.I.; Ali, M.; Mohan, S.; Hakeem, K.R.; Athar, M.T. An updated knowledge of Black seed (*Nigella sativa* Linn.): Review of phytochemical constituents and pharmacological properties. *J. Herb. Med.* **2021**, *25*, 100404. [CrossRef] [PubMed]
5. Barkah, N.N.; Wiryawan, K.G.; Retnani, Y.; Wibawan, W.T.; Wina, E. Physicochemical properties of products and waste of black seed produced by cold press method. *IOP Conf. Ser. Earth Environ. Sci.* **2021**, *756*, 012025. [CrossRef]
6. Abotbina, W.; Sapuan, S.M.; Ilyas, M.T.H.; Alkbir, M.F.M.; Ilyas, R.A.; Harussani, M.M. A Short Review on Black Seed Fibre (Nigella Sativa) Reinforced. In *Proceedings of the 8th Postgraduate Seminar on Natural Fibre Composites 2022*; Institute of Tropical Forest and Forest Products (INTROP), Universiti Putra Malaysia: Serdang, Malaysia, 2022; pp. 14–17.
7. Younus, H. *Molecular and Therapeutic Actions of Thymoquinone: Actions of Thymoquinone*; Springer: Berlin/Heidelberg, Germany, 2018; ISBN 9789811088001.
8. Ahmad, R.; Ahmad, N.; Shehzad, A. Solvent and temperature effects of accelerated solvent extraction (ASE) coupled with ultra-high pressure liquid chromatography (UHPLC-DAD) technique for determination of thymoquinone in commercial food samples of black seeds (*Nigella sativa*). *Food Chem.* **2019**, *309*, 125740. [CrossRef]
9. Vedove, T.M.; Maniglia, B.C.; Tadini, C.C. Production of sustainable smart packaging based on cassava starch and anthocyanin by an extrusion process. *J. Food Eng.* **2020**, *289*, 110274. [CrossRef]
0. Travalini, A.P.; Lamsal, B.; Magalhães, W.L.E.; Demiate, I.M. Cassava starch films reinforced with lignocellulose nanofibers from cassava bagasse. *Int. J. Biol. Macromol.* **2019**, *139*, 1151–1161. [CrossRef]
1. Farias, F.O.; Jasko, A.C.; Colman, T.A.D.; Pinheiro, L.A.; Schnitzler, E.; Barana, A.C.; Demiate, I.M. Characterisation of Cassava Bagasse and Composites Prepared by Blending with Low-Density Polyethylene. *Braz. Arch. Biol. Technol.* **2014**, *57*, 821–830. [CrossRef]
2. Shojaee-Aliabadi, S.; Hosseini, H.; Mohammadifar, M.A.; Mohammadi, A.; Ghasemlou, M.; Ojagh, S.M.; Hosseini, S.M.; Khaksar, R. Characterization of antioxidant-antimicrobial κ-carrageenan films containing *Satureja hortensis* essential oil. *Int. J. Biol. Macromol.* **2013**, *52*, 116–124. [CrossRef]
3. Ibrahim, M.I.J.; Sapuan, S.M.; Zainudin, E.S.; Zuhri, M.Y.M. Preparation and characterization of cornhusk/sugar palm fiber reinforced Cornstarch-based hybrid composites. *J. Mater. Res. Technol.* **2020**, *9*, 200–211. [CrossRef]
4. Kumaravel, S.; Hema, R.; Lakshmi, R. Production of Polyhydroxybutyrate (Bioplastic) and its Biodegradation by *Pseudomonas Lemoignei* and *Aspergillus Niger*. *E-J. Chem.* **2010**, *7*, S536–S542. [CrossRef]
5. Jumaidin, R.; Sapuan, S.M.; Jawaid, M.; Ishak, M.R.; Sahari, J. Thermal, mechanical, and physical properties of seaweed/sugar palm fibre reinforced thermoplastic sugar palm Starch/Agar hybrid composites. *Int. J. Biol. Macromol.* **2017**, *97*, 606–615. [CrossRef] [PubMed]
6. Lodha, P.; Netravali, A.N. Characterization of interfacial and mechanical properties of "green" composites with soy protein isolate and ramie fiber. *J. Mater. Sci.* **2002**, *37*, 3657–3665. [CrossRef]
7. Diyana, Z.; Jumaidin, R.; Selamat, M.; Ghazali, I.; Julmohammad, N.; Huda, N.; Ilyas, R. Physical Properties of Thermoplastic Starch Derived from Natural Resources and Its Blends: A Review. *Polymers* **2021**, *13*, 1396. [CrossRef]

38. Sangeetha, S.; Archit, R.; SathiaVelu, A. Phytochemical testing, antioxidant activity, HPTLC and FTIR analysis of antidiabeti plants *Nigella sativa, Eugenia jambolana, Andrographis paniculata* and *Gymnema sylvestre. Res. J. Biotechnol.* **2014**, *9*, 65–72.
39. Ma, X.; Cheng, Y.; Qin, X.; Guo, T.; Deng, J.; Liu, X. Hydrophilic modification of cellulose nanocrystals improves the physicochem ical properties of cassava starch-based nanocomposite films. *LWT* **2017**, *86*, 318–326. [CrossRef]
40. Silva, M.L.T.; Brinques, G.B.; Gurak, P.D. Desenvolvimento e caracterização de bioplásticos de amido de milho contendo farinh de subproduto de broto. *Braz. J. Food Technol.* **2020**, *23*. [CrossRef]
41. Huang, M.-F.; Yu, J.-G.; Ma, X.-F. Studies on the properties of Montmorillonite-reinforced thermoplastic starch composite *Polymer* **2004**, *45*, 7017–7023. [CrossRef]
42. Dias, A.B.; Müller, C.M.; Larotonda, F.D.; Laurindo, J.B. Mechanical and barrier properties of composite films based on rice flou and cellulose fibers. *LWT* **2011**, *44*, 535–542. [CrossRef]
43. Salaberria, A.M.; Labidi, J.; Fernandes, S.C. Chitin nanocrystals and nanofibers as nano-sized fillers into thermoplastic starch-base biocomposites processed by melt-mixing. *Chem. Eng. J.* **2014**, *256*, 356–364. [CrossRef]
44. da Rosa Zavareze, E.; Pinto, V.Z.; Klein, B.; El Halal, S.L.M.; Elias, M.C.; Prentice-Hernández, C.; Dias, A.R.G. Development c oxidised and heat–moisture treated potato starch film. *Food Chem.* **2012**, *132*, 344–350. [CrossRef] [PubMed]
45. Ilyas, R.A.; Sapuan, S.M.; Ishak, M.R.; Zainudin, E.S. Development and characterization of sugar palm nanocrystalline cellulos reinforced sugar palm starch bionanocomposites. *Carbohydr. Polym.* **2018**, *202*, 186–202. [CrossRef] [PubMed]
46. Ilyas, R.A.; Sapuan, S.M.; Ibrahim, R.; Abral, H.; Ishak, M.; Zainudin, E.; Asrofi, M.; Atikah, M.S.N.; Huzaifah, M.R.M.; Radz A.M.; et al. Sugar palm (*Arenga pinnata* (*Wurmb.*) Merr) cellulosic fibre hierarchy: A comprehensive approach from macro to nan scale. *J. Mater. Res. Technol.* **2019**, *8*, 2753–2766. [CrossRef]
47. Ilyas, R.; Sapuan, S.; Ibrahim, R.; Abral, H.; Ishak, M.; Zainudin, E.; Atikah, M.; Nurazzi, N.M.; Atiqah, A.; Ansari, M.; et al. Effec of sugar palm nanofibrillated cellulose concentrations on morphological, mechanical and physical properties of biodegradabl films based on agro-waste sugar palm (*Arenga pinnata* (*Wurmb.*) Merr) starch. *J. Mater. Res. Technol.* **2019**, *8*, 4819–4830. [CrossRef]
48. Syafri, E.; Yulianti, E.; Asrofi, M.; Abral, H.; Sapuan, S.M.; Ilyas, R.A.; Fudholi, A. Effect of sonication time on the thermal stabilit moisture absorption, and biodegradation of water hyacinth (*Eichhornia crassipes*) nanocellulose-filled bengkuang (*Pachyrhizu erosus*) starch biocomposites. *J. Mater. Res. Technol.* **2019**, *8*, 6223–6231. [CrossRef]
49. Edhirej, A.; Sapuan, S.M.; Jawaid, M.; Zahari, N.I. Preparation and characterization of cassava bagasse reinforced thermoplasti cassava starch. *Fibers Polym.* **2017**, *18*, 162–171. [CrossRef]
50. Ibrahim, M.I.J.; Sapuan, S.M.; Zainudin, E.S.; Zuhri, M.Y.M. Potential of using multiscale corn husk fiber as reinforcing filler i cornstarch-based biocomposites. *Int. J. Biol. Macromol.* **2019**, *139*, 596–604. [CrossRef]
51. Hazrol, M.D.; Sapuan, S.M.; Zainudin, E.S.; Wahab, N.I.A.; Ilyas, R.A. Effect of Kenaf Fibre as Reinforcing Fillers in Corn. *Polymer* **2022**, *14*, 1590. [CrossRef]
52. Othman, S.H.; Tarmiti, N.A.N.; Shapi'I, R.A.; Zahiruddin, S.M.M.; Tawakkal, I.S.M.A.; Basha, R.K. Starch/banana pseudosten biocomposite films for potential food packaging applications. *BioResources* **2020**, *15*, 3984–3998. [CrossRef]
53. Hazrati, K.; Sapuan, S.; Zuhri, M.; Jumaidin, R. Preparation and characterization of starch-based biocomposite films reinforced b *Dioscorea hispida* fibers. *J. Mater. Res. Technol.* **2021**, *15*, 1342–1355. [CrossRef]
54. Sanyang, M.L.; Sapuan, S.M.; Jawaid, M.; Ishak, M.R.; Sahari, J. Effect of Sugar Palm-derived Cellulose Reinforcement on th Mechanical and Water Barrier Properties of Sugar Palm Starch Biocomposite Films. *BioResources* **2016**, *11*, 4134–4145. [CrossRef]
55. López, O.; Versino, F.; Villar, M.; García, M. Agro-industrial residue from starch extraction of *Pachyrhizus ahipa* as filler c thermoplastic corn starch films. *Carbohydr. Polym.* **2015**, *134*, 324–332. [CrossRef] [PubMed]
56. Ilyas, R.; Sapuan, S.M.; Ibrahim, R.; Abral, H.; Ishak, M.R.; Zainudin, E.S.; Atiqah, A.; Atikah, M.S.N.; Syafri, E.; Asrofi, M.; et a Thermal, Biodegradability and Water Barrier Properties of Bio-Nanocomposites Based on Plasticised Sugar Palm Starch an Nanofibrillated Celluloses from Sugar Palm Fibres. *J. Biobased Mater. Bioenergy* **2020**, *14*, 234–248. [CrossRef]
57. López, O.V.; Ninago, M.; Lencina, M.M.S.; García, M.A.; Andreucetti, N.A.; Ciolino, A.E.; Villar, M.A. Thermoplastic starc plasticized with alginate–glycerol mixtures: Melt-processing evaluation and film properties. *Carbohydr. Polym.* **2015**, *126*, 83–9 [CrossRef]
58. Tavares, K.M.; de Campos, A.; Mitsuyuki, M.C.; Luchesi, B.R.; Marconcini, J.M. Corn and cassava starch with carboxymethy cellulose films and its mechanical and hydrophobic properties. *Carbohydr. Polym.* **2019**, *223*, 115055. [CrossRef]
59. Ramírez, M.G.L.; Satyanarayana, K.G.; Iwakiri, S.; de Muniz, G.I.B.; Tanobe, V.; Flores-Sahagun, T.S. Study of the properties c biocomposites. Part, I. Cassava starch-green coir fibers from Brazil. *Carbohydr. Polym.* **2011**, *86*, 1712–1722. [CrossRef]
60. Zoungranan, Y.; Lynda, E.; Dobi-Brice, K.K.; Tchirioua, E.; Bakary, C.; Yannick, D.D. Influence of natural factors on th biodegradation of simple and composite bioplastics based on cassava starch and corn starch. *J. Environ. Chem. Eng.* **202** *8*, 104396. [CrossRef]
61. Anugrahwidya, R.; Armynah, B.; Tahir, D. Bioplastics Starch-Based with Additional Fiber and Nanoparticle: Characteristics an Biodegradation Performance: A Review. *J. Polym. Environ.* **2021**, *29*, 3459–3476. [CrossRef]
62. Emadian, S.M.; Onay, T.T.; Demirel, B. Biodegradation of bioplastics in natural environments. *Waste Manag.* **2017**, *59*, 526–53 [CrossRef]
63. Sabbah, M.; Altamimi, M.; Di Pierro, P.; Schiraldi, C.; Cammarota, M.; Porta, R. Black Edible Films from Protein-Containin Defatted Cake of *Nigella sativa* Seeds. *Int. J. Mol. Sci.* **2020**, *21*, 832. [CrossRef]

sustainability

MDPI

Review

Recent Advances in Biomass Pyrolysis Processes for Bioenergy Production: Optimization of Operating Conditions

Dina Aboelela [1], Habibatallah Saleh [1], Attia M. Attia [1], Yasser Elhenawy [2,3,4], Thokozani Majozi [2] and Mohamed Bassyouni [4,5,6,*]

1 Faculty of Energy and Environmental Engineering, The British University in Egypt (BUE), El-Sherouk City 11837, Egypt; dina.aboelela@bue.edu.eg (D.A.); habibatallah184328@bue.edu.eg (H.S.); attia.attia@bue.edu.eg (A.M.A.)

2 School of Chemical and Metallurgical Engineering, University of the Witwatersrand, 1 Jan Smuts Avenue, Johannesburg 2000, South Africa; dr_yasser@eng.psu.edu.eg (Y.E.); thokozani.majozi@wits.ac.za (T.M.)

3 Mechanical Power Engineering Department, Port Said University, Port Said 42526, Egypt

4 Center of Excellence for Membrane Testing and Characterization (CEMTC), Port Said University, Port Said 42526, Egypt

5 Department of Chemical Engineering, Faculty of Engineering, Port Said University, Port Said 42526, Egypt

6 Faculty of Industry and Energy, East Port Said University of Technology, Saini, Port Said 45632, Egypt

* Correspondence: m.bassyouni@eng.psu.edu.eg

Abstract: Bioenergy has emerged to be among the primary choices for the short- and medium-term replacement of fossil fuels and the reduction in greenhouse gas (GHG) emissions. The most practical method for transforming biomass into biofuel is thermochemical conversion, which may be broken down into combustion, torrefaction, pyrolysis, hydrothermal liquefaction, and gasification. In this study, producing biofuels using a biomass pyrolysis process was investigated. This study explored the pyrolysis process and operating conditions to optimize the process parameters to maximize the desired product yields and quality. The pyrolysis process produces three main products, which are bio-oil, bio-char, and gas. There are three classifications for the pyrolysis method, with each of them producing a majority of a certain product. First, slow pyrolysis is conducted in the temperature range of 300–950 °C and residence time of 330–550 s. It produces around a 30% oil yield and 35% char yield, and thus, the majority yield of slow pyrolysis is char. Second, fast pyrolysis produces around 50% oil, 20% char, and 30% gas yields with a temperature range of 850–1250 °C and a residence time of 0.5–10 s. The average yield of flash pyrolysis was found to be 75% bio-oil, 12% bio-char, and 15% gas, which is conducted within less than 1 s. It was reported that the pyrolysis of biomass was simulated using ASPEN Plus, where the effects of several parameters, such as the temperature, heating rate, and residence time, on the product yield and composition were investigated. Pyrolysis was performed under different conditions ranging from 400 to 600 °C. The effects of different catalysts on the pyrolysis process were studied. It was found that the addition of a catalyst could increase the yield of bio-oil and improve the quality of the product. The optimal operating condition for the pyrolysis process was determined to be a temperature of 500 °C, which resulted in a higher bio-oil yield. It was found that the biofuel yield was enhanced by selecting appropriate raw materials, such as rice husk, along with the pyrolysis temperature (e.g., 450 °C) and particle size (350–800 μm), and using a low residence time and pressure.

Keywords: integrated system; pyrolysis methods; parameters; simulation

tation: Aboelela, D.; Saleh, H.; tia, A.M.; Elhenawy, Y.; Majozi, T.; assyouni, M. Recent Advances in omass Pyrolysis Processes for oenergy Production: Optimization Operating Conditions. *Sustainability* 23, *15*, 11238. https://doi.org/ .3390/su151411238

cademic Editors: Maurizio Volpe d Abu Naser Md Ahsanul Haque

eceived: 28 April 2023
evised: 1 July 2023
ccepted: 14 July 2023
iblished: 19 July 2023

1. Introduction

In contrast with past decades, which was a world where the affordable and limitless supply of fossil fuels could be relied upon, the present energy crisis has greatly increased uncertainty [1]. Hence, the worldwide scientific community has focused a lot of attention and efforts on developing a highly sustainable resource that might replace conventiona energy sources [2].

The only renewable resource that can be utilized to make solid, liquid, and gas fuels i biomass [3]. Also, biomass provides 14% of the energy used by the world [4]. Biofuels tha are liquid offer greater benefits than those that are solid or gaseous with respect to storag transportation, and high energy density [5]. They can also be incorporated into boiler motors, or turbines [6]. The biomass composition, along with the reaction mechanisn related to the size of the feed particles, temperature, reaction duration, and heating rat have an impact on the production of various fuels [7].

There are two kinds of biofuels: primary and secondary; both are produced usin biomass [8]. Also, they are divided into three generations. In the first generation o biofuels, starches or sugars from food crops, such as maize, sugarcane, and rapeseed, wer fermented to make bioethanol or biobutanol, and oil crops were trans-esterified to produc biodiesel [9]. An analysis of the first generation's life cycle showed that there was n net gain in energy. There was a debate going on among people that it is better to donat food to poor people rather than use it to produce fuel, and thus, researchers shifted to th second generation [10]. Agricultural waste (non-edible food) like lignocellulosic biomas was used to make second-generation biofuels. After shifting to the second generation, thi generation's life cycle assessment (LCA) showed that the net energy gain has increase Researchers then shifted to the third generation, which is a new technology to produc fuels from algae. Due to their significant lipid composition, ability to fix CO_2, and quic growth rate, microalgae are a favorable feedstock for the generation of biofuel. Bioethano biodiesel, and biohydrogen are the exemplification of third-generation biofuels that ar made from microalgae, seaweeds, and other microorganisms. The limits and shortcoming of first- and second-generation biofuels were resolved by third-generation biofuels. Larg scale neutral lipid buildup, high yield, CO_2 capture, and wastewater bioremediation ar further benefits [11].

Biomass can come from a diversity of sources: plant materials, microorganisms, an municipal solid waste. Forest timber feedstocks and agricultural waste are examples o plant resources [12]. Hard wood and soft wood are the two basic divisions of wood feedstocks. Sugar, starch, and oil seed crops are first-generation feedstocks. Sugarcan sugar beets, corn, sorghum, cassava, and wheat are examples of crops used to make suga and starch. Sunflower, oil palm, soy, coconut, jatropha, and rapeseed are examples of oi crops. Algae and lignocellulosic biomass from plants are utilized as plant resources fo second-generation biofuels. The biomass known as lignocellulosic comprises grasses, tree and crop wastes [4]. In recent years, the pyrolysis of biomass has attracted substantia interest as a potential source of bio-oil, gas, and bio-char [13]. There are a great number o articles that were published on this topic throughout the last decade, as shown in Figure 1

Several methods have been applied to convert biomass into biofuels, as shown i Figure 2. The most practical method for transforming biomass into biofuel is thermo chemical conversion, which may be broken down into combustion, torrefaction, pyrolysi hydrothermal liquefaction, and gasification [14]. By adjusting the process variables, th main goal of thermochemical conversion is to remove undesired by-products. Pyrolysi is a viable method for transforming biomass into biofuel at temperatures between 25 and 600 °C when it is inert. In fact, one new strategy is to use the pyrolysis method t manufacture bio-based chemicals and fuels from biomass [15]. The products of pyrolyzin biomass can be broken down into three categories: bio-oil, bio-char, and syngas. These thre byproducts have the ability to be used as sources of energy or in other applications, an they have several advantageous properties, including being friendly to the environmen having low costs, and degrading naturally [16].

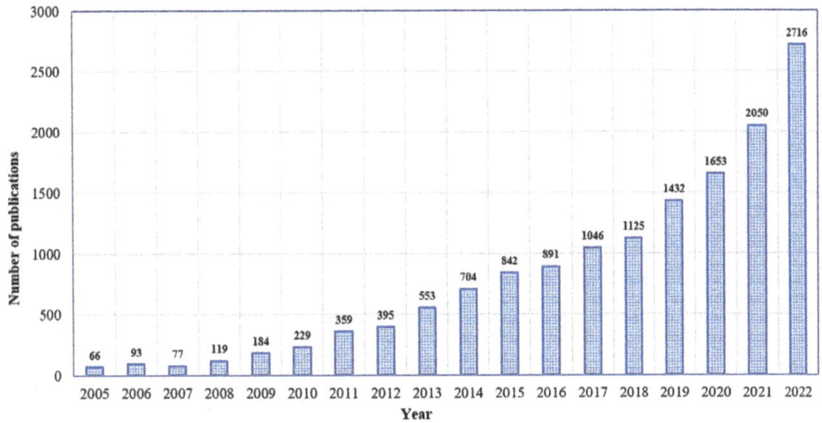

Figure 1. Number of publications per year on biomass pyrolysis.

Biofuels can be found in liquid form, gaseous form, or solid form. A variety of liquid biofuels are available. There is a possible carbon-neutral biofuel option in the form of bioethanol made from lignocellulosic biomass. Bioethanol's benefits as a biofuel include having a high octane number, low boiling point, increased heat of vaporization, and equivalent energy contents [17]. Vehicles may utilize fuel blends containing 85% (*v/v*) bioethanol without requiring any mechanical adjustments [18]. Both greenhouse gas (GHG) emissions and oil consumption may be substantially reduced by blending. There are three main processes that make up this system: pretreatment, enzymatic saccharification, and fermentation. Since 2008, a variety of work has been done to reduce costs by, for example, employing enzyme mixtures for improved saccharification, utilizing microbes for enhanced product yield, and manufacturing certain lofty value-added products to enhance the process economics. Biobutanol provides improved safety, lower hygroscopicity, lower igniting difficulty, more inter-solubility, higher viscosity, and better lubricity than its petroleum-based counterpart [19,20]. Features of bio-methanol, such as its great performance and lack of emissions, make it a safe fuel source [4]. The fact that methanol breaks down entirely into CO_2 and water in a steam environment makes it a useful component in fuel-cell-driven cars. 7-[3-(methylaminomethyl) phenoxy]methyl]quinoline-2-amine (M85) is composed of 85% methanol and 15% gasoline and may be utilized in most cars and trucks with just minor mechanical adjustments [4]. Ultrasonic processing of algal biomass can be used for biodiesel manufacturing [21]. To produce biodiesel, 51 percent of the lipids in the algal biomass are isolated and trans-esterified with CH_3OH (methanol) as a catalyst [22]. One of the available methods is fast pyrolysis, which is used to convert biomass into bio-oil [4]. Bio-oil is a liquid that has a distinct smokey aroma and its color is dark brown. Acids, alcohols, esters, ketones, phenols, aldehydes, and oligomers are all part of the complex combination that makes up these substances [23]. High water content, viscosity, ash content, oxygen concentration, and corrosiveness are all drawbacks of bio-oil as a fuel [4]. The high costs of manufacturing and low fuel quality of bio-oil are the main obstacles impeding its commercialization [24]. A method was devised to enhance the qualities of the bio-oil made from rice husk [25]. The enhanced oil's density was found to decrease from 1.24 to 0.95 g/cm^3, and its heating value rose from 16.0 to 27.2 MJ/kg. Additionally, the oil's pH decreased from 4.4 to 2.3 after the refinement process [4].

Regarding gaseous fuels, there are two kinds of gaseous biofuels, which are gas and biohydrogen. Last, but not least, there are solid biofuels, where the most significant type is bio-char. Bio-char is a carbon-dense byproduct of biomass breakdown at temperatures between 350 and 700 °C. Bio-char is useful both as a fertilizer and a soil conditioner. Reducing atmospheric CO_2 concentrations, increasing soil carbon stocks, and increasing soil carbon capture are all ways in which it contributes to climate change mitigation [26 There are nutrients in bio-char that help plants thrive. Soils conditioned with bio-cha made from Miscanthus by heating it to 400 °C for 10 min suppressed the growth of maiz seedlings. Soil conditioning using bio-char made from Miscanthus and collected afte heating at 600 °C for 60 min yielded contrasting results [4]. This suggests that improve yields from seedlings of different varieties and species need bio-char produced in differer ways [4,27].

While significant progress has been made in the area of biomass pyrolysis, there ar still several research gaps that need to be addressed to optimize the operating condition for bioenergy production, including the following: (i) There is a lack of selecting th standardized operating conditions for biomass pyrolysis. Different studies have use varying temperatures, residence times, heating rates, and catalysts, making it difficult t compare and generalize the results. The optimization of these operating conditions is cruci to maximizing the yield of desired bioenergy products and minimizing the formation c undesired by-products. (ii) Regarding the simulation of biomass pyrolysis, including th decomposition conditions, further research is needed to investigate the biomass pyrolysi simulation and identify the optimal conditions for maximizing the desired product yield (iii) The techno-economic analysis and life cycle assessment: The economic viability an environmental sustainability of biomass pyrolysis processes are essential for their successfu implementation. However, there is a lack of comprehensive techno-economic analyse that consider the optimization of the operating conditions. Further research is needed t evaluate the economic feasibility.

This study aimed to investigate the pyrolysis parameters to produce bio-oil, bio-cha and gases. The recent development methods were studied to provide a wide variety c sustainable bioenergy outputs with added value. There were several issues found wit the current oil/tar spraying techniques in anaerobic digestion (AD), including microbi toxicity and low production.

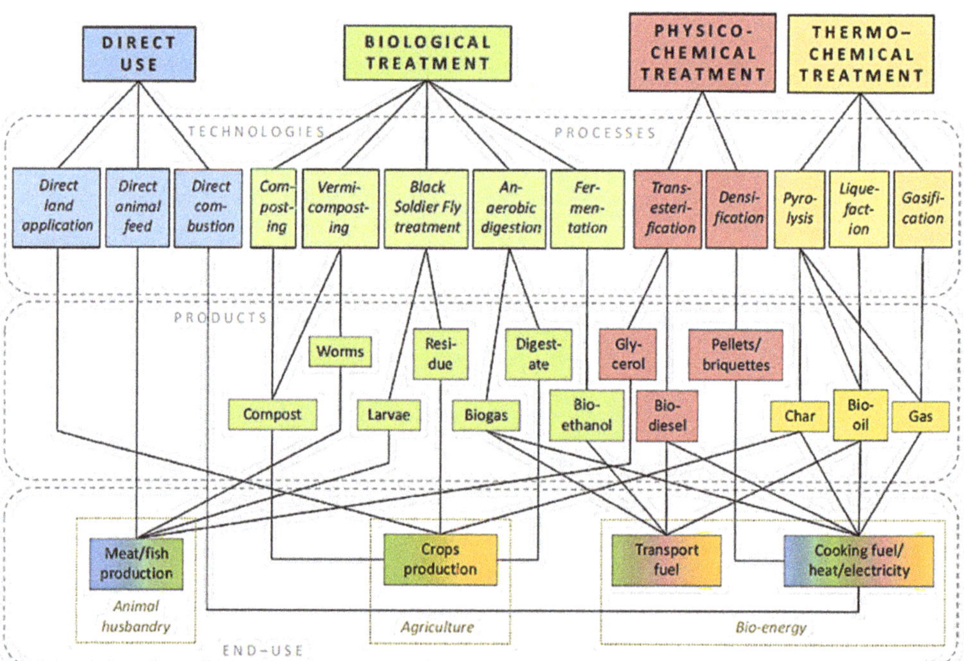

Figure 2. Treatment technologies for biomass [28].

2. Thermal Conversion

The three most common forms of thermal conversion are pyrolysis, gasification, and incineration. They are distinct from one another because of the ways in which they are used and, by extension, the results of their processes [29]. Charcoal, liquid, and gaseous byproducts can be produced from the pyrolysis of biomass in the absence of oxygen, as shown in Figure 3 [30].

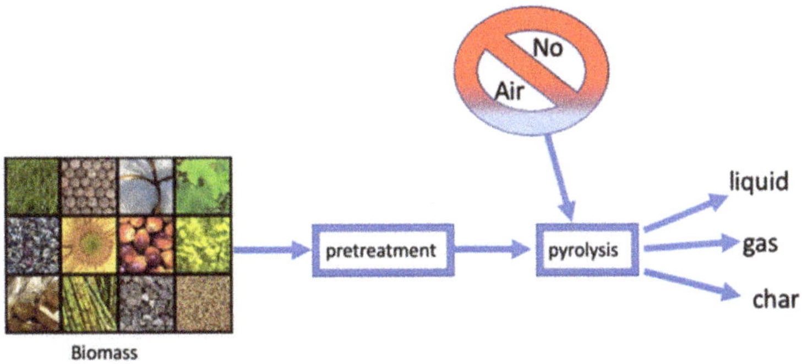

Figure 3. Schematic diagram for the pyrolysis method.

3. Pyrolysis Classifications

There are three distinct pyrolysis methods, namely, slow pyrolysis, fast pyrolysis, and flash pyrolysis [31,32]. Table 1 compares and contrasts these processes, highlighting the variations between them in terms of temperature, solid residence time, heating rate, biomass particle size, and product yield. The type of procedure and the process operating parameters determine the product distribution [33]. Figure 4 shows the majority percentages of each product, which are bio-oil, bio-char, and gas.

Table 1. Different parameters of the pyrolysis process.

Parameters	Slow Pyrolysis	Fast Pyrolysis	Flash Pyrolysis	Reference
Temperature (°C)	550–950	850–1250	900–1200	[33]
Heating rate (°C/s)	0.1–1.0	10–200	>1000	
Residence time (s)	300–550	0.5–10	<1	
Particle size (mm)	5–50	<1	<0.5	

3.1. Slow Pyrolysis

Since ancient times, slow pyrolysis has been utilized to increase char generation at low temperatures and low heating rates [31]. The generation of solid char and other liquids occurs in this procedure because the vapor residence time is excessively long (5 to 30 min) and elements in the vapor phase are inclined to react with one another. Slow pyrolysis, however, has several technological drawbacks that make it unsuitable for producing high-standard bio-oil [31]. Due to the prolonged residence time, the main product cracks, which may have a detrimental effect on the production and quality of the bio-oil. Additionally, considerable energy input is required because of the lengthy residence times and limited heat transmission [32].

3.2. Fast Pyrolysis

Fast pyrolysis entails rapidly heating biomass to temperatures between 850 and 1250 °C, with heating rates between 10 to 200 °C/s per second, over a period between 1 and 10 s [33]. Because the oil product yield in rapid pyrolysis is so much higher than the bio-char and syngas product yield, it is employed for bio-oil production [33]. Fast pyrolysis typically yields 60–75% liquid product, 15–25% charcoal product, and 10–20% non-condensable gaseous products [34]. Its goal is to heat the biomass to an appropriate temperature for thermal cracking to occur while minimizing the amount of time the liqui biomass is exposed to heat, which promotes the formation of char [35]. High temperature used in quick pyrolysis turn biomass into a liquid before it can react to generate char [33 These days, the fast pyrolysis process is used for more than only producing energy; it i also used in the food industries and for certain chemicals. Liquid fuels and a variety o specialty and chemical reactions may be produced using fast pyrolysis technology, whic has recently gained significant attention [36]. To separate the processing of solid biomas from its consumption, this liquid yield may be transported and stored conveniently an cheaply [36]. Further, it possesses the ability to supply a variety of useful compounds, eac of which has far greater value-added than fuels. In comparison with other processes, fas pyrolysis technology, particularly when used on a tiny scale, can offer cheap investmen costs and great energy efficiency [37]. There are several possible benefits to producin bio-oil via rapid pyrolysis, such as the storing and transporting of liquid fuels, its low cos the neutral CO_2 balance, and the utilization of second-generation materials, which has le to increased interest in this method in recent years [38]. Bio-oil raw materials and wast materials (forest residue, municipal and industrial waste, etc.) have a large energy densit compared with ambient gases [39].

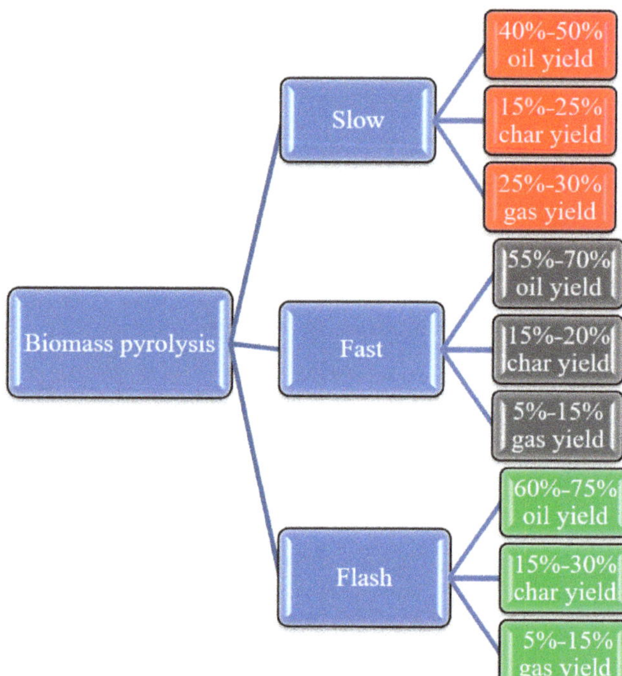

Figure 4. Yield of pyrolysis process at different conditions.

3.3. *Flash Pyrolysis*

A bio-oil output of up to 75% may be attained using the flash pyrolysis of biomass [40 making it a potential technique to produce solid, liquid, and gaseous fuel from biomass. I this process, the particles are heated quickly, the reaction temperature is high (between 45 and 1000 °C), and the gas is released within less than one second [41]. However, there ar some technological constraints to be considered, such as (i) the low thermal stability an corrosive properties of produced bio-oil; (ii) the presence of solids in the oil; (iii) furthe polymerization and condensation reactions can lead to an increase in the viscosity of th bio-oil over time, making it more challenging to handle, pump, and process; (iv) biomas used in flash pyrolysis often contains alkali metals, such as potassium and sodium, where these alkali metals can be released and dissolved in the bio-oil during pyrolysis; and (v) during flash pyrolysis, small amounts of water can be generated as a byproduct, where this pyrolytic water can mix with the bio-oil and affect its stability, quality, and compatibility with downstream processes [42].

4. Parameters That Affect the Pyrolysis Process

Temperature, feed particle size, residence time, biomass type, catalyst, heating rate, and pressure are all important parameters that affect the pyrolysis process. Both the primary and secondary processes involved in the degradation of biomass during pyrolysis need heat and mass transport. Primary degradation occurs at low temperatures (about 200–300 °C) and involves the breakdown of the complex biomass structure. This process produces intermediate products, such as volatiles, char, and bio-oil. The principal products differ based on the biomass mix and the pyrolysis conditions. Volatiles are gases and vapors that are produced during pyrolysis. Char is the solid residue that is left over after the volatiles have been discharged. It is a carbon-rich, high-surface-area substance that can be utilized as a solid fuel or as a precursor for the synthesis of activated carbon. Bio-oil, commonly known as pyrolysis oil or biomass oil, is a black, viscous liquid produced via volatile condensation. To obtain higher-value products, bio-oil can be improved further using techniques such as hydrotreating or fractional distillation. Secondary degradation occurs at higher temperatures (usually above 500 °C) and is characterized by the thermal cracking of the primary products formed during primary degradation. This process produces more gases, such as light hydrocarbons (e.g., ethylene and propylene), and boosts the overall production of gases and liquids. Secondary decomposition reactions are often faster and more exothermic than primary decomposition events. Syngas, also known as synthesis gas, is a carbon monoxide (CO) and hydrogen (H_2) mixture that can be created during the secondary breakdown of biomass. Secondary biomass breakdown can generate several additional gases and vapors, including higher hydrocarbons, tars, and light oxygenates. These products can be developed and treated further for specialized uses. Several parameters influence both the primary and secondary degradation processes in pyrolysis, including the temperature, heating rate, residence duration, and composition of the biomass feedstock. It is feasible to modify the pyrolysis process to produce desired product yields and qualities for a variety of applications, including bioenergy, biofuels, and bio-based compounds, by optimizing these parameters. The breakdown of lignin, cellulose, and hemicellulose into simpler compounds is an example of a primary reaction. The decomposition of intermediates is a key feature of secondary reactions. First, the major products must be broken down into smaller molecules so that cellulose may be converted into sugars; second, the transformation of the main products into huge molecules and char takes place [43].

4.1. Temperature

The pyrolysis process relies on several factors, one of which is temperature. A variety of temperature values are required for the breakdown and devolatilization of biomass constituents. Heavily tarred substances result from the breakdown of hemicellulose and non-condensable gases at temperatures below 300 °C. When biomass is heated to temperatures above 550 °C, it breaks down into its parts and releases several different chemical substances [43]. Acetic acid, levoglucosan, hydroxy-acetaldehyde, and 5-hydromerthylfurfural are all cellulose-derived chemicals, whereas phenolics were derived from lignin [35]. The literature contains several studies that addressed the importance of temperature in the pyrolysis process and bio-oil yield, including the ones listed in Table 2.

Table 2. Effect of temperature on bio-oil yield products.

Biomass	Temperature (°C)	Oil Yield (wt. %)	References
Wheat straw	600	34	[44–47]
Rice husk	450	70	
Coal	500	42	
Sunflower cake	550	41	
Hardwood samples	532	66.89	
Soybean cake	530	41	
Bagasse	500	66.1	

The ideal temperature for producing bio-oils in high yields during fast pyrolysi was reported to be between 400 and 500 °C. The char yield decreased as the pyrolysi temperature increased. Additionally, it was found that additional devolatilization o the primary char balances the generation of secondary char at higher temperatures. A temperatures between 300 and 400 °C, around 80 to 90% of the entire bulk conversion wa achieved [48]. The final pyrolysis temperature has a considerable impact on the conten and oil of the liquid effluent. Another study showed that a higher temperature resulted i higher biomass conversion efficiencies because more energy was needed to break down th cellulose bonds at higher temperatures. The rate of the solid cake's decomposition varie with temperature. When the temperature of the soybean cake was raised from 400 °C t 700 °C, an additional 11.82 wt.% of the material was decomposed [49]. Sunflower cake' yield at 450–700 °C was 10.7 wt.% [50]. The highest liquid yield was achieved durin the pyrolysis process between 500 and 550 °C. Secondary reactions, such as the rate o production of gasses increased when the temperature increased to 600 °C. However, the oi yield increased at around 570 °C, whereas the gas yield increased when the temperatur values increased from 430 to 730 °C [51]. Stabler species were generated during secondar breaking when the ultimate pyrolysis temperature was raised, where the functional-grour containing compounds were found. Polycyclic aromatic hydrocarbons, such as pyren phenanthrene, and others, were formed and accumulated at higher temperatures. Sinc dehydration and decarboxylation take place at high temperatures, the bio-oil content increased while the oxygenated concentration decreased [43].

4.2. *Size of Feed Particles*

Oil production and quality are both affected by the size of the feed particles used i the process. A common trend in biomass pyrolysis is the preference for smaller particl sizes due to the ease and uniformity with which they heat up. For rotating cone pyrolysis, i was recommended that particles be no more than 20 mm in diameter; for fluid bed system this number should be no more than 2 mm; and for a circulating fluid bed, it should be les than 6 mm. Some of the materials and their respective optimal particle size are summarize in Table 3 [52–54].

Table 3. Particle sizes used in the pyrolysis process with different sources of biomass.

Substrate	Optimal Particle Size	Reference
Wood particles	350–800 μm	[52]
Hazelnut	0.225 < dp < 0.425	[52]
Municipal solid waste (MSW)	Uncrushed, 1–2 cm	[53]
Rapeseed	<0.4 mm–>1.8 mm	[54]

Particles of wood with a diameter of 350 μm were fully pyrolyzed, whereas particle with a diameter of 800 μm were converted at a height of 0.9 m [52]. The maximum yiel was achieved at 0.45–0.6 mm for hazelnuts and municipal solid waste (MSW) [53]. At a siz of 0.6 to 0.85 mm, rapeseed produced the highest yield [54]. In the pyrolysis examinatio of orange trash, researchers found three distinct particle size ranges (300–180, 180–150 μm and <150 μm) [55]. For particles less than 150 μm in size, it was found that the thermal behaviors of the beginning and end of the pyrolysis process were varied. Because of variations in surface area, this variation was the result of heat and mass transport processes. However, the liquid output dropped because the larger biomass particles demand more heat and have a low heat transfer coefficient. Limitations in heat transport resulted in higher activation energy for large particles [56]. A high liquid yield was attained from large particles if these particles differ from one another in terms of characteristics like bulk density and oxygen content of the oil. Liquid products were generated when there was less obstruction. There were fewer reactive species and less energy in the liquid because of the increased oxygen level. If the particle size were to be reduced, it would cost more to complete the pyrolysis process since grinding equipment would be required. The technique is more expensive since it uses energy to reduce the size of a particle from its original, larger form. It was noted that different types of biomasses and pyrolizers produced different particle sizes. Optimizing the liquid product yield required careful consideration of both the pyrolizer and biomass sources selection [43].

4.3. Residence Time

One of the most crucial factors in the production of liquid fuels is the residence time. In pyrolysis, a fraction of a minute or less of residence time is optimal for maximizing liquid production. Low residence times are often favored for producing high-quality bio-oil. In pyrolysis, the yield of the liquid products is increased as time goes on due to a secondary reaction. More time in the reactor could be necessary for complete conversion; however, the best potential yield from the liquid is obtained after a short time. The yields of the liquid products increase due to the short residence period at a lower pressure. By varying the residence period from 15 to 40 min with fir sawdust, the effect of residence time was observed [57]. At 30 min, the product liquid yield was the highest (21.22%). It was noticed that bio-oil production increased with the biomass heating rate, even at durations above 40 s. The quantity of oxygen present during pyrolysis was also a factor in the final quality of the oil produced. Two-step pyrolysis, consisting of pyrolysis and oil generation, was used to reduce the oxygen levels.

To produce liquid products with good quality and yield, the residence time must be optimized [43]. When pyrolyzing raw sorghum bagasse at 525 °C, raising the vapor residence time from 0.2 s to 0.9 s lowered the bio-oil yields from 75% to 57% while simultaneously boosting char and gas yields [58]. Similar to this study, the oil yield from pyrolysis of sweet gum hardwood at 700 °C dropped from 22 wt.% to 15 wt.% when the vapor residence time was increased from 0.7 s to 1.7 s [59]. Product distributions as a function of vapor residence time were studied; however, the relationship between the vapor residence time and pyrolysis temperature, and its effect on yields and quality, needs further investigation [60].

4.4. Types of Biomass

Biomass can be classified into two main groups, namely, vegetable-derived and animal-derived, as shown in Figure 5. Lignocellulosic material is made up of three different components: cellulose, hemicellulose, and lignin, all of which can be found in different proportions. Hemicellulose degrades between 470 and 530 °C, cellulose between 510 and 770 °C, and lignin between 550 and 770 °C. Toward the end, ash is produced from the biomass, which contains trace quantities of inorganic substances, like potassium, sodium, phosphorus, calcium, and magnesium. The final product's elements are extremely sensitive to their constituents. More cellulose and hemicellulose are present at the outset, leading to greater oil output. All three components have distinct temperatures at which they degrade. Cellulose is crystalline and breaks down faster, whereas lignin is complicated and has a greater degree of polymerization [43]. Stronger structural integrity makes decomposing the lignin more challenging, but it results in a bigger char yield [61]. However, the deterioration of this material may be aided by the application of a high heating rate and temperatures, leading to a greater liquid output. Because of the high volatility and reactivity caused by th presence of such a huge quantity of volatile material, bio-oil production is stimulated [62 In their study of bio-oil production from rice straw and bamboo sawdust, the biomass wit a higher volatile material concentration produced a higher yield [63].

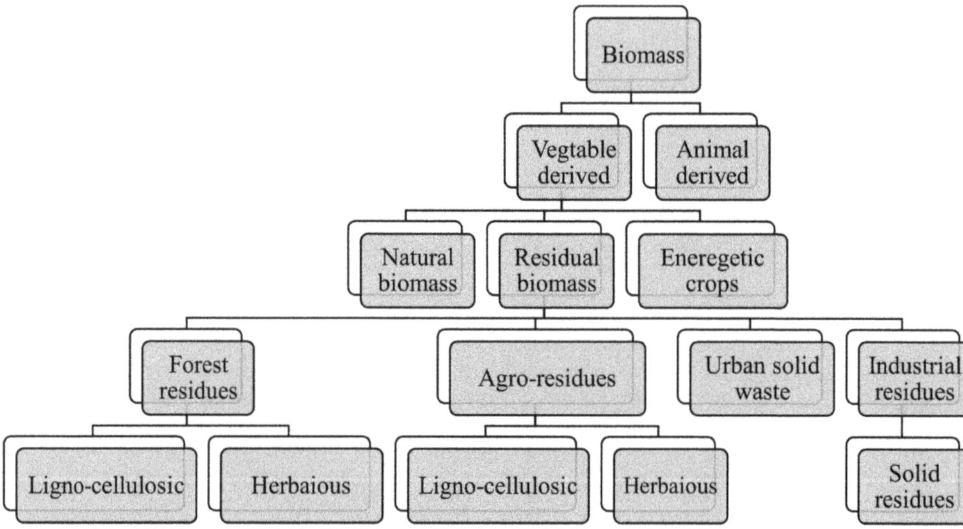

Figure 5. Schematic diagram of biomass types.

The percentages of cellulose, hemicellulose, and lignin in hazelnut shells were 30 wt.% 23% wt.%, and 38% wt.%, respectively. The percentages of cellulose, hemicellulose, an lignin in farm waste were 17 wt.%, 7 wt.% and 18 wt.%, respectively. Therefore, when ligni levels increased, the bio-oil yields declined but the bio-char yields increased. Agricultura byproducts showed a higher bio-oil output than hazelnut shells.

4.5. Catalyst

Biomass pyrolysis can be conducted with or without catalysts [64]. Catalysts hav been utilized to enhance several characteristics of bio-oils, including their ability to b repolymerized, their total acid number, their corrosivity, and their compatibility wit petroleum products. Certain pyrolysis processes require specific catalysts [65]. Fluidize and fixed-bed reactors are commonly used in catalytic pyrolysis. These catalysts ca be either provided in a solid or vapor phase. The results of the two procedures wer distinct because of the differences in contact time and response mechanism. In situ an ex situ pyrolysis improvements of beetle-killed trees were performed in the presenc of an HZSM-5 catalyst [66]. More benzene and toluene were produced in the ex sit upgrade, but the specificity for xylenes and aromatics containing carbon 9 was highe in the in situ upgrade. In the presence of a ZrO_2-FeOx catalyst, woodchips made fro Japanese cedar have a catalytic effect [67]. The ratio of catalyst loading to the feed rate o pyroligneous acid production determined the amount of feed that was converted to keton Rapeseed cake was converted gradually between 150 °C and 550 °C, depending on th kind of catalyst utilized. Noncatalytic test (34.06 wt.%) > Na_2CO_3 (27.10 wt.%) > HZSM 5 (26.43 wt.%) > Al_2CO_3 (21.64 wt.%) [68] in terms of total organic compounds. Usin rapid pyrolysis, desilicated ZSM-5 zeolite was employed as a catalyst for lignocellulosi biomass [69]. The conversion rate and the amount of unwanted coke may both be optimize by carefully regulating the amount of desilicated ZSM-5 utilized in the process. To acquir a large liquid yield, the catalyst choice is crucial. As the catalytic utilization increase coke generation is reduced, leading to a greater yield of aromatics [43]. By facilitatin the dissociation of complex biomass molecules and promoting the production of desired products, catalysts improved the pyrolysis reactions. Catalysts can lower the activation energy required for the pyrolysis reactions, thereby reducing the operating temperature and enhancing the reaction rate. This led to faster and more efficient conversion of biomass into desired products. Catalysts help with increasing the yield of bio-oil, which is a valuable product obtained from biomass pyrolysis. They can promote the cracking of larger molecules, resulting in higher production of lighter hydrocarbons that make up bio-oil. Using certain catalysts can influence the distribution of pyrolysis products, favoring the formation of specific compounds. For example, certain catalysts can promote the production of valuable chemicals, such as furans or phenolic compounds, which have various applications in the chemical industry. Biomass pyrolysis can sometimes lead to the formation of coke, which is a solid carbonaceous residue. Catalysts can suppress coke formation by catalyzing secondary reactions that consume coke precursors or by promoting the gasification of carbonaceous species. Another undesired product that can form during biomass pyrolysis is tar. Tar is a complex mixture of high-molecular-weight compounds that can be problematic in bio-oil production and utilization. Catalysts can help with reducing tar formation during pyrolysis by catalyzing tar-cracking reactions or by promoting secondary reactions that convert tar into more desirable products. Some catalysts can be regenerated and reused, allowing for multiple cycles of biomass pyrolysis. This can contribute to the economic feasibility of the process by reducing catalyst costs and improving the overall process sustainability. Catalysts can enhance the pyrolysis reactions by facilitating the breakdown of complex biomass molecules and promoting desirable product formation. Cracking compounds with greater molecular weights into lighter hydrocarbon products is a common method utilized with catalysts to improve the kinetics of pyrolysis reactions [70]. However, the distributions of products produced by various catalysts vary depending on the parameters in which they were used. Pyrolysis catalysts were divided into three categories according to their intended use. The first class was combined with biomass just before it was introduced to the reactor [71]. The second group was introduced into the reactor, allowing for direct interaction with the vapors, solids, and tars [72]. The third set was transferred to a secondary reactor that followed the pyrolysis reactor.

4.6. Heating Rate

The degradation of biomass into products relies heavily on the heating rate. The rapid breakdown of biomass into its parts in fast pyrolysis calls for a very high heating rate. Also, the highest yields of liquid products are achieved with the shortest residence time and highest heating rate. Thus, fewer undesirable chemicals are generated as a result of the shorter contact duration of the secondary reaction. It was proposed that a heating rate of up to 1000 °C/s might be used. Increases in the production of aliphatic and carbonyl chemicals were observed during rapid pyrolysis of coconut biomass [73]. The optimal temperature for maximizing oil production was observed to be influenced by the heating rate. Different heating rates (50 °C/s, 150 °C/s, and 250 °C/s) were applied to the esparto biomass, and the results showed that at 500 °C, the liquid yield was 45 wt.% for the 50 °C/s and 150 °C min^{-1} conditions, and 57 wt.% for the 250 °C min^{-1} condition. The ideal temperature increased from 500 °C to 550 °C at a heating rate of 250 °C/s [43]. The formation of volatiles was increased by increasing the heating rate. If the heating rate is high enough, the temperature will rise to its maximum, but if the heating rate is low enough, the temperature will remain at its minimum [74]. High heating rates are associated with high-quality final products because they reduce the amount of water present, stop secondary reactions from occurring, and create less oxygen [43].

4.7. Pressure

In most cases, atmospheric pressure is used for pyrolysis [43]. Researchers concluded that when completing pyrolysis, the pressure is greater than that of the atmosphere, which results in a greater bio-char yield [33]. Char is formed when the pressure is raised, causing the vapors to remain exposed to the carbon-based substance for a longer duration and for secondary carbon to be produced via decomposition [75]. The amount of carbon in bio-char can be affected by the high pressure within the reactor. Increases in the energy density of bio-char result from high-pressure pyrolysis of biomass, which increases the bio-char's carbon content [76]. Gases, including nitrogen and argon, as well as water vapor are employed in the pyrolysis process. Nitrogen gas (N_2) is a common inert gas. It was found that the yield of liquid oil was unaffected by the existence of inert gas. By doubling the flow rate from 50 cm^3 min^{-1} to 100 cm^3 min^{-1} [77], the liquid yield was increased by 3 wt.%. Nonetheless, a high liquid product yield may be achieved with a minimal gas flow. The higher the gas flow rate, the more gases are produced, as more volatile substances are evaporated. Also, steam can be utilized as a sweep gas. It was noted that the liquid yield improved when steam was used as the sweep gas [78]. If the oxygen content of the gas is processed, it is reduced by decreasing the gas flow rate, and the bulk density of the gas is increased, and thus, more liquid product is produced.

5. Pretreatment of Biomass

Microwave pretreatment is a promising technology for the conversion of lignocellulosic biomass into sustainable biofuels due to its ability to enhance the efficiency of the conversion process. On rare occasions, microwave pretreatment was investigated. Based on a few studies, here are some recent advances in microwave pretreatment [79]. Improvement of enzymatic hydrolysis: Microwave pretreatment improves the efficiency of enzymatic hydrolysis of lignocellulosic biomass. A study showed that microwave pretreatment of corn stover at 180 °C for 10 min resulted in a 70% increase in glucose yield during enzymatic hydrolysis compared with untreated corn stover. Reduction in energy consumption: Microwave pretreatment can reduce the energy consumption required for the conversion of lignocellulosic biomass into biofuels [80]. A study showed that microwave pretreatment of corn stover at 160 °C for 10 min reduced the energy consumption of the subsequent hydrolysis and fermentation processes by 25%. Enhancement of biofuel production: Microwave pretreatment can also enhance the production of biofuels from lignocellulosic biomass [81]. A study showed that microwave pretreatment of corn straw at 200 °C for 10 min resulted in a 26.7% increase in ethanol yield during fermentation compared with untreated corn straw. Optimization of process parameters: The efficiency of microwave pretreatment can be further improved through the optimization of process parameters such as temperature, time, and power [82]. Another study showed that optimization of the microwave pretreatment parameters for corn stover, such as a temperature of 180 °C, a time of 10 min, and a power of 800 W, resulted in a 72.3 wt.% increase in glucose yield during enzymatic hydrolysis [83].

6. Pyrolysis Product Properties

The pyrolysis of biomass yields three main byproducts: char, gases, and oil, which, when cooled to room temperature, condenses into a dark brown viscous liquid. It is between 350 °C and 500 °C where the most liquid is produced [84]. Due to differences in pyrolysis operations, various reactions take place at varying temperatures. At increased temperatures, molecules in the liquid and residual solid were broken down into tiny ones, enriching the gaseous component [85]. The production of charcoal is obtained at a low temperature with a low heating rate approach. The production of liquid products requires low temperature, high heating rate, and short gas residence time procedure. The production of fuel gas is achieved at a high temperature with a low heating rate and long gas residence time [31].

6.1. Bio-Oil

The condensed vapor of a pyrolysis process is known as bio-oil, which is a liquid. An alternative fuel oil use is possible. Compared with the heat content of hydrocarbon fuels, bio-oils are only about 40% as effective [31]. Since it is a liquid, it can be readily transported and stored, and its energy density is higher than that of biomass gasification fuel [82]. Pyrolysis oil, known as "bio-oil" or "bio-crude", has a wide variety of oxygenated molecules. Carbonyls, carboxyls, and phenolics are just a few of the chemical functional groups found in bio-oil that present opportunities for its use [86]. In spite of this, the thermo-physical characteristics of pyrolysis bio-oil are affected by a wide variety of elements that are now poorly understood [87]. An overview of some of the physical oil's features and characteristics is given in Table 4.

Table 4. Physical properties and characteristics of bio-oil.

Properties	Oil Characteristics	Interpretation	Ref.
Form	Free-flowing, organic liquid with a dark reddish-brown color	Oil's chemical composition and the presence of micro-carbon	[31]
Odor	Unique scent: a sharp, smokey odor.	Acids and aldehydes with smaller molecular weights	[31]
Density	Extremely high in comparison to fossil fuel Bio-oil from pyrolysis: 1.2 kg/L 0.85 kg/L for fossil fuels	High levels of moisture and significant molecular contamination	[31]
Viscosity	40–100 cP	Various feedstock types, water content, and the gathering of several non-heavy ends	[88]
Heat value	26.7 MJ/kg	High oxygen content	[89]
Aging	With time, there is an increase in viscosity, a reduction in volatility, phase separation, and gum deposition	A high pH value and complex structure	[31]
Miscibility	Petroleum fuel is completely immiscible in non-polar solvents, yet miscible with polar solvents	Polar in nature	[31]

Approximately 300 to 400 different chemicals make up pyrolysis oil [88]. The pyrolysis oil changes physically and chemically during storage, as several reactions happen and volatiles are released as a result of aging [89]. According to several studies, aging effects and reactions are sped up at higher temperatures, but if pyrolysis oil is preserved in a cold environment, the impacts can be mitigated [90]. Scientists found that the thermal efficiency of pyrolysis oils in combustion engine operations is comparable to that of diesel fuel; however, they showed severe ignition delay [91]. In order to reliably ignite pyrolysis oil, a modest amount of preheated combustion air is required. Heating rate, pyrolysis temperature, and residence time are only a few of the process factors that might affect pyrolysis oil yields, quality, and stability [92,93]. The ash amount and composition of the pyrolysis oil can be changed based on the reactor type (ablative/fixed), particle size, and char formation, all of which influence the rate, efficiency, and mechanism of biomass pyrolysis. Early interest in bio-oil was motivated by worries about crude oil shortages, yet recently, the environmentally friendly benefits of biomass fuels have grown to become an even more compelling and significant element. Due to the issues encountered when using bio-oils as fuel in conventional machinery designed for use with petroleum-derived fuels, such as boilers, engines, and gas turbines, the industry has yet to adopt them as a commercial standard. The major causes of this include bio-oil's viscosity, coking, and corrosion [31].

To produce energy, heat, and chemicals, bio-oil will be substituted for fossil fuels. Some applications for bio-oil are illustrated in Figure 6.

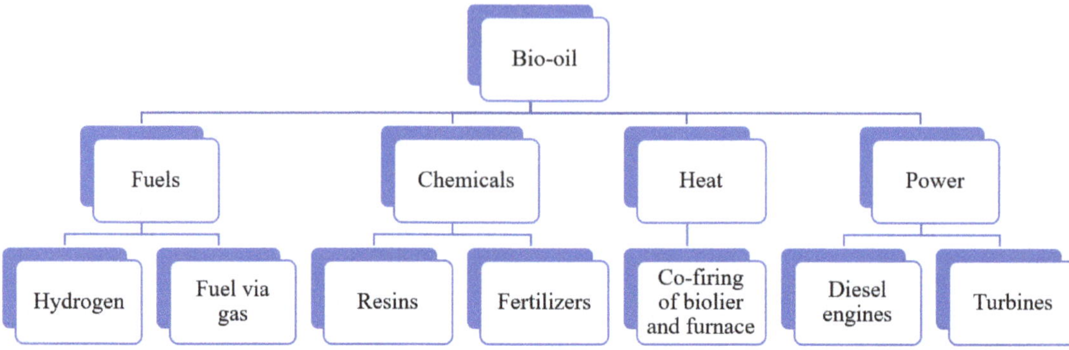

Figure 6. Applications of bio-oil.

6.2. Bio-Char

Significant mass loss in the form of volatiles occurs during the thermal breakdown o lignin and hemicellulose to produce a hard amorphous carbon matrix known as bio-cha The main solid yield is bio-char (also known as charcoal), which consists of unconverte organic solids and carbonaceous wastes created from the partial or total breakdown o biomass components, in addition to a mineral fraction [58]. The type of feedstock an the pyrolysis conditions determine the char's physical and chemical characteristics [58 Biomass and pyrolysis conditions determine the percentage of bio-char generated (10–35% Figure 7 displays the variations in bio-char yields across three temperature ranges durin pyrolysis [31,94]. The quantity of bio-char is significant in the low-temperature (450–500 °C zone due to weaker devolatilization rates and low carbon conversions. The generation o bio-char is drastically cut back in the second zone of average temperature (550–650 °C). was found that between 18–17 wt.% bio-char was the greatest production percentage i this temperature range. Extremely poor bio-char output was observed in the region wit temperatures greater than 650 °C [95]. The impact of varying temperatures on bio-cha production is shown in Figure 7.

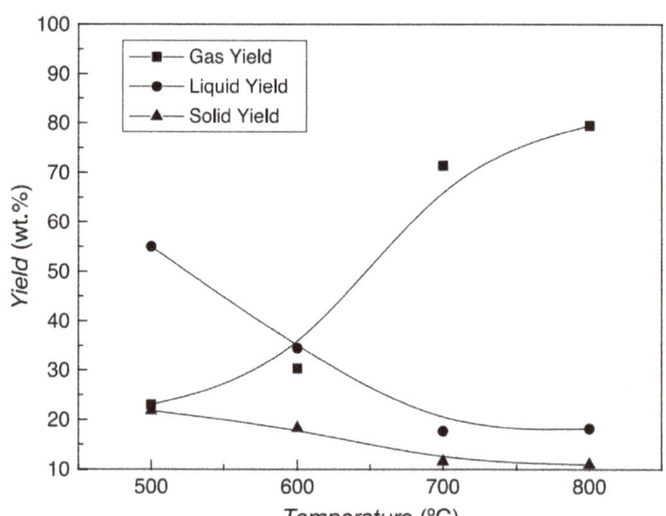

Figure 7. Different temperatures with different bio-char yields [31].

The physical features of bio-char may be considerably altered by adjusting some variables [96], including the reactor type and form, biomass type and drying treatment, platform molecules, particle size, chemical activity, heating rate, residence time, pressure, and inert gas flow rate [97]. Higher heating rates (up to 105–500 °C/s), shorter residence times, and finer platform molecules all contribute to the production of finer bio-char, while slower pyrolysis operating conditions and larger particle size of the feedstock provide coarser bio-char [98]. Further, bio-char made from wood-based biomass is often more coarsely granular [99]. In contrast, pyrolysis procedures produce finer and more brittle organized bio-char from agricultural wastes [100].

6.3. Gaseous (Syngas)

Potential byproducts of biomass pyrolysis include carbon dioxide, carbon monoxide, hydrogen, and hydrocarbons with reduced carbon numbers like methane and ethane. Certain gases like propane, ammonia, nitrogen oxide, sulfur oxide, and alcohols with minimal carbon numbers are obtained [58]. In slow pyrolysis operations, bio-gas is also generated, ranging from roughly 10 wt.% to 35 wt.%. Nevertheless, flash pyrolysis at high temperatures can produce more syngas. For instance, using calcined dolomite as a catalyst, it was reported that syngas was generated from the pyrolysis of municipal solid waste in a bench-scale downstream fixed-bed reactor at temperatures between 750 °C and 900 °C [101]. A temperature of 900 °C resulted in a gas output of 78.87 wt.% [102]. Secondary processes, including decarboxylation, decarbonylating, dehydrogenation, deoxygenation, and cracking, are subsequently performed on the volatile molecules and tar to produce the various components of syngas [103]. Tar decomposition and thermal cracking favor higher temperatures, leading to a greater proportion of syngas but lower oil and char production [104]. Moisture content affects heat transmission during pyrolysis, which is bad for producing syngas. Gaseous products are drastically reduced due to the high moisture content, which aids in the separation of water-soluble components from the gaseous phase [105]. When compared with wet biomass, the early stages of pyrolysis create the most gas for a specific temperature when using dry biomass. This happens because higher relative humidity lengthens the time required for a surface to dry [106]. Pyrolysis gas can be used for a variety of purposes, including the generation of heat or electricity through gas burning and compression ignition engines [107], whether alone or in combination with coal; the generation of specific gas elements, including (CH_4, H_2) or other volatiles; and the synthesizing of liquid biofuels. The heated pyrolytic gas can be recycled back into the pyrolysis reactor as a carrier gas, or it can be utilized to warm the inert sweep gas [58].

7. Simulation

Software simulation is required to determine the impact of pyrolysis temperature on the production of bio-char, as it is a time- and energy-efficient approach. The simulation process of biomass using Aspen Plus is shown in Figure 8 [107].

Using Aspen Plus, users can create detailed models of the pyrolysis process, including the complex chemical reactions and heat transfer mechanisms involved [108]. It allows users to define the reaction kinetics of the pyrolysis process based on experimental data or theoretical models. This includes specifying the rate of primary and secondary reactions during the decomposition of the feedstock. The software provides a vast database of thermodynamic properties for various substances, which can be used to accurately model the behavior of different species during pyrolysis [109]. Also, the tool is efficient in maximizing the pyrolysis process's operational parameters, including the heating rate, temperature, solid residence time, and feedstock size. Numerous experimental pyrolysis studies on fruit wastes, including the hulls of the Karanja (*Pongamia pinnata*) fruit [110], fruit bunches of the oil palm (*Elaeis guineensis*) [111], empty fruit bunches of the sweet lime (*Citrus limetta*) [112], peels of lemon [113], pomegranate (*Punica granatum* L.) [114], watermelon peel (*Citrullus lanatus*) [115], jackfruit (*Artocarpus heterphyllus*) peel [116], pine (*Pinus*) fruit shell [117], casuarina (*Casuarina equisetifolia*) fruit waste [118], coconut (*Cocos nucifera*) shell, and longan (*Dimocarpus longan*) fruit seed [118], were carried out. There were even a few studies on the pyrolysis of orange peel, banana peel, mango endocarp, apricot kernel shell, and date seed. There have not been many studies that used Aspen Plus in pyrolysis experiments despite the fact that it is used in practice. Ismail et al. used Aspen Plus to conduct a pyrolysis investigation on used tires [119]. The Aspen Plus model was effectively used to estimate the pyrolysis yield production, as well as to look into how temperature affected the pyrolysis product output. A scientist used Aspen Plus to carry out a microwave pyrolysis investigation on four biomass wastes, including *Calophyllum inophyllum* seed, *Acacia nilotica*, Bael shell, and rice husk [120–122]. The goal of the simulation research was to pinpoint the biomass that produced the highest output of bio-oil. According to the Aspen Plus findings, *C. inophyllum* biomass had the greatest bio-oil output (48 weight percent). Another researcher investigated the pyrolysis of municipal green trash using modeling and experiments [121]. The study aimed to validate the experimental results using Aspen Plus. The goal of the research was to maximize the output of bio-oil by optimizing the pyrolysis process operational variables, including the feedstock size, temperature, moisture, and air-to-fuel ratio. The simulation and experimental findings were found to be in good agreement, demonstrating that the Aspen Plus simulator is an efficient tool for predicting pyrolysis products. The simulation findings also showed that Aspen Plus can be utilized to sucwtcessfully optimize pyrolysis operating parameters, and the findings may be used for experimental analysis. Utilizing an Aspen simulation model and typical pyrolysis settings, the pyrolysis products of five different fruit wastes were examined. These wastes were orange peel, banana peel, mango endocarp, apricot kernel shell, and date pits [123,124]. Before using simulation to predict the yields of fruit waste pyrolysis, the model was first used to validate it by utilizing published data. The simulation's findings showed that all of the fruit wastes had high syngas yields of 46–55 wt.%. and high bio-char yields of 39–51 wt.%., but a poor bio-oil yield of 11 wt.%. The high volatile content of 50–78 wt.% of all the fruit wastes was responsible for their significant syngas output. Date pits had the greatest bio-char output of 50.92 wt.%. and mango endocarp had the greatest syngas production of 54.23 wt.%. among the fruit wastes. Date pits have a high elemental carbon concentration and a medium cellulose composition, which contributed to their strong char output. Mango endocarp's large elemental oxygen and hydrogen concentrations and extremely high holocellulose concentration were likely responsible for the fruit's high syngas output. According to the research, all of the selected fruit wastes were utilized as pyrolysis feedstocks to produce syngas. The research further predicted the bio-char from date pits pyrolysis.

Aspen Plus software has been widely utilized in other domains, including the manufacture of biodiesel and ethanol [125], coal or biomass gasification [126], and flue gas pollution control [127]. There is less research on utilizing it to model the process of biomass pyrolysis to create bio-char. Several studies demonstrated that Aspen Plus software can accurately and realistically model chemical manufacturing processes. An appropriate model was selected to investigate the impact of temperature variations on bio-char generation based on various pyrolysis settings and reactions. In addition, software simulation was used to determine the heat duty of the reactor at various pyrolysis temperatures, allowing for the analysis of the reaction process from the viewpoint of heat balance. Second, the biomass was pyrolyzed in a nitrogen stream using a tube furnace, and the yield of bio-char was determined by dividing the mass of the resulting product by the mass of the raw materials. The effect of temperature on the synthesis of bio-char was examined using various pyrolysis temperature values.

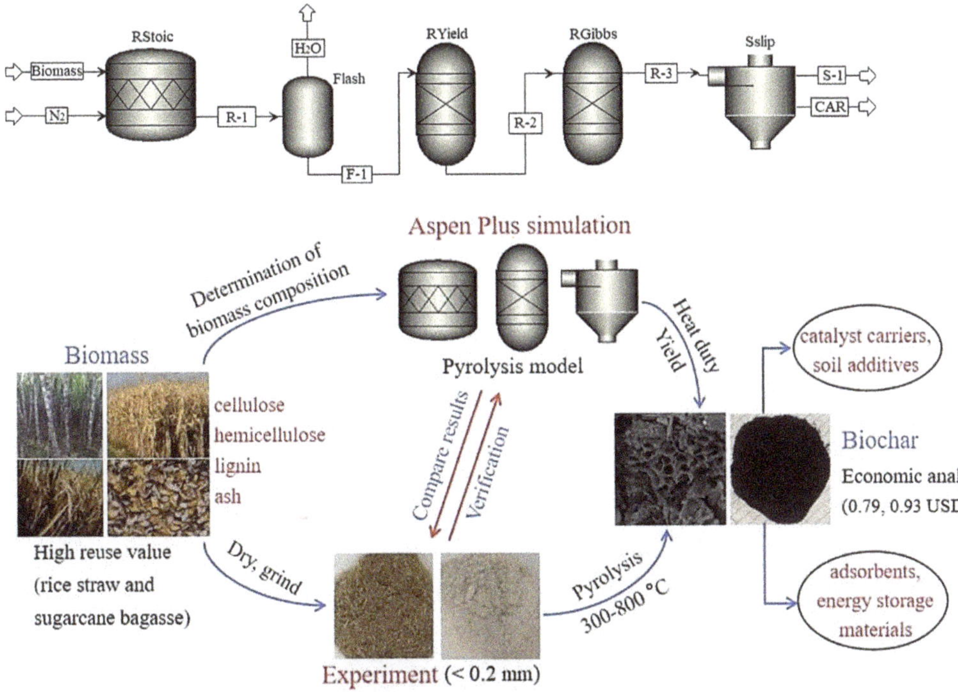

Figure 8. Effect of pyrolysis temperature on the yield of bio-char, as examined using a software simulation [107].

8. Product Treatment

8.1. Bio-Oil

The acidity, viscosity, chemical instability, and poor heating value [128] of bio-oil prevent it from being utilized as a "drop-in" transportation fuel [129]. The chemical discrepancies between bio-oil and fuel-grade hydrocarbons suggest that oxygen extraction is necessary; nevertheless, this process is costly [88]. There are two ways to get rid of the oxygen: either as H_2O (requiring the addition of hydrogen) or as CO_2 (reducing the fuel output) [130]. The bio-oil can be deoxygenated in one of two major ways: catalytic cracking or catalytic hydrodeoxygenation. Acidic zeolites and high temperatures (773–823 K) are required for catalytic cracking. This reduces biofuel production while increasing bio-char formation (>20 wt.%) [131]. Hydrogen unit activities in chemical engineering can be defined by the term "hydrotreatment" [132]. Hydrodeoxygenation, hydrocracking, and hydrogenation are all terms used to describe the same process when hydrogen is present with the intention of deoxygenating, cracking, or hydrogenating a molecule or complex mixture [133]. Two-stage hydrotreatment of bio-oil is a popular method [134]. At certain temperatures (100 and 300 °C), the active carbonyl and carboxyl functional groups must be converted into alcohols in the initial phase, known as stabilization [135]. However, the reactivity of certain acids (such as acetic acid) can be affected by factors such as the reaction temperature, residence time, and catalyst type [136]. A second process, namely, cracking and hydrodeoxygenation (HDO), is conducted around (350 °C and 400 °C) [137].

8.2. Bio-Char

Water quenching or leaching, heat treatments/aeration, aging/weathering, activation, particle size reduction, and palletization/granulation are used for post-treatment bio-char processing. Advantages of treatment methods are listed in Table 5.

Table 5. Types of bio-char treatments.

Treatment	Advantages	Reference
Water quenching or leaching	- Bio-char has easily soluble components that need to be removed. - Essential procedure for shutting down pyrolysis in many low-tech setups, such as conical kiln "fame-curtain" pyrolyzers.	[138]
Heat treatments/aeration	- It was proposed that off-gassing ethylene can be achieved by the very straightforward aeration of bio-char after its manufacture. - Toxic naphthalene is removed.	[139]
Aging/weathering	- It enhances bio-char properties through surface oxidation and increases the porosity.	[140]
Activation	- Increases the surface area. - Increases the pore volume.	[141]
Particle size reduction	- Particle fusion occurs during pyrolysis at high pressures, leading to bio-char that is more dense, and hence, smaller in size.	[142]

8.3. Gas Treatments and Applications

An electric furnace can be utilized to heat feedstock before it is fed into a fixed-bed reactor. An inert gas, like nitrogen, is compressed into the reactor (1). Throughout the operation, gases and vapors are released, and bio-char is gathered. Figure 9 shows the possible treatment process of pyrolysis products.

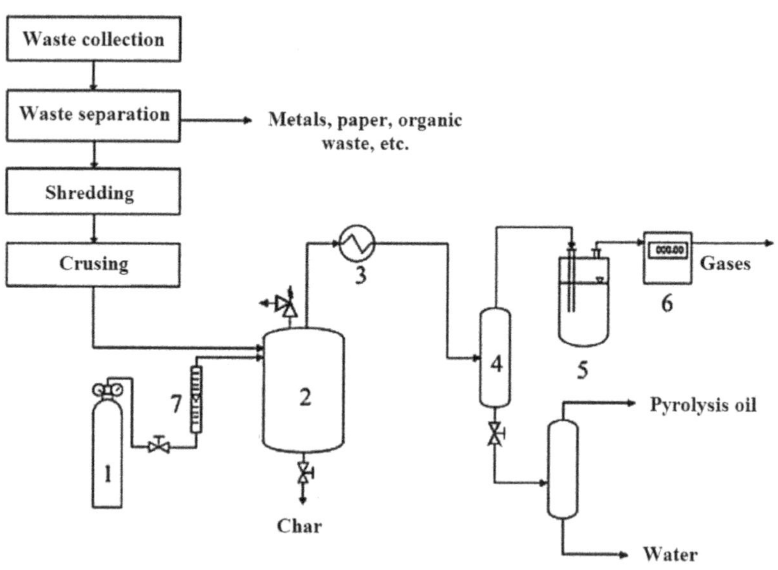

Figure 9. Waste treatment and pyrolysis products. (1—N2 bottle; 2—reactor; 3—heat exchanger; 4—separation unit, 5—water trap; 6—gas flow meter; 7—rotameter) [143].

The gas is released from the reactor (2), then cooled in a heat exchanger (3). Then, the gas mixture is fed into a separation unit (4) to separate gases from the stream. After the gas is separated, it is further purified by passing it through a water trap (5) to remove the leftover water; after the water is trapped, the stream is passed through a gas flow meter (6) to measure the production yield of gases. Flow rate of nitrogen gas is measured using rotameter (7) [143].

With no free oxygen present, waste is heated to a temperature of about 500 °C using an external heat source. To generate electricity, the feedstock is broken down into its volatile components to generate syngas, which is then fed into a boiler to generate steam, with the exhaust gas being treated in an emission control unit [144]. As shown in Figure 10, waste is first pretreated, which involves screening to remove contaminants (metals, glasses, and stones), shredding to reduce the particle size, and drying to remove any water. Then, pretreated waste is moved through a pyrolysis kiln to produce char and gas. The gas produced is then transferred to a combustion chamber. A boiler is used to produce gas and steam. Char produced from the pyrolysis kiln is treated in a char treatment unit to produce syngas, increasing the gas yield to produce electricity. A flue gas stream is passed through a flow gas treatment to make it safer for the environment. Syngas is collected to increase the gas yield. Steam is either used directly in domestic heating or used to produce electricity.

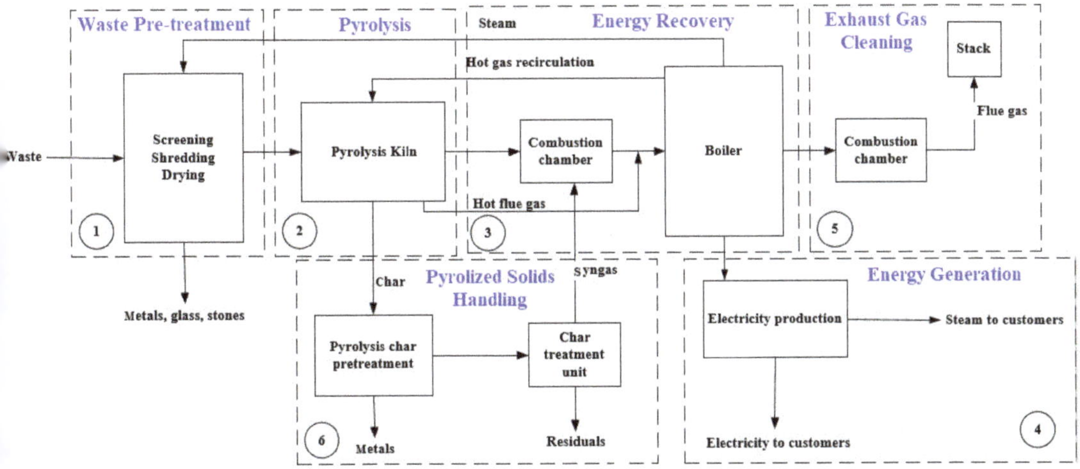

Figure 10. The use of pyrolysis waste for energy recovery and generation.

9. Recent Developments

The unsustainable treatment and disposal of food waste (FW) is a global environmental crisis, along with the recalcitrant organic residues (RORs) produced by FW treatment plant processes. There is an urgent need to fully harness FW and ROR through the introduction of recycling renewable technology. Two potential methods for degrading FW and its leftovers are anaerobic digestion (AD) and pyrolysis, which together provide a wide variety of sustainable bioenergy outputs with added value. There are a few issues with the current oil/tar spraying techniques used in anaerobic digestion (AD), including their microbial toxicity and low production [145].

9.1. Pyrolysis and Anaerobic Digestion Integrated Process

Pyrolysis, which can produce gas, oil, and a solid residue (char) that can all be further recycled, is an alternative approach to increase the energy and economic value of recalcitrant organic residue (ROR) usage [146–148]. Scientists throughout the world are increasingly interested in this method for its potential to valorize a wide range of waste products and yield useful products, as indicated in Table 6 [149–152]. A pyrolysis and AD coupling arrangement may be broken down into three distinct types: AD and pyrolysis, pyrolysis and AD, and AD–pyrolysis–AD.

Table 6. Waste products and coupling technology process.

Waste	Coupling Technology Process	Remarks	Ref.
Lignocellulosic biomass	- AD–pyrolysis - Pyrolysis–AD - AD–pyrolysis–AD	When comparing the AD–pyrolysis process with the standard AD process, the electricity benefit may be increased by around 42%. Few efforts have been made to combine pyrolysis with AD, and there is a pressing need for more research into the hazardous substances produced during pyrolysis. The decomposition of biomass using AD–pyrolysis–AD is feasible, and the resulting sludge and residues may be put to good use.	[149]
Paper mill sludge	AD combined with pyrolysis	The integrated method increases energy independence throughout the treatment phase.	[150]
Food waste	Combining AD and pyrolysis	This research shed light on the evaluation of AD pyrolysis for FW treatment and its subsequent transformation into gas, oil, and solid yields for energy generation. This integrated method allows for the efficient concentration of nutrient elements optimized for use in soil conditioning and agronomy.	[151]
Recalcitrant organic residues (ROR)	Two-stage pyrolysis coupled with AD	During bio-methanation, ROR's high H_2 to CO ratio (60:20 vol.%) in syngas produces almost 100% more CH_4 than the control. With a high CO content, the breakdown rate of H_2 is slowed down because of the higher concentration of H_2. Conventional ROR treatments have limitations that can be avoided by combining second-stage pyrolysis with the AD process. By using AD to process FW, a byproduct rich in hydrogen (H_2) is produced, as well as improved bio-methanation.	[152–155]

9.2. Challenges and Disposal of Recalcitrant Organic Residues Using the Anaerobic Digestion of Food Waste

Composting and the AD process are widely used in the process of food waste valorization. Anaerobic digestion (AD) involves the biological breakdown of organic materials in the absence of oxygen. Despite the fact that treating FW can help with AD, roughly 30% of the total food waste materials end up as ROR after screening and biological treatment as shown in Figure 11 [156]. Plastics and high lignin matters are the main types of solid residual matter produced during the AD process, and neither can be completely degraded without biochemical treatment. Plastics, lignin, and biomass all present significant treatment issues; nevertheless, they may be used for other purposes, such as resource recovery and power generation [157–159]. There are many treatment challenges that arise regarding ROR disposal. Landfilling, open burning, and incinerating these waste products are all inefficient ways to dispose of them. More environmental challenges (such as soil deterioration, GHG emissions, and water pollution) are being brought about by the disposal methods now in use. The rapid increase in ROR production rates necessitates immediate action in the form of environmentally responsible waste management strategies [160–163].

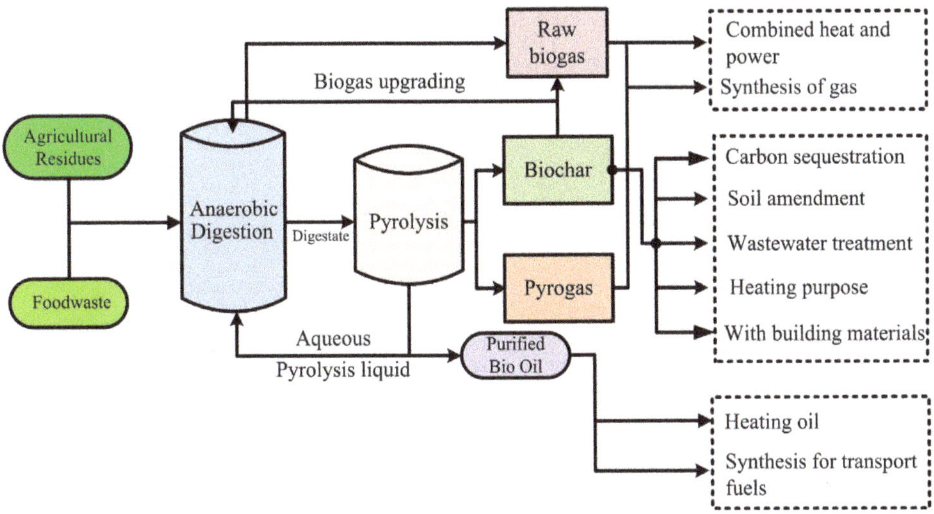

Figure 11. Pyrolysis and anaerobic digestion integrated process [156].

10. Bio-Oil Using Microalgae

The fast growth and high lipid content of microalgae make them a great candidate as a source of bio-oil, which can be extracted from microalgae through two common methods: hydrothermal liquefication (HTL) and pyrolysis [164]. Using these methods, microalgae are heated in the absence of oxygen to produce bio-oil. The biomass is rapidly heated to a high temperature, causing the microalgae cells to break down and release volatile organic compounds, which are then condensed into bio-oil [165,166]. The characteristics of the bio-oil produced include a high energy content, high viscosity, and high acidity. The composition of bio-oil depends on the type of microalgae, as well as the pyrolysis conditions, such as temperature, heating rate, and residence time. A study showed that pyrolysis of Nannochloropsis microalgae at 450 °C for 30 min yielded bio-oil with a heat value of 35.7 MJ/kg, a viscosity of 79.9 cP, and an acidity of 4.6 mg KOH/g [167].

Hydrothermal liquification (HTL) is a process in which heat and water are used under high pressure to cause degradation; it involves four stages: heating, pressurization, depressurization, and separation [168]. During the heating stage, the microalgae are heated to a high temperature in the presence of water, causing the biomass to break down and released organic compounds, which are condensed into bio-oil [169]. The characteristics of bio-oil produced showed a high energy content, low viscosity, and low acidity [170]. The composition of bio-oil also depends on the type of microalgae, as well as the operating conditions, such as temperature, pressure, and residence time. It was shown in a study that the HTL of Chlorella vulgaris microalgae at 300 °C and 20 MPa for 30 min resulted in bio-oil with a heat value of 31.5 MJ/kg, a viscosity of 2.87 cP, and an acidity of 0.35 mg KOH/g [171].

11. Production of Bio-Char from Crop Residues and Its Application for Anaerobic Digestion

Bio-char has attracted substantial interest as a potential material to improve soil fertility and boost plant growth. Bio-char has also been utilized as a feedstock in anaerobic digestion, which turns organic matter into gases and provides a renewable energy source.

Several studies were conducted to examine the formation of bio-char from crop leftovers and its use in anaerobic digestion. For example, one study looked into the production of bio-char from corn stover and its use as a co-substrate for anaerobic digestion. The results showed that adding bio-char boosted gas yield and improved anaerobic digestion process stability [172]. A similar study investigated the production of bio-char from wheat straw and its use in anaerobic digestion. The results showed that adding bio-char boosted methane output and improved substrate biodegradability [173]. Another study looked into the production of bio-char from rice straw and its use in anaerobic digestion. The addition of bio-char enhanced the gas yield while decreasing the ammonia and hydrogen sulfide contents in the gas [174]. These studies showed the potential of bio-char as a feedstock for anaerobic digestion, as well as its capacity to improve gas production and quality. However, more research is needed to optimize the generation of bio-char from crop residues and its use in anaerobic digestion under various conditions.

12. Economic Studies

An economic study of the pyrolysis process involves evaluating the costs and potential revenues associated with implementing and operating a pyrolysis plant. The financial viability and profitability of the process are assessed. Some key factors typically considered in an economic study of pyrolysis are the capital cost, feedstock cost, operating cost, product revenues, energy generation and utilization, by-product handling and disposal, and revenue streams and incentives [175]. The capital cost includes the cost of acquiring or leasing land and preparing the site for construction, purchasing or installing pyrolysis reactors, feedstock handling systems, product collection and storage equipment, auxiliary systems, and other required infrastructure. Second, the feedstock cost includes costs associated with acquiring the biomass feedstock, including harvesting, collection, and transportation to the pyrolysis plant, and a consideration of the potential fluctuations in feedstock prices due to market conditions, seasonality, and availability. Third, operating costs include personnel salaries and training; employee benefits; electricity and fuel required for operating the pyrolysis process, including heating the reactors and powering various equipment; ongoing costs for maintaining and repairing equipment and infrastructure; and expenses associated with laboratory analysis and quality control [176,177]. Then, product revenues include assessing the market demand and price for bio-oil, evaluating potential markets and applications for bio-char, and identifying potential uses for syngas, including on-site energy generation, heat production, or conversion into other value-added chemicals. Energy generation and utilization include assessing the feasibility and potential revenue from generating electricity, heat, or steam from the excess energy produced during the pyrolysis process, and evaluating energy costs and potential savings from utilizing the generated energy internally, reducing reliance on external sources. Then, by-product handling and disposal involves analyzing costs associated with by-product handling, such as transportation, storage, and potential treatment or disposal methods for residues or waste streams, and exploring opportunities for by-product utilization, such as selling bio-char for agricultural applications or exploring synergies with other industries. Revenue streams and incentives include investigating potential revenue streams from renewable energy credits, carbon credits, or government incentives that promote renewable energy and sustainable practices, and identifying tax benefits, grants, or subsidies that are available for biomass-based projects. In Figure 12, a Sankey diagram illustrates the heat pathways [178]. The heat input, including electric energy, is analyzed. Electric energy is transformed into thermal energy and electric energy losses, and energy from biomass and charcoal is changed to product energy and its losses. It was found that heat losses from the bottom to the top of the reactor were about 28–35% [175]. This amount of losses directly affects the product yields [176]. The amount of liquid products is slightly lower than the theoretical and designed values.

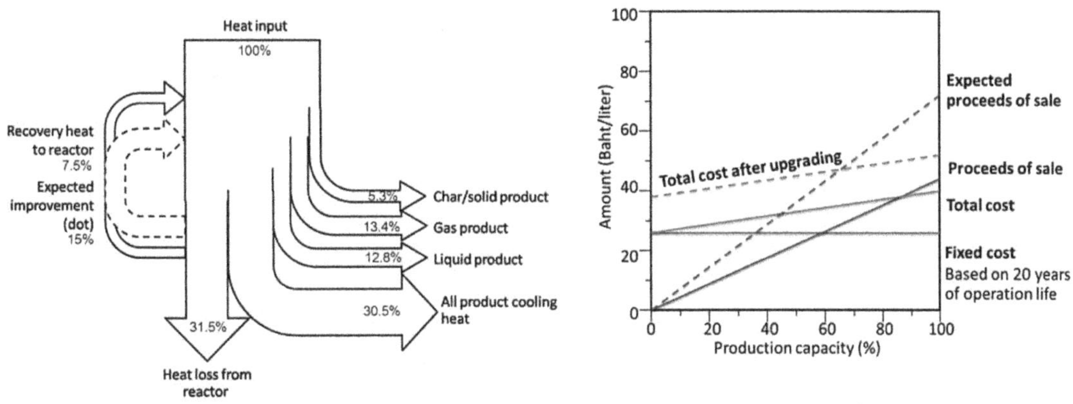

Figure 12. Sankey diagram showing the heat energy [178].

13. Conclusions

Thermochemical conversion is the most practical method for transforming biomass into biofuel. The pyrolysis process produces three main products, which are bio-oil, bio-char, and gas. The optimal operating temperature for a high yield of bio-oil from biomass was found to be in the range of 450–550 °C. A high bio-char production was found below 500 °C. The addition of a catalyst can increase the yield of bio-oil and improve the quality of the product. Biofuel yield is enhanced by selecting raw materials, such as using rice husk, along with an appropriate pyrolysis temperature (e.g., 450 °C) and particle size (350–800 μm), and using a low residence time and pressure. The paper concludes that bioenergy has emerged as one of the primary choices for the short- and medium-term replacement of fossil fuels and the reduction in greenhouse gas (GHG) emissions.

Author Contributions: Conceptualization, D.A., H.S., A.M.A., Y.E., T.M. and M.B.; methodology, D.A., H.S., A.M.A., Y.E., T.M. and M.B.; software, D.A., H.S., A.M.A., Y.E., T.M. and M.B.; validation, D.A., H.S., A.M.A., Y.E., T.M. and M.B.; investigation, D.A., H.S., A.M.A., Y.E., T.M. and M.B.; resources, D.A., H.S., A.M.A., Y.E., T.M. and M.B.; data curation, D.A., H.S., A.M.A., Y.E., T.M. and M.B.; writing—original draft preparation, D.A., H.S., A.M.A., Y.E., T.M. and M.B.; writing—review and editing, D.A., H.S., A.M.A., Y.E., T.M. and M.B.; visualization, D.A., H.S., A.M.A., Y.E., T.M. and M.B.; funding acquisition, Y.E., T.M. and M.B. All authors have read and agreed to the published version of the manuscript.

Funding: This project has received funding from the European Union's Horizon 2020 Research and Innovation Programme under Grant Agreement 963530.

Institutional Review Board Statement: Not applicable.

Informed Consent Statement: Not applicable.

Data Availability Statement: The data presented in this study are available on request from the corresponding author.

Acknowledgments: The researchers would like to acknowledge the assistance provided by the South African national energy development institute (sanedi) for funding the project. This project has received funding from the European Union's Horizon 2020 Research and Innovation Programme under Grant Agreement 963530.

Conflicts of Interest: The authors declare no conflict of interest.

References

1. Haque, A.N.M.A.; Naebe, M. Sustainable biodegradable denim waste composites for potential single-use packaging. *Sci. Tota Environ.* **2022**, *809*, 152239. [CrossRef] [PubMed]
2. Hosen, M.D.; Hossain, M.S.; Islam, M.A.; Haque, A.N.M.A.; Naebe, M. Utilisation of natural wastes: Water-resistant sem transparent paper for food packaging. *J. Clean. Prod.* **2022**, *364*, 132665. [CrossRef]
3. Haque, A.N.M.A.; Naebe, M.; Mielewski, D.; Kiziltas, A. Waste wool/polycaprolactone filament towards sustainable use in 3I printing. *J. Clean. Prod.* **2023**, *386*, 135781. [CrossRef]
4. Alhathal Alanezi, A.; Bassyouni, M.; Abdel-Hamid, S.M.S.; Ahmed, H.S.; Abdel-Aziz, M.H.; Zoromba, M.S.; Elhenawy, Theoretical Investigation of Vapor Transport Mechanism Using Tubular Membrane Distillation Module. *Membranes* **2021**, *11*, 56 [CrossRef]
5. Ionel, I.O.A.N.A.; Cioabla, A.E. Biogas production based on agricultural residues. From history to results and perspective *WSEAS Trans. Environ. Dev.* **2010**, *8*, 591–603.
6. Faaij, A. Modern biomass conversion technologies. *Mitig. Adapt. Strateg. Glob. Chang.* **2006**, *11*, 343–375. [CrossRef]
7. Vamvuka, D. Bio-oil, solid and gaseous biofuels from biomass pyrolysis processes—An overview. *Int. J. Energy Res.* **2011**, *3* 835–862. [CrossRef]
8. Verma, M.; Godbout, S.; Brar, S.K.; Solomatnikova, O.; Lemay, S.P.; Larouche, J.P. Biofuels production from biomass by therm chemical conversion technologies. *Int. J. Chem. Eng.* **2012**, *2012*, 542426. [CrossRef]
9. Brennan, L.; Owende, P. Biofuels from microalgae—A review of technologies for production, processing, and extractions o biofuels and co-products. *Renew. Sustain. Energy Rev.* **2010**, *14*, 557–577. [CrossRef]
10. Mata, T.M.; Martins, A.A.; Caetano, N.S. Microalgae for biodiesel production and other applications: A review. *Renew. Sustai Energy Rev.* **2010**, *14*, 217–232. [CrossRef]
11. Bassyouni, M. Dynamic mechanical properties and characterization of chemically treated sisal fiber-reinforced polypropylen biocomposites. *J. Reinf. Plast. Compos.* **2018**, *37*, 1402–1417. [CrossRef]
12. Ahmad, H.; Farooq, A.; Bassyouni, M.I.; Sait, H.H.; El-Wafa, M.A.; Hasan, S.W.; Ani, F.N. Pyrolysis of Saudi Arabian date paln waste: A viable option for converting waste into wealth. *Life Sci. J.* **2014**, *11*, 667–671.
13. Sait, H.H.; Hussain, A.; Bassyouni, M.; Ali, I.; Kanthasamy, R.; Ayodele, B.V.; Elhenawy, Y. Hydrogen-rich syngas and biocha production by non-catalytic valorization of date palm seeds. *Energies* **2022**, *15*, 2727. [CrossRef]
14. Bassyouni, M.; Iqbal, N.; Iqbal, S.S.; Abdel-Hamid, S.S.; Abdel-Aziz, M.H.; Javaid, U.; Khan, M.B. Ablation and thermo-mechanica investigation of short carbon fiber impregnated elastomeric ablatives for ultrahigh temperature applications. *Polym. Degrad. Sta* **2014**, *110*, 195–202. [CrossRef]
15. Bassyouni, M.; ul Hasan, S.W.; Abdel-Aziz, M.H.; Abdel-hamid, S.S.; Naveed, S.; Hussain, A.; Ani, F.N. Date palm wast gasification in downdraft gasifier and simulation using ASPEN HYSYS. *Energy Convers. Manag.* **2014**, *88*, 693–699. [CrossRef]
16. Jeong, J.Y.; Yang, C.W.; Lee, U.D.; Jeong, S.H. Characteristics of the pyrolytic products from the fast pyrolysis of palm kernel cak in a bench-scale fluidized bed reactor. *J. Anal. Appl. Pyrolysis* **2020**, *145*, 104708. [CrossRef]
17. Lynd, L.R.; Cushman, J.H.; Nichols, R.J.; Wyman, C.E. Fuel ethanol from cellulosic biomass. *Science* **1991**, *251*, 1318–132 [CrossRef]
18. Balat, M.; Balat, H.; Öz, C. Progress in bioethanol processing. *Prog. Energy Combust. Sci.* **2008**, *34*, 551–573. [CrossRef]
19. da Silva Trindade, W.R.; dos Santos, R.G. Review on the characteristics of butanol, its production and use as fuel in interna combustion engines. *Renew. Sustain. Energy Rev.* **2017**, *69*, 642–651. [CrossRef]
20. Nanda, S.; Golemi-Kotra, D.; McDermott, J.C.; Dalai, A.K.; Gökalp, I.; Kozinski, J.A. Fermentative production of butano Perspectives on synthetic biology. *New Biotechnol.* **2017**, *37*, 210–221. [CrossRef]
21. Yadav, P.; Varma, A.K.; Mondal, P. Production of biodiesel from algal biomass collected from Solani River using Ultrasoni Technique. *Int. J. Renew. Energy Res.* **2014**, *4*, 714–724.
22. Levine, R.B.; Pinnarat, T.; Savage, P.E. Biodiesel production from wet algal biomass through in situ lipid hydrolysis an supercritical transesterification. *Energy Fuels* **2010**, *24*, 5235–5243. [CrossRef]
23. Zhao, Z.; Jiang, Z.; Xu, H.; Yan, K. Selective production of phenol-rich bio-oil from corn straw waste by direct microwave pyrolysi without extra catalyst. *Front. Chem.* **2021**, *9*, 700887. [CrossRef]
24. Solantausta, Y.; Oasmaa, A.; Sipila, K.; Lindfors, C.; Lehto, J.; Autio, J.; Jokela, P.; Alin, J.; Heiskanen, J. Bio-oil production fro biomass: Steps toward demonstration. *Energy Fuels* **2012**, *26*, 233–240. [CrossRef]
25. Sutrisno, B.; Hidayat, A. Upgrading of bio-oil from the pyrolysis of biomass over the rice husk ash catalysts. *IOP Conf. Ser. Mate Sci. Eng.* **2016**, *162*, 012014. [CrossRef]
26. Glaser, B.; Haumaier, L.; Guggenberger, G.; Zech, W. The 'Terra Preta' phenomenon: A model for sustainable agriculture in th humid tropics. *Naturwissenschaften* **2001**, *88*, 37–41. [CrossRef] [PubMed]
27. Ławińska, O.; Korombel, A.; Zajemska, M. Pyrolysis-Based Municipal Solid Waste Management in Poland—SWOT Analysi *Energies* **2022**, *15*, 510. [CrossRef]
28. Lohri, C.R.; Diener, S.; Zabaleta, I.; Mertenat, A.; Zurbrügg, C. Treatment technologies for urban solid biowaste to create valu products: A review with focus on low-and middle-income settings. *Rev. Environ. Sci. Bio/Technol.* **2017**, *16*, 81–130. [CrossRef]

9. Galvagno, S.; Casu, S.; Casabianca, T.; Calabrese, A.; Cornacchia, G. Pyrolysis process for the treatment of scrap tyres: Preliminary experimental results. *Waste Manag.* **2002**, *22*, 917–923. [CrossRef]
0. Demirbas, A.; Arin, G. An overview of biomass pyrolysis. *Energy Sources* **2002**, *24*, 471–482. [CrossRef]
1. Li, S.; Xu, S.; Liu, S.; Yang, C.; Lu, Q. Fast pyrolysis of biomass in free-fall reactor for hydrogen-rich gas. *Fuel Process. Technol.* **2004**, *85*, 1201–1211. [CrossRef]
2. Tippayawong, N.; Kinorn, J.; Thavornun, S. Yields and gaseous composition from slow pyrolysis of refuse-derived fuels. *Energy Sources Part A* **2008**, *30*, 1572–1580. [CrossRef]
3. Tripathi, M.; Sahu, J.N.; Ganesan, P. Effect of process parameters on production of biochar from biomass waste through pyrolysis: A review. *Renew. Sustain. Energy Rev.* **2016**, *55*, 467–481. [CrossRef]
4. Bridgwater, A.V. Renewable fuels and chemicals by thermal processing of biomass. *Chem. Eng. J.* **2003**, *91*, 87–102. [CrossRef]
5. Mohan, D.; Pittman, C.U., Jr.; Steele, P.H. Pyrolysis of wood/biomass for bio-oil: A critical review. *Energy Fuels* **2006**, *20*, 848–889. [CrossRef]
6. Brammer, J.G.; Lauer, M.; Bridgwater, A.V. Opportunities for biomass-derived "bio-oil" in European heat and power markets. *Energy Policy* **2006**, *34*, 2871–2880. [CrossRef]
7. Al-Rumaihi, A.; Shahbaz, M.; Mckay, G.; Mackey, H.; Al-Ansari, T. A review of pyrolysis technologies and feedstock: A blending approach for plastic and biomass towards optimum biochar yield. *Renew. Sustain. Energy Rev.* **2022**, *167*, 112715. [CrossRef]
8. Chiaramonti, D.; Oasmaa, A.; Solantausta, Y. Power generation using fast pyrolysis liquids from biomass. *Renew. Sustain. Energy Rev.* **2007**, *11*, 1056–1086. [CrossRef]
9. Meier, D.; Oasmaa, A.; Peacocke, C.; Piskorz, J.; Radlein, D. *Fast Pyrolysis of Biomass: A Handbook*; U.S. Department of Energy: Washington, DC, USA, 1999.
0. Demirbas, A. Recent advances in biomass conversion technologies. *Energy Edu Sci. Technol.* **2000**, *6*, 19–41.
1. Aguado, R.; Olazar, M.; Gaisán, B.; Prieto, R.; Bilbao, J. Kinetic study of polyolefin pyrolysis in a conical spouted bed reactor. *Ind. Eng. Chem. Res.* **2002**, *41*, 4559–4566. [CrossRef]
2. Cornelissen, T.; Yperman, J.; Reggers, G.; Schreurs, S.; Carleer, R. Flash co-pyrolysis of biomass with polylactic acid. Part 1: Influence on bio-oil yield and heating value. *Fuel* **2008**, *87*, 1031–1041. [CrossRef]
3. Kaur, R.; Gera, P.; Jha, M.K. Study on effects of different operating parameters on the pyrolysis of biomass: A review. *J. Biofuels Bioenergy* **2015**, *1*, 135. [CrossRef]
4. Luo, Z.; Wang, S.; Liao, Y.; Zhou, J.; Gu, Y.; Cen, K. Research on biomass fast pyrolysis for liquid fuel. *Biomass Bioenergy* **2004**, *26*, 455–462. [CrossRef]
5. Zhang, S.; Yan, Y.; Li, T.; Ren, Z. Upgrading of liquid fuel from the pyrolysis of biomass. *Bioresour. Technol.* **2005**, *96*, 545–550. [CrossRef]
6. Ozbay, N.; Pütün, A.E.; Pütün, E. Bio-oil production from rapid pyrolysis of cottonseed cake: Product yields and compositions. *Int. J. Energy Res.* **2006**, *30*, 501–510. [CrossRef]
7. Demirbas, A. The influence of temperature on the yields of compounds existing in bio-oils obtained from biomass samples via pyrolysis. *Fuel Process. Technol.* **2007**, *88*, 591–597. [CrossRef]
8. Biagini, E.; Barontini, F.; Tognotti, L. Devolatilization of biomass fuels and biomass components studied by TG/FTIR technique. *Ind. Eng. Chem. Res.* **2006**, *45*, 4486–4493. [CrossRef]
9. Uzun, B.B.; Pütün, A.E.; Pütün, E. Fast pyrolysis of soybean cake: Product yields and compositions. *Bioresour. Technol.* **2006**, *97*, 569–576. [CrossRef]
0. Yorgun, S.A.İ.T.; Şensöz, S.; Koçkar, Ö.M. Flash pyrolysis of sunflower oil cake for production of liquid fuels. *J. Anal. Appl. Pyrolysis* **2001**, *60*, 1–12. [CrossRef]
1. Shen, D.K.; Gu, S. The mechanism for thermal decomposition of cellulose and its main products. *Bioresour. Technol.* **2009**, *100*, 6496–6504. [CrossRef]
2. Chen, L.; Dupont, C.; Salvador, S.; Grateau, M.; Boissonnet, G.; Schweich, D. Experimental study on fast pyrolysis of free-falling millimetric biomass particles between 800 C and 1000 C. *Fuel* **2013**, *106*, 61–66. [CrossRef]
3. Demiral, İ.; Şensöz, S. Fixed-bed pyrolysis of hazelnut (*Corylus avellana* L.) bagasse: Influence of pyrolysis parameters on product yields. *Energy Sources Part A Recovery Util. Environ. Eff.* **2006**, *28*, 1149–1158. [CrossRef]
4. Onay, O.; Koçkar, O.M. Fixed-bed pyrolysis of rapeseed (*Brassica napus* L.). *Biomass Bioenergy* **2004**, *26*, 289–299. [CrossRef]
5. Lopez-Velazquez, M.A.; Santes, V.; Balmaseda, J.; Torres-Garcia, E. Pyrolysis of orange waste: A thermo-kinetic study. *J. Anal. Appl. Pyrolysis* **2013**, *99*, 170–177. [CrossRef]
6. Haykiri-Acma, H. The role of particle size in the non-isothermal pyrolysis of hazelnut shell. *J. Anal. Appl. Pyrolysis* **2006**, *75*, 211–216. [CrossRef]
7. Guo, X.; Li, S.; Zheng, Y.; Ci, B. Preparation and characterization of bio-oil by microwave pyrolysis of biomass. *Energy Sources Part A Recovery Util. Environ. Eff.* **2016**, *38*, 133–139. [CrossRef]
8. Kan, T.; Strezov, V.; Evans, T.J. Lignocellulosic biomass pyrolysis: A review of product properties and effects of pyrolysis parameters. *Renew. Sustain. Energy Rev.* **2016**, *57*, 1126–1140. [CrossRef]
9. Boroson, M.L.; Howard, J.B.; Longwell, J.P.; Peters, W.A. Product yields and kinetics from the vapor phase cracking of wood pyrolysis tars. *AIChE J.* **1989**, *35*, 120–128. [CrossRef]

60. Uddin, M.N.; Daud, W.W.; Abbas, H.F. Effects of pyrolysis parameters on hydrogen formations from biomass: A review. *Rsc Ad* **2014**, *4*, 10467–10490. [CrossRef]
61. Guedes, R.E.; Luna, A.S.; Torres, A.R. Operating parameters for bio-oil production in biomass pyrolysis: A review. *J. Anal. App Pyrolysis* **2018**, *129*, 134–149. [CrossRef]
62. Omar, R.; Idris, A.; Yunus, R.; Khalid, K.; Isma, M.A. Characterization of empty fruit bunch for microwave-assisted pyrolysi *Fuel* **2011**, *90*, 1536–1544. [CrossRef]
63. Jung, S.H.; Kang, B.S.; Kim, J.S. Production of bio-oil from rice straw and bamboo sawdust under various reaction condition in a fast pyrolysis plant equipped with a fluidized bed and a char separation system. *J. Anal. Appl. Pyrolysis* **2008**, *82*, 240–24 [CrossRef]
64. Cardoso, C.R.; Ataide, C.H. Analytical pyrolysis of tobacco residue: Effect of temperature and inorganic additives. *J. Anal. App Pyrolysis* **2013**, *99*, 49–57. [CrossRef]
65. Aysu, T. Catalytic pyrolysis of Alcea pallida stems in a fixed-bed reactor for production of liquid bio-fuels. *Bioresour. Techno* **2015**, *191*, 253–262. [CrossRef]
66. Luo, G.; Resende, F.L. In-situ and ex-situ upgrading of pyrolysis vapors from beetle-killed trees. *Fuel* **2016**, *166*, 367–37 [CrossRef]
67. Mansur, D.; Yoshikawa, T.; Norinaga, K.; Hayashi, J.I.; Tago, T.; Masuda, T. Production of ketones from pyroligneous acid o woody biomass pyrolysis over an iron-oxide catalyst. *Fuel* **2013**, *103*, 130–134. [CrossRef]
68. Smets, K.; Roukaerts, A.; Czech, J.; Reggers, G.; Schreurs, S.; Carleer, R.; Yperman, J. Slow catalytic pyrolysis of rapeseed cak Product yield and characterization of the pyrolysis liquid. *Biomass Bioenergy* **2013**, *57*, 180–190. [CrossRef]
69. Li, J.; Li, X.; Zhou, G.; Wang, W.; Wang, C.; Komarneni, S.; Wang, Y. Catalytic fast pyrolysis of biomass with mesoporous ZSM- zeolites prepared by desilication with NaOH solutions. *Appl. Catal. A Gen.* **2014**, *470*, 115–122. [CrossRef]
70. Siengchum, T.; Isenberg, M.; Chuang, S.S. Fast pyrolysis of coconut biomass–An FTIR study. *Fuel* **2013**, *105*, 559–565. [CrossRef
71. Şensöz, S.; Angın, D. Pyrolysis of safflower (*Charthamus tinctorius* L.) seed press cake in a fixed-bed reactor: Part 2. Structura characterization of pyrolysis bio-oils. *Bioresour. Technol.* **2008**, *99*, 5498–5504.
72. Antal Jr, M.J.; Allen, S.G.; Dai, X.; Shimizu, B.; Tam, M.S.; Grønli, M. Attainment of the theoretical yield of carbon from biomas *Ind. Eng. Chem. Res.* **2000**, *39*, 4024–4031. [CrossRef]
73. Antal, M.J.; Grønli, M. The art, science, and technology of charcoal production. *Ind. Eng. Chem. Res.* **2003**, *42*, 1619–164 [CrossRef]
74. Acıkgoz, C.; Onay, O.; Kockar, O.M. Fast pyrolysis of linseed: Product yields and compositions. *J. Anal. Appl. Pyrolysis* **2004**, *7* 417–429. [CrossRef]
75. Pütün, A.E.; Apaydın, E.; Pütün, E. Rice straw as a bio-oil source via pyrolysis and steam pyrolysis. *Energy* **2004**, *29*, 2171–218 [CrossRef]
76. Ahmed, I.I.; Gupta, A.K. Pyrolysis and gasification of food waste: Syngas characteristics and char gasification kinetics. *App Energy* **2010**, *87*, 101–108. [CrossRef]
77. Demiral, I.; Şensöz, S. The effects of different catalysts on the pyrolysis of industrial wastes (olive and hazelnut bagasse). *Bioresou Technol.* **2008**, *99*, 8002–8007. [CrossRef]
78. He, M.; Hu, Z.; Xiao, B.; Li, J.; Guo, X.; Luo, S.; Yang, F.; Feng, Y.; Yang, G.; Liu, S. Hydrogen-rich gas from catalytic stean gasification of municipal solid waste (MSW): Influence of catalyst and temperature on yield and product composition. *Int. Hydrogen Energy* **2009**, *34*, 195–203. [CrossRef]
79. Hao, X.H.; Guo, L.J.; Mao, X.A.; Zhang, X.M.; Chen, X.J. Hydrogen production from glucose used as a model compound o biomass gasified in supercritical water. *Int. J. Hydrogen Energy* **2003**, *28*, 55–64. [CrossRef]
80. Zhang, H.; Yu, J.; Li, L.; Ma, Y. Microwave-assisted pretreatment of corn stover for enhancing enzymatic hydrolysis an fermentation. *Fuel Process. Technol.* **2021**, *210*, 106624. [CrossRef]
81. Lu, Y.; Wang, Y.; Wang, Y.; Li, Z. Microwave-assisted pretreatment of corn stover for the production of biofuels: A review. *Renew Sustain. Energy Rev.* **2020**, *119*, 109592. [CrossRef]
82. Gao, Z.; Li, Y.; Wang, Y.; Liu, Y.; Ma, X. Microwave-assisted pretreatment of corn straw for efficient bioethanol productio *Bioresour. Technol.* **2021**, *335*, 125246. [CrossRef]
83. Zhang, H.; Xu, Y.; Ma, Y.; Li, L.; Yu, J. Optimization of microwave-assisted alkali pretreatment for enhancing enzymatic hydrolysi of corn stover. *Fuel* **2020**, *266*, 117080. [CrossRef]
84. Fahmi, R.; Bridgwater, A.V.; Donnison, I.; Yates, N.; Jones, J.M. The effect of lignin and inorganic species in biomass on pyrolysi oil yields, quality and stability. *Fuel* **2008**, *87*, 1230–1240. [CrossRef]
85. Demirbas, A. Production of gasoline and diesel fuels from bio-materials. *Energy Sources Part A* **2007**, *29*, 753–760. [CrossRef]
86. Balat, M.; Balat, M.; Kırtay, E.; Balat, H. Main routes for the thermo-conversion of biomass into fuels and chemicals. Part 1 Pyrolysis systems. *Energy Convers. Manag.* **2009**, *50*, 3147–3157. [CrossRef]
87. Jahirul, M.I.; Rasul, M.G.; Chowdhury, A.A.; Ashwath, N. Biofuels production through biomass pyrolysis—A technologica review. *Energies* **2012**, *5*, 4952–5001. [CrossRef]
88. Adarsh, K.; Kumar, J.; Bhaskar, T. Utilization of lignin: A sustainable and eco-friendly approach. *J. Energy Inst.* **2020**, *93*, 235–27

9. Çulcuoglu, E.; Ünay, E.; Karaosmanoglu, F.; Angin, D.; Şensöz, S. Characterization of the bio-oil of rapeseed cake. *Energy Sources* **2005**, *27*, 1217–1223. [CrossRef]
0. Oasmaa, A.; Kuoppala, E. Fast pyrolysis of forestry residue. 3. Storage stability of liquid fuel. *Energy Fuels* **2003**, *17*, 1075–1084. [CrossRef]
1. Oasmaa, A.; Meier, D. Norms and standards for fast pyrolysis liquids: 1. Round robin test. *J. Anal. Appl. Pyrolysis* **2005**, *73*, 323–334. [CrossRef]
2. Shihadeh, A.; Hochgreb, S. Diesel engine combustion of biomass pyrolysis oils. *Energy Fuels* **2000**, *14*, 260–274. [CrossRef]
3. Abdullah, N. An Assessment of Pyrolysis for Processing Empty Fruit Bunches. Ph.D. Dissertation, Aston University, Birmingham, UK, 2005.
4. Rocha, J.; Gomez, E.; Mesa Perez, J.; Cortez, L.; Seye, O.; Brossard Gonzalez, L. The demonstration fast pyrolysis plant to biomass conversion in brazil. In Proceedings of the VII World Renew Energy Congress, Cologne, Germany, 29 June–5 July 2002.
5. González, J.F.; Román, S.; Encinar, J.M.; Martínez, G. Pyrolysis of various biomass residues and char utilization for the production of activated carbons. *J. Anal. Appl. Pyrolysis* **2009**, *85*, 134–141. [CrossRef]
6. Downie, A.; Crosky, A.; Munroe, P. Physical properties of biochar. In *Biochar for Environmental Management*; Lehmann, J., Joseph, S., Eds.; Earthscan Publications Ltd.: London, UK, 2009; Volume 2, pp. 154–160.
7. Cetin, E.; Moghtaderi, B.; Gupta, R.; Wall, T.F. Influence of pyrolysis conditions on the structure and gasification reactivity of biomass chars. *Fuel* **2004**, *83*, 2139–2150. [CrossRef]
8. Lua, A.C.; Yang, T.; Guo, J. Effects of pyrolysis conditions on the properties of activated carbons prepared from pistachio-nut shells. *J. Anal. Appl. Pyrolysis* **2004**, *72*, 279–287. [CrossRef]
9. Sohi, S.; Lopez-Capel, E.; Krull, E.; Bol, R. Biochar, climate change and soil: A review to guide future research. *CSIRO Land Water Sci. Rep.* **2009**, *5*, 17–31.
00. Kantarelis, E.; Zabaniotou, A. Valorization of cotton stalks by fast pyrolysis and fixed bed air gasification for syngas production as precursor of second generation biofuels and sustainable agriculture. *Bioresour. Technol.* **2009**, *100*, 942–947. [CrossRef]
01. Tang, L.; Huang, H. Plasma pyrolysis of biomass for production of syngas and carbon adsorbent. *Energy Fuels* **2005**, *19*, 1174–1178. [CrossRef]
02. He, M.; Xiao, B.; Liu, S.; Hu, Z.; Guo, X.; Luo, S.; Yang, F. Syngas production from pyrolysis of municipal solid waste (MSW) with dolomite as downstream catalysts. *J. Anal. Appl. Pyrolysis* **2010**, *87*, 181–187. [CrossRef]
03. Wei, L.; Thomasson, J.A.; Bricka, R.M.; Batchelor, W.D.; Columbus, E.P.; Wooten, J.R. *Experimental Study of a Downdraft Gratifier*; ASABE Meeting Paper No. 066029; American Society of Agricultural and Biological Engineers: St. Joseph, MI, USA, 2006.
04. Dasappa, S.; Paul, P.J.; Mukunda, H.S.; Rajan, N.K.S.; Sridhar, G.; Sridhar, H.V. Biomass gasification technology–a route to meet energy needs. *Curr. Sci.* **2004**, *87*, 908–916.
05. Al Arni, S.; Bosio, B.; Arato, E. Syngas from sugarcane pyrolysis: An experimental study for fuel cell applications. *Renew. Energy* **2010**, *35*, 29–35. [CrossRef]
06. Hossain, A.K.; Davies, P.A. Pyrolysis liquids and gases as alternative fuels in internal combustion engines–A review. *Renew. Sustain. Energy Rev.* **2013**, *21*, 165–189. [CrossRef]
07. Liu, Y.; Yang, X.; Zhang, J.; Zhu, Z. Process simulation of preparing biochar by biomass pyrolysis via aspen plus and its economic evaluation. *Waste Biomass Valorization* **2022**, *13*, 2609–2622. [CrossRef]
08. Elkhalifa, S.; AlNouss, A.; Al-Ansari, T.; Mackey, H.R.; Parthasarathy, P.; Mckay, G. Simulation of food waste pyrolysis for the production of biochar: A Qatar case study. In *Computer Aided Chemical Engineering*; Elsevier: Amsterdam, The Netherlands, 2019; Volume 46, pp. 901–906.
09. Tan, Y.L.; Abdullah, A.Z.; Hameed, B.H. Product distribution of the thermal and catalytic fast pyrolysis of karanja (*Pongamia pinnata*) fruit hulls over a reusable silica-alumina catalyst. *Fuel* **2019**, *245*, 89–95. [CrossRef]
10. Lee, X.J.; Lee, L.Y.; Hiew, B.Y.Z.; Gan, S.; Thangalazhy-Gopakumar, S.; Ng, H.K. Valorisation of oil palm wastes into high yield and energy content biochars via slow pyrolysis: Multivariate process optimisation and combustion kinetic studies. *Mater. Sci. Energy Technol.* **2020**, *3*, 601–610. [CrossRef]
11. Sukumar, V.; Manieniyan, V.; Senthilkumar, R.; Sivaprakasam, S. Production of bio oil from sweet lime empty fruit bunch by pyrolysis. *Renew. Energy* **2020**, *146*, 309–315. [CrossRef]
12. Selvarajoo, A.; Wong, Y.L.; Khoo, K.S.; Chen, W.H.; Show, P.L. Biochar production via pyrolysis of citrus peel fruit waste as a potential usage as solid biofuel. *Chemosphere* **2022**, *294*, 133671. [CrossRef] [PubMed]
13. Saadi, W.; Rodríguez-Sánchez, S.; Ruiz, B.; Souissi-Najar, S.; Ouederni, A.; Fuente, E. Pyrolysis technologies for pomegranate (*Punica granatum* L.) peel wastes. Prospects in the bioenergy sector. *Renew. Energy* **2019**, *136*, 373–382. [CrossRef]
14. Lam, S.S.; Liew, R.K.; Lim, X.Y.; Ani, F.N.; Jusoh, A. Fruit waste as feedstock for recovery by pyrolysis technique. *Int. Biodeterior. Biodegrad.* **2016**, *113*, 325–333. [CrossRef]
15. Adeniyi, A.G.; Otoikhian, K.S.; Ighalo, J.O.; Mohammed, I.A. Pyrolysis of different fruit peel waste via a thermodynamic model. *ABUAD J. Eng. Res. Dev.* **2019**, *2*, 16–24.
16. Alves, J.L.F.; Da Silva, J.C.G.; da Silva Filho, V.F.; Alves, R.F.; de Araujo Galdino, W.V.; Andersen, S.L.F.; De Sena, R.F. Determination of the bioenergy potential of Brazilian pine-fruit shell via pyrolysis kinetics, thermodynamic study, and evolved gas analysis. *BioEnergy Res.* **2019**, *12*, 168–183. [CrossRef]

117. Parthasarathy, P.; Sheeba, K.N. Combined slow pyrolysis and steam gasification of biomass for hydrogen generation—A reviev *Int. J. Energy Res.* **2015**, *39*, 147–164. [CrossRef]
118. Castilla-Caballero, D.; Barraza-Burgos, J.; Gunasekaran, S.; Roa-Espinosa, A.; Colina-Márquez, J.; Machuca-Martínez, F Hernández-Ramírez, A.; Vázquez-Rodríguez, S. Experimental data on the production and characterization of biochars de rived from coconut-shell wastes obtained from the Colombian Pacific Coast at low temperature pyrolysis. *Data Brief* **202** *28*, 104855. [CrossRef] [PubMed]
119. Yang, J.; Yu, M.; Chen, W. Adsorption of hexavalent chromium from aqueous solution by activated carbon prepared from longa seed: Kinetics, equilibrium and thermodynamics. *J. Ind. Eng. Chem.* **2015**, *21*, 414–422. [CrossRef]
120. Ismail, H.Y.; Abbas, A.; Azizi, F.; Zeaiter, J. Pyrolysis of waste tires: A modeling and parameter estimation study using Aspe Plus®. *Waste Manag.* **2017**, *60*, 482–493. [CrossRef] [PubMed]
121. Dineshkumar, M.; Shrikar, B.; Ramanathan, A. Development of computer aided modelling and optimization of microwav pyrolysis of biomass by using aspen plus. *IOP Conf. Ser. Earth Environ. Sci.* **2019**, *312*, 012006. [CrossRef]
122. Kabir, M.J.; Chowdhury, A.A.; Rasul, M.G. Pyrolysis of municipal green waste: A modelling, simulation and experimenta analysis. *Energies* **2015**, *8*, 7522–7541. [CrossRef]
123. AlNouss, A.; Parthasarathy, P.; Mackey, H.R.; Al-Ansari, T.; McKay, G. Pyrolysis study of different fruit wastes using an Aspe Plus model. *Front. Sustain. Food Syst.* **2021**, *5*, 604001. [CrossRef]
124. Treeyawetchakul, C. Preliminary Modified Biodiesel Production by Coupling Reactive distillation with a Steam Reformer vi Aspen Plus®. *IOP Conf. Ser. Mater. Sci. Eng.* **2020**, *778*, 012064. [CrossRef]
125. Quintero, J.A.; Cardona, C.A. Process simulation of fuel ethanol production from lignocellulosics using aspen plus. *Ind. Eng Chem. Res.* **2011**, *50*, 6205–6212. [CrossRef]
126. Oladokun, O.; Nyakuma, B.B.; Ivase, T.J. Gasification of Nigerian Lignite Coals under Air-Steam Conditions using ASPEN Plu for the Production of Hydrogen and Syngas. *Pet. Coal* **2021**, *63*, 332–339.
127. Pilar González-Vázquez, M.; Rubiera, F.; Pevida, C.; Pio, D.T.; Tarelho, L.A. Thermodynamic analysis of biomass gasificatio using aspen plus: Comparison of stoichiometric and non-stoichiometric models. *Energies* **2021**, *14*, 189. [CrossRef]
128. Flagiello, D.; Di Natale, F.; Erto, A.; Lancia, A. Wet oxidation scrubbing (WOS) for flue-gas desulphurization using sodiun chlorite seawater solutions. *Fuel* **2020**, *277*, 118055. [CrossRef]
129. Diebold, J.P. *Review of the Chemical and Physical Mechanisms of the Storage and Stability of Fast Pyrolysis Bio-Oils*; NREL 1; Therma chemie, Inc.: Lakewood, CO, USA, 2000.
130. Evans, R.J.; Milne, T.A. Molecular characterization of the pyrolysis of biomass. *Energy Fuels* **1987**, *1*, 123–137. [CrossRef]
131. Stankovikj, F.; McDonald, A.G.; Helms, G.L.; Olarte, M.V.; Garcia-Perez, M. Characterization of the water-soluble fraction o woody biomass pyrolysis oils. *Energy Fuels* **2017**, *31*, 1650–1664. [CrossRef]
132. Koike, N.; Hosokai, S.; Takagaki, A.; Nishimura, S.; Kikuchi, R.; Ebitani, K.; Suzuki, Y.; Oyama, S.T. Upgrading of pyrolysis bio-o using nickel phosphide catalysts. *J. Catal.* **2016**, *333*, 115–126. [CrossRef]
133. Kadarwati, S.; Oudenhoven, S.; Schagen, M.; Hu, X.; Garcia-Perez, M.; Kersten, S.; Li, C.Z.; Westerhof, R. Polymerizatio and cracking during the hydrotreatment of bio-oil and heavy fractions obtained by fractional condensation using Ru/C an NiMo/Al2O3 catalyst. *J. Anal. Appl. Pyrolysis* **2016**, *118*, 136–143.
134. Han, Y.; Gholizadeh, M.; Tran, C.C.; Kaliaguine, S.; Li, C.Z.; Olarte, M.; Garcia-Perez, M. Hydrotreatment of pyrolysis bio-oil: A review. *Fuel Process. Technol.* **2019**, *195*, 106140.
135. Hew, K.L.; Tamidi, A.M.; Yusup, S.; Lee, K.T.; Ahmad, M.M. Catalytic cracking of bio-oil to organic liquid product (OLP). *Bioresou Technol.* **2010**, *101*, 8855–8858. [CrossRef]
136. Elliott, D.C.; Hart, T.R.; Neuenschwander, G.G.; Rotness, L.J.; Olarte, M.V.; Zacher, A.H.; Solantausta, Y. Catalytic hydroprocessin of fast pyrolysis bio-oil from pine sawdust. *Energy Fuels* **2012**, *26*, 3891–3896. [CrossRef]
137. Elliott, D.C.; Hart, T.R. Catalytic hydroprocessing of chemical models for bio-oil. *Energy Fuels* **2009**, *23*, 631–637. [CrossRef]
138. Elliott, D.C.; Lee, S.J.; Hart, T.R. *Stabilization of Fast Pyrolysis Oil: Post Processing Final Report (No. PNNL-21549)*; Pacific Northwes National Lab. (PNNL): Richland, WA, USA, 2012.
139. Zacher, A.H.; Olarte, M.V.; Santosa, D.M.; Elliott, D.C.; Jones, S.B. A review and perspective of recent bio-oil hydrotreatin research. *Green Chem.* **2014**, *16*, 491–515.
140. de Miguel Mercader, F.; Groeneveld, M.J.; Kersten, S.R.; Geantet, C.; Toussaint, G.; Way, N.W.; Schaverien, C.J.; Hogendoorn, K. Hydrodeoxygenation of pyrolysis oil fractions: Process understanding and quality assessment through co-processing in refiner units. *Energy Environ. Sci.* **2011**, *4*, 985–997. [CrossRef]
141. Stankovikj, F.; Tran, C.C.; Kaliaguine, S.; Olarte, M.V.; Garcia-Perez, M. Evolution of functional groups during pyrolysis o upgrading. *Energy Fuels* **2017**, *31*, 8300–8316. [CrossRef]
142. De Miguel Mercader, F.; Koehorst, P.J.J.; Heeres, H.J.; Kersten, S.R.; Hogendoorn, J.A. Competition between hydrotreating an polymerization reactions during pyrolysis oil hydrodeoxygenation. *AIChE J.* **2011**, *57*, 3160–3170. [CrossRef]
143. Ateş, F.; Miskolczi, N.; Borsodi, N. Comparision of real waste (MSW and MPW) pyrolysis in batch reactor over different catalyst Part I: Product yields, gas and pyrolysis oil properties. *Bioresour. Technol.* **2013**, *133*, 443–454. [CrossRef] [PubMed]
144. Cornelissen, G.; Pandit, N.R.; Taylor, P.; Pandit, B.H.; Sparrevik, M.; Schmidt, H.P. Emissions and char quality of flame-curtai "Kon Tiki" kilns for farmer-scale charcoal/biochar production. *PLoS ONE* **2016**, *11*, e0154617. [CrossRef] [PubMed]

45. Fulton, W.; Gray, M.; Prahl, F.; Kleber, M. A simple technique to eliminate ethylene emissions from biochar amendment in agriculture. *Agron. Sustain. Dev.* **2013**, *33*, 469–474. [CrossRef]
46. Fabbri, D.; Rombolà, A.G.; Torri, C.; Spokas, K.A. Determination of polycyclic aromatic hydrocarbons in biochar and biochar amended soil. *J. Anal. Appl. Pyrolysis* **2013**, *103*, 60–67. [CrossRef]
47. Miskolczi, N.; Ates, F.; Borsodi, N. Comparison of real waste (MSW and MPW) pyrolysis in batch reactor over different catalysts. Part II: Contaminants, char and pyrolysis oil properties. *Bioresour. Technol.* **2013**, *144*, 370–379. [CrossRef]
48. Pyrolysis. 23 March 2014. Available online: https://proreds.eu/technology/pyrolysis/ (accessed on 5 February 2023).
49. Giwa, A.S.; Xu, H.; Chang, F.; Zhang, X.; Ali, N.; Yuan, J.; Wang, K. Pyrolysis coupled anaerobic digestion process for food waste and recalcitrant residues: Fundamentals, challenges, and considerations. *Energy Sci. Eng.* **2019**, *7*, 2250–2264. [CrossRef]
50. Giwa, A.S.; Chang, F.; Xu, H.; Zhang, X.; Huang, B.; Li, Y.; Wu, J.; Wang, B.; Vakili, M.; Wang, K. Pyrolysis of difficult biodegradable fractions and the real syngas bio-methanation performance. *J. Clean. Prod.* **2019**, *233*, 711–719. [CrossRef]
51. Luo, S.; Xiao, B.; Hu, Z.; Liu, S. Effect of particle size on pyrolysis of single-component municipal solid waste in fixed bed reactor. *Int. J. Hydrogen Energy* **2010**, *35*, 93–97. [CrossRef]
52. Feng, Q.; Lin, Y. Integrated processes of anaerobic digestion and pyrolysis for higher bioenergy recovery from lignocellulosic biomass: A brief review. *Renew. Sustain. Energy Rev.* **2017**, *77*, 1272–1287. [CrossRef]
53. Song, X.D.; Chen, D.Z.; Zhang, J.; Dai, X.H.; Qi, Y.Y. Anaerobic digestion combined pyrolysis for paper mill sludge disposal and its influence on char characteristics. *J. Mater. Cycles Waste Manag.* **2017**, *19*, 332–341. [CrossRef]
54. Opatokun, S.A.; Strezov, V.; Kan, T. Product based evaluation of pyrolysis of food waste and its digestate. *Energy* **2015**, *92*, 349–354. [CrossRef]
55. Giwa, A.S.; Xu, H.; Wu, J.; Li, Y.; Chang, F.; Zhang, X.; Jin, Z.; Huang, B.; Wang, K. Sustainable recycling of residues from the food waste (FW) composting plant via pyrolysis: Thermal characterization and kinetic studies. *J. Clean. Prod.* **2018**, *180*, 43–49. [CrossRef]
56. Rickwinder, S.; Paritosh, K.; Pareek, N.; Vivekanand, V. Integrated system of anaerobic digestion and pyrolysis for valorization of agricultural and food waste towards circular bioeconomy. *Bioresour. Technol.* **2022**, *360*, 127596.
57. Tang, Z.; Chen, W.; Chen, Y.; Yang, H.; Chen, H. Co-pyrolysis of microalgae and plastic: Characteristics and interaction effects. *Bioresour. Technol.* **2019**, *274*, 145–152. [CrossRef]
58. Zakaria, M.; Sharaky, A.M.; Al-Sherbini, A.-S.; Bassyouni, M.; Rezakazemi, M.; Elhenawy, Y. Water desalination using solar thermal collectors enhanced by nanofluids. *Chem. Eng. Tech.* **2022**, *45*, 15–25. [CrossRef]
59. Pattiya, A.; Titiloye, J.O. Pyrolysis of biomass for biofuels: A review. *Sustainability* **2019**, *11*, 2250.
60. Zhang, S.; Li, J.; Yu, H. Catalytic pyrolysis of biomass for biofuels production: A review. *Sustainability* **2020**, *12*, 8939.
61. Chen, H.; Chen, Y.; Li, M.; Zheng, Y. Fast pyrolysis of biomass for biofuels: A review. *Sustainability* **2019**, *11*, 6017.
62. Xu, J.; Wang, Y.; Ma, X. Catalytic pyrolysis of lignocellulosic biomass for biofuels production: A review. *Sustainability* **2020**, *12*, 4435.
63. Nanda, S.; Dalai, A.K. Catalytic pyrolysis of biomass for biofuels production. *Sustainability* **2017**, *9*, 2023.
64. Chen, S.; Li, Q. Bio-oil production via fast pyrolysis of biomass residues: A review. *Sustainability* **2016**, *8*, 1302.
65. Xiu, S.; Shahbazi, A. Bio-oil production and upgrading research: A review. *Sustainability* **2012**, *4*, 1577–1618. [CrossRef]
66. Anwar, S.; Zaini, M.A.A.; Kumar, R.; Rashid, U. Pyrolysis of microalgae biomass for bio-oil production. *Energy Convers. Manag.* **2017**, *141*, 159–168.
67. Chia, S.R.; Chew, K.W.; Show, P.L.; Yap, Y.J. Recent advancements in pyrolysis technology for sustainable production of biofuels from microalgae. *Renew. Sustain. Energy Rev.* **2018**, *81*, 1853–1872.
68. Shafeeq, A.; Ahmad, M.; Javed, M.R. Pyrolysis of microalgae biomass for bio-oil production: A critical review. *Biomass Convers. Biorefin.* **2021**, *11*, 1095–1132.
69. Kumar, M.; Sharma, S.; Singh, J.; Kumar, S. Pyrolysis of microalgae for bio-oil production: A critical review. *Renew. Sustain. Energy Rev.* **2020**, *118*, 109512. [CrossRef]
70. Chen, J.; Duan, P.; Li, X.; Ma, X. Hydrothermal liquefaction of microalgae for bio-oil production: A review. *Algal Res.* **2019**, *42*, 101614.
71. Gao, F.; Li, C.; Zhang, X.; Zhang, Y. Hydrothermal liquefaction of microalgae: A review. *Renew. Sustain. Energy Rev.* **2019**, *109*, 272–287.
72. Yen, H.W.; Hu, I.C.; Chen, C.Y. Hydrothermal liquefaction of microalgae for bio-oil production: A review. *Bioresour. Technol.* **2019**, *289*, 121639.
73. Li, S.; Yang, B.; Zhang, Y.; Zhu, Y.; Zhang, Y. Hydrothermal liquefaction of Chlorella vulgaris for bio-oil production: Effects of reaction parameters and kinetic models. *Energy* **2019**, *169*, 470–478. [CrossRef]
74. Wang, X.; Liu, D.; Zhao, X.; Li, S.; Li, J.; Zhang, R. Effect of biochar addition on the performance of anaerobic digestion of corn stover: A comparative study. *Bioresour. Technol.* **2020**, *297*, 122469.
75. Liu, T.; Zhang, L.; Zhang, W.; Li, Y.; Bian, Y. Effects of wheat straw biochar on anaerobic digestion of corn stover and cow dung: Methane production, biodegradability, and microbial community structure. *Bioresour. Technol.* **2021**, *325*, 124678.
76. Chen, Y.; Xie, Y.; Wang, H.; Wei, L.; Chen, Y. Effect of biochar addition on the anaerobic digestion of rice straw: Methane production and carbon transformation. *Energy* **2019**, *168*, 1169–1177.

177. Chawannat, J.; Tippayawong, N. Technical and economic analysis of a biomass pyrolysis plant. *Energy Procedia* **2015**, *79*, 950–95
178. Karittha, I.-O.; Wiyaratn, W.; Arpornwichanop, A. Technical and economic assessment of the pyrolysis and gasification integrate process for biomass conversion. *Energy* **2018**, *153*, 592–603.

sustainability

eview

;ustainable Adsorbents from Plant-Derived Agricultural Vastes for Anionic Dye Removal: A Review

.bu Naser Md Ahsanul Haque [1,*], Nigar Sultana [2], Abu Sadat Muhammad Sayem [3] nd Shamima Akter Smriti [4]

1 Institute for Frontier Materials, Deakin University, Waurn Ponds, VIC 3216, Australia
2 Department of Chemistry, Jagannath University, Dhaka 1100, Bangladesh
3 Manchester Fashion Institute, Manchester Metropolitan University, Manchester M15 6BG, UK
4 Department of Fabric Engineering, Bangladesh University of Textiles, Dhaka 1208, Bangladesh
* Correspondence: a.haque@deakin.edu.au

Abstract: The extensive use of dyes in numerous industries results in massive dye discharge in the wastewater, which is a major cause of water pollution. Globally, the consumption of dyes is near seven hundred thousand tons across different sectors, of which around 10–15% goes into the wastewater. Among the dye kinds, anionic dyes make up the main proportion, having a 32–90% share in the wastewater. Different plant-derived wastes, which are sustainable given their natural abundance, effectiveness, and low cost, are frequently proposed for dye separation. However, these adsorbents are inherently more suitable for cationic dyes than anionic dyes. In recent years, the modification of these wastes has been progressively considered to suit them to anionic dye removal. These modifications involve mechanical, thermal, or chemical treatments, or combinations. These attempts propose two-way benefits, as one abundant waste is being used to cure another severe problem, and eventually both could be diminished. This review has a key focus on the evaluation of plant-derived adsorbents and their modifications, and particularly for anionic dye adsorption. Overall, the mechanism of adsorption and the suitability of the current methods are discussed, and their future potential is explored.

Keywords: lignocellulose; natural biomass; dye wastewater; adsorption isotherm; Congo Red; diffusion and kinetics

itation: Haque, A.N.M.A.; Sultana, .; Sayem, A.S.M.; Smriti, S.A. ustainable Adsorbents from lant-Derived Agricultural Wastes r Anionic Dye Removal: A Review. *ustainability* **2022**, *14*, 11098. ttps://doi.org/10.3390/ 141711098

cademic Editor: Dino Musmarra

eceived: 4 August 2022
ccepted: 31 August 2022
ublished: 5 September 2022

1. Introduction

Dyes are comprehensively used in food, pharmaceutical, textile, plastic, metal, and many other manufacturing industries to impart the colors of choice to the end products. Currently, more than seven hundred thousand tons of dyes are required each year across different sectors, of which at least 10–15% are discarded into the wastewater [1]. Although the use of dyes is unavoidable, it is also responsible for massive water pollution, which is a major concern. Anionic dyes are the known main fraction (nearly 32–90%), compared with all other dye kinds (such as cationic or nonionic dye), in the wastewater. These include the acid, direct, reactive, and soluble forms of some nonionic dyes, which are difficult to eliminate due to their high solubility in water [2]. These dyes not only alter the color of water, but they also inhibit light penetration and reduce the rate of photosynthesis and the oxygen level, causing damage to the entire aquatic ecosystem [2]. Often, these dyes are carcinogenic and initiate various diseases in humans, such as the dysfunction of the kidneys, reproductive system, liver, brain, and central nervous system, and they bring hazards to other living organisms [3]. Therefore, it is essential to remove these dyes from wastewater to attain a sustainable environment. In the recent past, there has been a growing interest in the removal of anionic dyes from wastewater, and particularly in the last few years. For instance, a search of the keyword 'anionic dye removal' in the Web of Science database (title, abstract, keywords only) recalled 4981 articles starting from 1975 (this number is 3637 from

Scopus): 3938 have been published from 2015 to the present (~79%), and 2457 have been published from just 2019 to the present date (~49%). These studies have been performed across varied disciplines, such as from the engineering, environmental, chemistry, and materials perspectives, which shows the urge to solve the problem from multiple sectors (Figure 1).

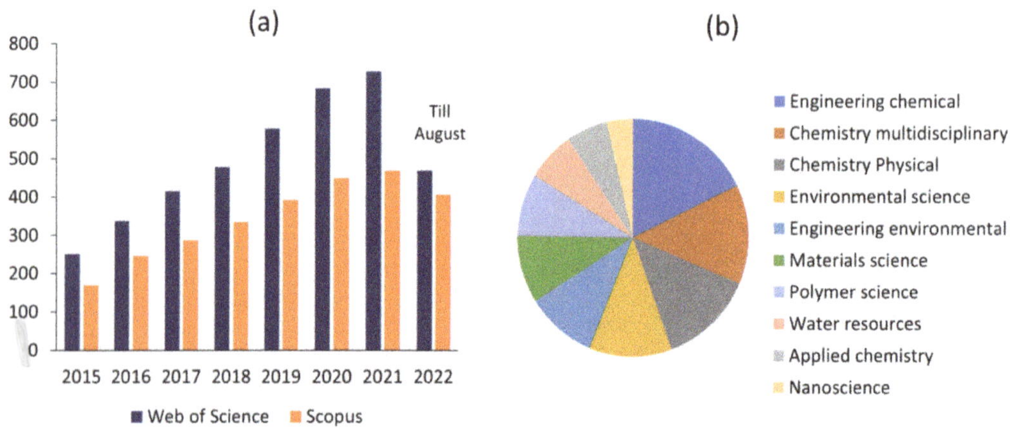

Figure 1. The number of studies retrieved from the Web of Science and Scopus databases: (**a**) on 'anionic dye removal' since 2015, and (**b**) their distribution across ten key subject categories.

Many techniques are proposed for removing anionic dyes from wastewater, such as physical (membrane filtration, ion exchange, irradiation, coagulation), chemical (oxidation, ozonation, photochemical, electrochemical destruction), and biological (aerobic and anaerobic microbial degradation) treatments [4]. However, the adsorption method is still a preferred method of dye removal due to its simplicity and efficiency [5]. There is an ongoing interest in preparing more sustainable adsorbents from agricultural and other natural plant-derived wastes, as they are widely available at a low cost. These resources are chemically known as lignocellulose, and they are often collectible at a negligible cost. The composition of lignocellulose biomasses (such as wheat straw, rice straw, oat straw, bagasse, cotton gin trash, and so on) is mostly cellulose (40–50%), intimately associated with hemicellulose (20–30%) and lignin (10–25%) [6]. However, due to their chemical nature, they are more prone to adsorb cationic dyes than anionic dyes. This is because the lignocellulose structure is chemically composed of different anions as its active sites, such as the hydroxyl and carbonyl groups, and it generally shows a negative surface charge [7]. Thus, it is important to modify them to take advantage of their abundance and utilize them for anionic dye separation. Over the years, several modification methods of these wastes have been proposed, including combinations of mechanical, thermal, or chemical treatments, to enhance the adsorption ability of these resources.

Because the use of agricultural wastes for dye separation is a massive area of research and the interest is increasing, currently, there are several reviews available that summarize these findings. However, these are mostly based on all classes of dyes [4,8–10], rather than with any particular focus on the anionicdye class. Due to the natural attraction of cationic dyes to the lignocellulosic structure, the removal of cationic dyes has received more interest compared with the removal of anionic dyes, which limits the discussions for the anionic dye adsorption in reviews in which both classes are discussed. However, as mentioned above, anionic dyes are the major proportion of all other dyes, and the separation of these dyes needs specific attention. The adsorption of anionic dyes is essential, but more challenging compared with cationic dyes. In recent years, numerous common lignocellulosic wastes, such as rice husk [11], wheat straw [12], sawdust [13], cotton waste [7], banana waste [14], corn stalks [15], orange peel [16], eucalyptus bark [17], coffee waste [18], and many other resources, have been proposed for the successful separation of anionic dyes, showing the variety and possibility of these raw materials for anionicdye removal. Therefore, this review aims to evaluate the findings reported for anionicdye adsorption from plant-derived agricultural wastes, and to explore the suitability of the proposed methods and reveal their future potential.

2. Types of Anionic Dyes

Based on their ionic nature, dyes can be classified into two categories: ionic and nonionic (Figure 2). Anionic dyes are a type of ionic dye in which anions are the key active groups. Some nonionic dyes, such as vat and disperse dyes, need to be converted into soluble anionic forms before being applied on a substrate [7]. Anionic dyes can be classified into reactive, direct, and acid dyes based on their application process.

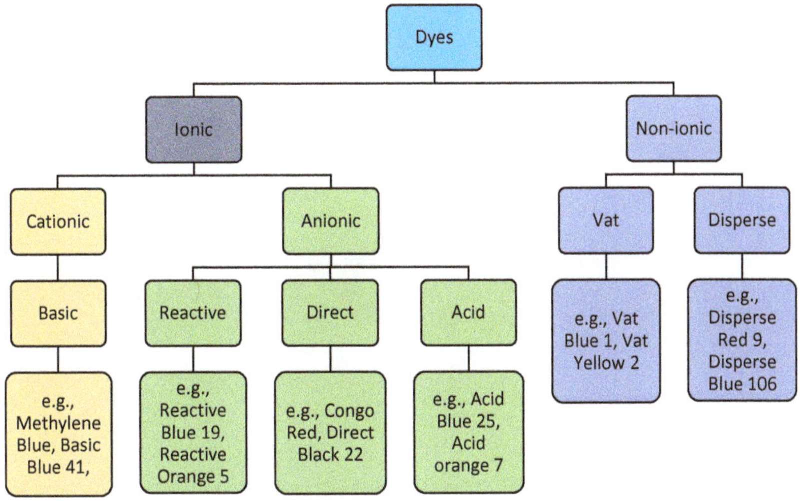

Figure 2. Ion-based-dye classification with examples.

Historically, the chemistry of dyes and pigments has mostly been studied in the context of textile materials, as the coloration of textiles demands that extensive quality parameters be met, which are not relevant to other industries. Understanding their mechanism of interaction with fibers (such as cellulose) can also provide an indication of their interaction ability when adsorbents from agricultural wastes are used for dye removal, as the chemical nature of these wastes is also mostly cellulosic or lignocellulosic.

2.1. Reactive Dye

Reactive dyes can form a covalent bond with certain fiber active sites, such as the hydroxyl group (-OH) or amino group (-NH_2) [19]. Lignocellulosic biomasses contain plenty of hydroxyl groups around their structures, thus making them accessible to interact with reactive dyes. Commercial reactive dyes commonly contain the dichlorotriazine, trichloropyrimidine, aminochloro-s-triazine, sulphatoethylsulphone, dichloroquinoxaline, aminofluoro-s-triazine, and difluorochloropyrimide groups [20]. Because reactive dyes and the surface of cellulosic fibers become negatively charged in water, electrolytes (such as sodium chloride or Glauber salt) are often used to neutralize the fiber surface. No matter how reactive the chemical groups of these dyes are, it is hard to achieve over 80% dye fixation in actual dyeing, which causes the obvious discharge of unfixed dyes as hazardous effluent [21].

2.2. Direct Dye

Direct dyes are named from their high affinity towards cellulosic fiber, and they are thus said to be directly appliable onto cellulose-based fiber systems during dyeing. The major chromophores in the structure of direct dyes are the azo, anthraquinoid, phthalocyanine and metal complexes [19]. These dyes are highly substantive, although not reactive. The use of electrolytes is also required for the successful attachment of direct dyes to cellulosic fibers.

2.3. Acid Dye

Acid dyes have acidic groups in their structures and are mainly applied on noncellulosic fibers, such as protein and polyamides in acidic conditions. The chromophoric systems in acid dyes are sulphonated azo, anthraquinone, nitrodiphenylamine, triphenylmethane and xanthenes [22]. Cationized fibers can produce a bond with an anion of acid dyes through an electrostatic force [23].

3. Chemical Nature of Plant-Derived Agricultural Wastes

Common plant-based agricultural wastes are a complex combination of cellulose, hemicellulose, and lignin, and are thus also widely known as lignocellulosic biomass. As shown in Figure 3, these constituents are complexly attached in the lignocellulosic structure. Cellulose is the major component that is linearly formed from the D-glucose subunits, and it is commonly known for its higher crystallinity [24]. However, the hemicellulose part (such as pentose and hexose) is randomly oriented within cellulose chains and is known to impact negatively on the crystallinity. Their chain length is much shorter than the typical chain length of cellulose. The other part, lignin, is the outer layer that holds together both cellulose and hemicellulose [25]. It is a complex structure of cross-linked phenolic subunits, although its actual structure is still to be identified [26]. Throughout the lignocellulosic structure, there are abundant hydroxyl groups, which make this material highly hydrophilic and ideal for dye adsorption from water. For the same reason, their surface is electronegative, and they attract cationic dyes more as the cations of dye become easily attached to the anionic hydroxyl groups. Therefore, their properties need to be altered for efficient anionic dye removal, which has frequently been attempted by researchers over the years.

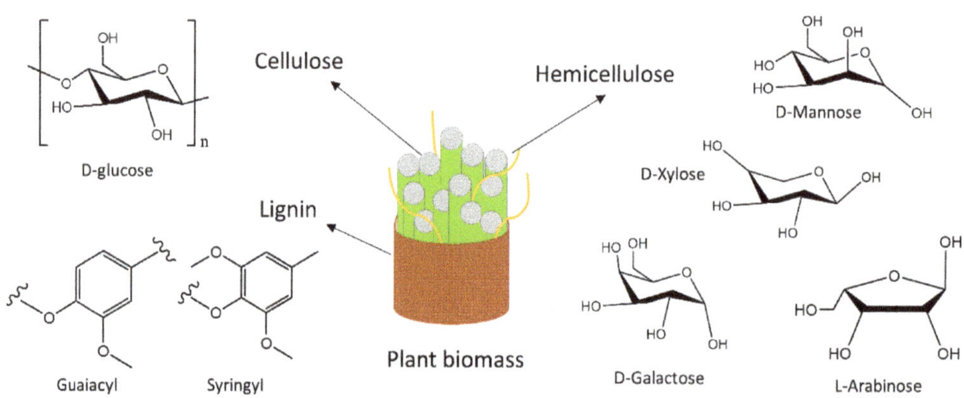

Figure 3. Common chemical groups in plant-derived agricultural waste.

4. Adsorbents for Anionic Dyes

As mentioned above, due to the electronegative nature of the surface of agricultural wastes (i.e., lignocellulosic structure), anionic dyes show a repulsive effect during the attachment through ionic interactions, if the adsorbent surface is not modified [7]. The modification of these materials can be divided into three main broad concepts: (1) mechanical, (2) chemical, and (3) thermal modifications. The concept of mechanical modification is mainly to reduce the particle size of the initial material by cutting or mechanical milling [27]. This may involve some chemical treatment (e.g., for cleaning purposes), but it will not necessarily alter the chemical groups or ionic behavior. However, the principle of the second concept (i.e., chemical modification) is exactly the opposite, where the modification of the chemical groups is performed to make them chemically more interactive with anionic dyes [16]. This method may involve mechanical milling as a preliminary step, although chemical treatments are the main driving factor for the dye adsorption. The other concept is thermal modification, where both mechanical milling and chemical treatment may be present, although the main principle is to convert the biomass into a carbonaceous material (such as activated carbon) by high-temperature carbonization processes [28], which often result in a high surface area, which leads to the key improvement in the adsorption property. All three methods are promising from different aspects and have their benefits and drawbacks.

4.1. Mechanical Modifications

The mechanical modification of biomass mainly involves the mechanical-milling operation, where the materials are crushed into fine or coarse powder (Figure 4 and Table 1). Although the chemical properties are not altered, the adsorption capacity largely depends on bath conditions and is often found as a complex synergistic influence from many parameters. Although anionic dyes have a general repulsive effect on the lignocellulosic surface, at the initial stage, the dye adsorption largely depends on the force generated by the dye-concentration gradient from the solution [29]. At the beginning of adsorption, this force helps the dye molecules to reach the adsorbent surface, overcoming the repulsive effect. However, after a certain level, when the concentration gradient is reduced by some of the dyes adsorbed onto the adsorbent, the repulsive effect starts to take the active role and limits the adsorption rate. Hence, even with a similar surface area, mechanically modified biomass can often adsorb more cationic dyes than anionic dyes [30]. This indicates that there is the possibility of remaining vacant sites inside the adsorbent (not completely saturated), even when the equilibrium is reached.

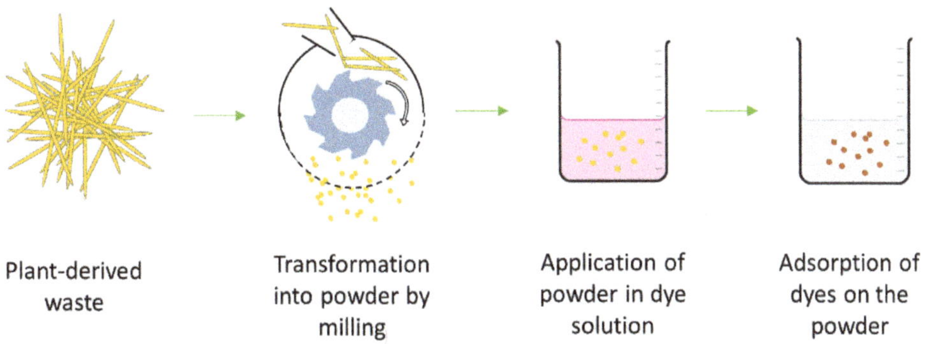

Figure 4. Schematic of the key concept of the mechanical modification of plant-derived agricultural waste for anionic dye removal.

Table 1. Maximum adsorption capacities (q_{max}) of anionic dye adsorbents derived from mechanicall modified agricultural wastes.

Resources	Particle Size (μm)	Dye	q_{max} (mg/g)	Reference
Waste tea residue	0.3475	Acid Blue 25	127.14	[31]
Date stones and jujube shells	50–100	Congo Red	45.08–59.55	[32]
Waste banana pith	>53	Direct Red	5.92	[33]
Waste banana pith	>53	Acid Brilliant Blue	4.42	[33]
Jujuba seeds	53–150	Congo Red	55.56	[34]
Cotton plant wastes	75–500	Remazol Black B	35.7–50.9	[35]
Lotus	<100	Congo Red	0.783–1.179	[36]
Almond shell	100–500	Eriochrome Black T	123.92	[37]
Waste of corn silk	250–500	Reactive Blue 19	71.6	[27]
Waste of corn silk	250–500	Reactive Red 218	63.3	[27]
Banana peel, cucumber peel, and potato peel	250–500	Orange G	20.9–40.5	[30]
Eucalyptus bark	250–700	Solar Red BA	43.5	[17]
Eucalyptus bark	250–700	Solar Brittle Blue A	49	[17]
Saccharomyces cerevisiae (yeast)	315–400	Acid Red 14	18–23	[38]
Mushroom waste	<400	Direct Red 5B	18	[39]
Mushroom waste	<400	Direct Black 22	15.46	[39]
Mushroom waste	<400	Direct Black 71	20.19	[39]
Mushroom waste	<400	Reactive Black 5	14.62	[39]
Ash seed	≤1000	Cibacron Blue	67.114	[40]
Bean peel	≤1000	Cibacron Blue	28.490	[41]
Jute processing waste	10,000	Congo Red	13.18	[39,42]
Corn stigmata	Ground (size not mentioned)	Indigo Carmine	63.7	[39,43]
Cotton gin trash	Film form	Acid Blue 25	35.46	[7]

It is almost a rule of thumb that a smaller-size particle possesses more surface area and hence, it can hold a greater amount of dye. This has been found to be true in countless studies, and it has been proven for anionic dye adsorption as well. For example, the adsorption capacity of eucalyptus bark towards Solar Red BA (a direct dye) was below 10 mg/g when a particle size of 0.51–0.71 mm was used, although it reached 18.15 mg/g when a <0.255 mm size was used [17]. A similar observation was also found in another report in which the adsorption of Eriochrome Black T was significantly improved (~22.8%) when the particle size of an almond-shell adsorbent was reduced from >500 μm to <100 μm [37].

4.2. Chemical Modifications

Due to the poor interaction of lignocellulosic biomass with anionic dyes [1], the chemical modification of their surface has received keen attention. In most cases, the main idea was to bring cationic groups to the surface, which facilitated the dye adsorption. Several methods of achieving the cationized adsorbent are reported in varied dimensions. As shown in Figure 5, these can be divided into four main categories: treatment with amine-based chemicals, cationic surfactants, quaternary ammonium compounds, and acid solutions.

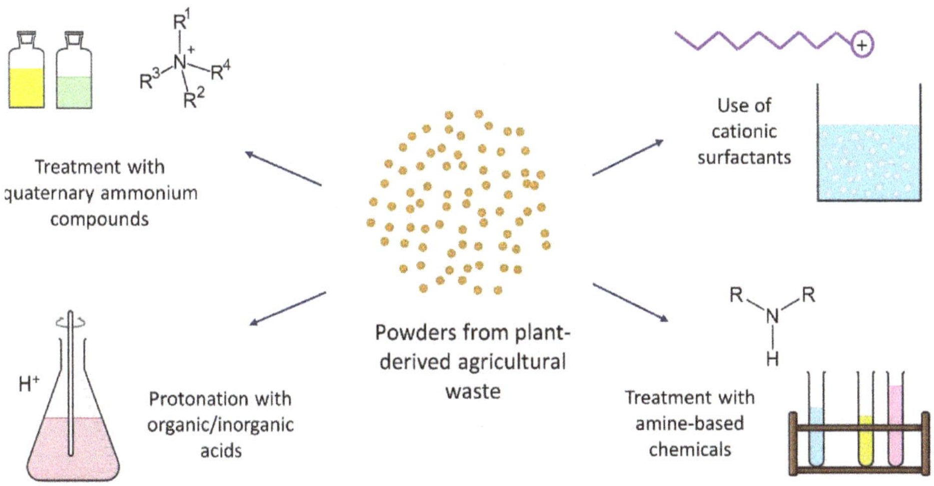

Figure 5. Different chemical-treatment methods are proposed for modifying plant-derived agricultural waste.

Different amine-rich compounds have mostly been chosen by researchers for the chemical modification of lignocellulosic biomass because they contain a high number of amines that easily become attached to lignocellulose and attract anionic dyes. However, these methods often require multiple steps of chemical treatment to effectively modify the structure. For instance, Wong et al. used polyethylenimine to treat coffee waste for the adsorption of reactive and acid dye [18]. The method included the impregnation of coffee-waste powder in a polyethylenimine solution at a moderate temperature (65 °C) for 6 h, and then cross-linking using glutaraldehyde. This delivered 77.5 mg/g and 34.4 mg/g maximum adsorptions of Reactive Black 5 and Congo Red dyes, respectively. Nevertheless, a comparison with the unmodified coffee waste was not reported to validate the true influence of the cationization. In another study, nanocrystalline cellulose extracted from hardwood bleached pulp was amino-functionalized by ethylenediamine [44]. The process involved the treatment of ethylenediamine at 30 °C for 6 h, and an in situ reduction by $NaBH_4$ at room temperature for 3 h, which resulted in the 555.6 mg/g maximum adsorption of Acid Red GR, although no comparison was shown with unmodified lignocellulose. In a different study, wheat straw was treated with epichlorohydrin and N,N-dimethylformamide at 85 °C for 1 h. Furthermore, an ethylenediamine treatment (45 min at 8 °C) and trimethylamine treatments (120 min at 85 °C) were applied to achieve the final amino-modified wheat-straw powder, which showed maximum adsorption capacities of 714.3 mg/g for anionic dye (Acid Red 73) and 285.7 mg/g for reactive dye (Reactive Red 24) [12].

Coating with amine-based polymer was also performed, where sawdust coated with polyaniline nanofibers removed up to 212.97 mg/g of Acid Red G [45]. These nanofibers were prepared by the polymerization of an aniline monomer and oxidant (($FeCl_3$). The inclusion of polyaniline nanofibers enhanced the surface area of the biomass (from 1.22 to 16.26 m^2/g), which was useful for the adsorption. In one study, an amine-rich biopolymer, (chitosan) was used to modify lignocellulose for anionic dye adsorption [7]. The chitosan was dissolved in an acetic acid solution, and the cotton-gin-trash film was impregnated for 1 h at room temperature. The resultant modified adsorbent showed a significant increase in the adsorption capacity of Acid Blue 25 (151.5 mg/g) compared with that of the unmodified biomass (35.5 mg/g). The amine-based adsorbent was also prepared as a form of a nanocomposite. For instance, banana-peel cellulose was combined with chitosan and silver nanoparticles to investigate its suitability to adsorb Reactive Orange 5, reaching a maximum adsorption capacity of 125 mg/g [14].

In the second category, treatment with cationic surfactants commonly increases th aliphatic carbon content in biomass [15], but most importantly, they induce the cationi groups on the surface, which help anionic dye removal. Several studies have been pe formed to modify lignocellulosic biomass using different cationic surfactants. These includ cetylpyridinium bromide [15], hexadecylpyridinium bromide [46,47], hexadecylpyridiniun chloride [48], hexadecylpyridinium chloride monohydrate [49], cetyltrimethylammoniun bromide [50], and hexadecyltrimethylammonium bromide [51]. This technique commonl involves the continuous shaking of the crushed lignocellulose particles in the surfactan solutions, and mostly at room temperature. However, the maximum adsorption achieve through this technique is somewhat lower than the amine-based methods. For example, few of the adsorption capacities reported for this method are 70 mg/g dye by modifie wheat straw [47], 30.8–31.1 mg/g dye by modified corn stalks [15], 55–89 mg/g dye b modified coir pith [51], and 146.2 $\pm$ 2.4 mg/g dye by modified orange-peel powder [16 which are comparatively lower than most of those reported for amine-based modification (Table 3). This is probably related to the abundance of amide groups [18] in the amine-base compounds, which are more prone to attach to the lignocellulosic structure, as well as wit the anionic dyes, by electrostatic attraction [7,18].

There are few studies on the modification of lignocellulose by quaternary ammo nium compounds for anionic dye adsorption, which mostly showed high efficiency. Fo example, Jiang and Hu reported hydroxypropyloctadecyldimethylammonium-modifie rice husk cellulose for reactive (Diamine Green B) and acid-dye (Congo Red and Aci Black 24) adsorption. This modification required multiple steps, which included alkalin and ultrasonic treatments, and treatments with epichlorohydrin and N,N-dimethyl-1 octadecylamine/isopropanol. The resultant adsorbent was able to adsorb up to ~580 mg/ of the anionic dye (Congo Red) [52]. In another study, biomass from palm kernel she was quaternized using N-(3-chloro-2-hydroxypropyl)trimethylammonium chloride in a alkaline aqueous mixture, which resulted in 207.5 mg/g of maximum adsorption [53].

The fourth category of the chemical modification of the lignocellulose surface i through acidic treatment. For instance, it was claimed that a phosphoric acid treatmen worked by increasing the surface area and led to the grafting of phosphate onto the ligno cellulose structure. However, the adsorption capacity reported in the proposed metho was quite low (15.96 mg/g) [54]. A similar result (adsorption capacity of 13.39 mg/g of Re active Red 196) was reported when concentrated HCl was used for sawdust treatment [13 Slightly higher adsorption has also been reported, such as 40.98 mg/g of Drimarine Blac CL-B removal by HCl-modified peanut husk [55]. The only exception (a higher remova efficiency) was the treatment of fermentation wastes with nitric acid, where the raw biomas structure was reported to be protonated, replacing the naturally available ionic species, an thereby resulting in a 185.2 mg/g adsorption of Reactive Black 5 [56]. However, there wa no following study on the nitric acid treatment.

A combination of amine-based chemicals and acid treatment, such as cationizatio by dichloroethane and methyl amine, followed by protonation using acetic acid, was als reported in one study to modify the biomass surface [16]. The resultant orange-peel adso bent showed a maximum adsorption capacity of 163 mg/g of Congo Red. Furthermor the silylation of biomass was also found to be effective for anionic dye adsorption, wher a maximum adsorption of 208.33 mg/g was reported [57]. Moreover, Lin et al. propose an amphoteric-modification method of delignified wheat straw by a monochloroaceti acid treatment, and then grafting with 2-dimethylamino ethyl methacrylate. The resultan adsorbent could be used to adsorb both anionic and cationic dyes by switching betwee acidic or alkaline conditions ranging from pH 2 to 10 [58].

Aside from the four major kinds of treatments, some other methods, such as mod fications of biomass using oxides and metal salts, have also been reported. For exampl the modification of wood biowaste using Al_2O_3 by the solvo-reaction method showed high adsorption capacity of Reactive Blue 19 (441.9 mg/g) [59]. Furthermore, the modifica tion of sawdust using $FeCl_3$ (maximum 212.97 mg/g Acid Red G removal) [45], the us of $ZnCl_2$ on pulp waste (maximum 285.7 mg/g Methyl Orange removal) [60], hazelnut bagasse (maximum 450 mg/g Acid Blue 350 removal) [61], and pine cone (maximum 118.06 mg/g Alizarin Red S removal) [62], and the use of $CaCl_2$ with peanut husk (maximum 40.98 mg/g Drimarine Black CL-B removal) [55], have also been reported, showing considerable efficiency.

Overall, the trend in Table 3 shows that effective dye removal by chemically modified adsorbents is more related to the modification methods, rather than to the type of dye. For example, the acidic modification of lignocellulose often produced a lower efficiency either for acid or reactive dyes, although the adsorption capacity was reported to be higher for both when amine-based chemicals were used. When the same adsorbent was used to adsorb acid and reactive dyes, mostly the acid dyes showed a higher removal efficiency [12,51,52], although an exception was also seen [48]. Modification with different amine-based chemicals was found to be more effective for anionic dye adsorption, which was probably related to the abundance of cationic amine groups in these compounds.

Table 2. Maximum adsorption capacities (q_{max}) of anionic dye adsorbents derived from chemically modified agricultural wastes.

Resources	Chemical Modification	Modification Type	Dye	q_{max} (mg/g)	Reference
Wheat straw	Grafting with 2-dimethylamino ethyl methacrylate monomer	Amine-based	Orange II	506	[58]
Wheat straw	Treatment with hexadecylpyridinium bromide	Cationic surfactant	Light Green	70.01 ± 3.39	[47]
Peanut husk	Modified by hexadecylpyridinium bromide	Cationic surfactant	Light Green	60.5	[46]
Wheat straw	Treatment with epichlorohydrin, N,N-dimethylformamide, ethylenediamine, and trimethylamine	Amine-based	Acid Red 73	714.3	[12]
Wheat straw	Treatment with epichlorohydrin, N,N-dimethylformamide, ethylenediamine, and trimethylamine	Amine-based	Reactive Red 24	285.7	[12]
Barley straw	Treatment with hexadecylpyridinium chloride monohydrate	Cationic surfactant	Acid Blue 40	51.95	[49]
Barley straw	Treatment with hexadecylpyridinium chloride monohydrate	Cationic surfactant	Reactive Blue 4	31.5	[49]
Corn stalks	Treatment by cetylpyridinium bromide	Cationic surfactant	Acid Red	30.77	[15]
Corn stalks	Treatment by cetylpyridinium bromide	Cationic surfactant	Acid Orange	31.06	[15]
Banana peel	Reinforcement with nanoparticles and chitosan	Amine-based	Reactive Orange 5	125	[14]
Sawdust	Coating with polyaniline	Amine-based	Acid Red G	212.97	[45]
Sawdust	Treatment with cetyltrimethylammonium bromide	Cationic surfactant	Congo Red	9.1	[50]
Sawdust	Treatment with concentrate HCl	Acid treatment	Reactive Red 196	13.39	[13]
Rice husk	Treatment with hydroxypropyloctadecyldimethylammonium	Quaternary ammonium compounds	Diamine Green B	207.15	[52]
Rice husk	Treatment with hydroxypropyloctadecyldimethylammonium	Quaternary ammonium compounds	Acid Black 24	268.88	[52]
Rice husk	Treatment with hydroxypropyloctadecyldimethylammonium	Quaternary ammonium compounds	Congo Red	580.09	[52]
Peanut husk	Treatment with alginate and $CaCl_2$	Metal salt	Drimarine Black CL-B	40.98	[55]
Peanut husk	Treatment by hexadecylpyridinium bromide in batch mode	Cationic surfactant	Light Green	146.2 ± 2.4	[63]

Table 3. Maximum adsorption capacities (q_{max}) of anionic dye adsorbents derived from chemicall modified agricultural wastes.

Resources	Chemical Modification	Modification Type	Dye	q_{max} (mg/g)	Reference
Peanut husk	Hydrochloric acid treatment	Acid treatment	Drimarine Black CL-B	51.02	[55]
Oil palm empty fruit bunches	Silylation	Amine-based	Procion Red	208.33	[57]
Orange peel	Treatment with dichloroethane, methyl amine, and acetic acid	Amine-based	Congo Red	163	[16]
Cotton gin trash	Cationized by chitosan	Amine-based	Acid Blue 25	151.52	[7]
Aquatic plant	Phosphoric acid treatment and low-temperature activation	Acid treatment	Direct Red 89	15.96	[54]
Coffee waste	Treatment with polyethylenimine	Amine-based	Reactive Black 5	77.52	[18]
Coffee waste	Treatment with polyethylenimine	Amine-based	Congo Red	34.36	[18]
Hardwood kraft pulp	Grafting cellulose nanocrystals with ethylenediamine	Amine-based	Acid Red GR	555.6	[44]
Palm kernel shell	Quaternized by N-(3-chloro-2-hydroxypropyl) trimethylammonium chloride	Quaternary ammonium compounds	Reactive Black 5	207.5	[53]
Fermentation waste	Protonated by nitric acid	Acid treatment	Reactive Black 5	185.2	[56]
Waste coir pith	Treatment with hexadecyltrimethylammonium solution	Cationic surfactant	Acid Brilliant Blue	159	[51]
Waste coir pith	Treatment with hexadecyltrimethylammonium solution	Cationic surfactant	Procion Orange	89	[51]
Wood residue	Aluminum oxide modification	Metal oxide	Reactive Blue 19	29.83	[64]
Wood biowaste	Aluminum oxide modification	Metal oxide	Reactive Blue 19	441.9	[59]

4.3. *Thermal Modifications*

The thermal modification of plant-derived wastes is mainly performed to prepare activated carbon (AC). AC is an advanced form of charcoal that is basically a porous carbonaceous structure with an extended surface area [65]. It has been accepted as an effective adsorbent and has been applied widely in water and wastewater treatment, air purification, and solvent recovery, as well as in the pharmaceutical and medical industries and industrial processes. According to a market research report, the market demand for AC was estimated to be USD 5.7 billion worldwide in 2021, and it is predicted to reach USD 8.9 billion by 2026 [66]. A variety of plant-derived wastes have been used as raw materials for the preparation of AC for anionic dye adsorption, as these resources are rich in carbon content, cost effective, renewable, and abundant [67].

Carbonization and activation are basically the two steps that are involved in preparing AC [68]. Activation can be performed by different methods, such as mechanical or thermal, chemical, or a combination of both (i.e., physiochemical) [67], and it can also be biological [69]. These include several methods and their combinations, such as chemical treatment, air oxidation, electrochemical oxidation, and plasma, microwave, and ozone treatment for enhancing the adsorption performance of AC [70–74]. However, for AC prepared for anionic dye adsorption, the chemical-activation technique (using activating agents, such as zinc chloride or phosphoric acid) [75] is mostly preferred (Figure 6) because of its faster activation time, higher carbon yield, simplicity, lower operating temperature and well-developed pore structure. The proper selection of materials can enhance the adsorption potentiality of AC through the development of its pore structure (total pore volume, pore size distribution) and surface functionalities [76].

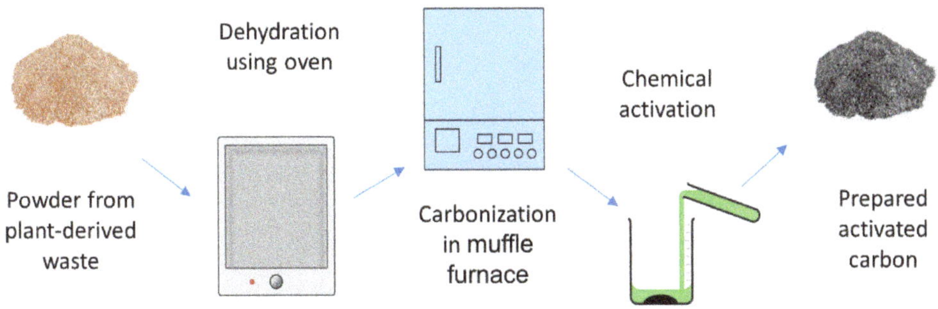

Figure 6. Schematic of thermal-modification steps for plant-derived agricultural waste towards the preparation of activated carbon for anionic dye removal.

The main functional groups on the surface of AC that are responsible for removing dyes are carboxyl, carbonyl, phenols, lactones, and quinones. A positive surface charge on AC can be attained by applying an alkaline treatment, which can accelerate the adsorption of anionic dyes. The porous carbon of alkali-treated AC can enhance the reaction with acid-dye molecules by dipole–dipole, H-bonding, and covalent bonding [69]. As listed in Table 4, ACs prepared from many agricultural byproducts, such as peanut hull, coir pith, rice husk, coffee husk, and water hyacinth, have successfully been used for eliminating anionic dyes from wastewater.

Table 4. Maximum anionic dye adsorptions (q_{max}) reported in different studies onto activated carbon prepared from agricultural waste.

Resources	Surface Area (m^2/g)	Anionic Dye	q_{max} (mg/g)	Reference
Java citronella	Not reported	Congo Red	4.29	[77]
Palm tree waste	648.90	Congo Red	10.4	[78]
Water hyacinth	Not reported	Congo Red	14.367	[79]
Wastewater sludge	98.8	Amaranth	19.6	[80]
Coffee husk	613 ± 14	Indigo Carmine	36.63	[81]
Carob waste	921.07	Reactive Black 5	36.90	[82]
Rick husk	272	Acid Yellow 36	86.9	[83]
Pinecone	878.07	Alizarin Red S	118.06	[62]
Vegetable waste	Not reported	Eriochrome Black T	120.50	[28]
Date palm fronds	431.82	Methyl Orange	163.132	[84]
Sawdust	516.3	Acid Yellow 36	183.8	[83]
Hazelnut bagasse	1489	Acid Blue 350	450	[61]
Psyllium stalks	Not reported	Coomassie Brilliant Blue	237.2	[83,85]
Waste potato peels	Not reported	Cibacron Blue	270.3	[83,86]
Pulp waste	1022.46	Methyl Orange	285.71	[60,85]
Date palm fronds	431.82	Eriochrome Black T	309.59	[84]
Grape waste	1455	Acid yellow 36	386	[84,87]
Pink shower	283.4	Congo Red	970	[75]

The dye-adsorption capacity of different ACs is substantially controlled by the pH of the dye solution, possibly because the electrostatic interaction between ionized dye molecules and carbon surface functionalities that change the solution pH and rate of adsorption also varies. In the case of anionic dye, high dye uptake is favored at a lower pH or acidic medium, as the complete protonation of the carbon surface functionalities enhances the electrostatic attraction with anionic dyes, thereby increasing the dye uptake. From previous studies, it was observed that the OH group in the solution increased with an increasing pH, producing electrostatic repulsion with anionic dye, and thus decreasing the dye uptake. Therefore, the maximum adsorption of anionic dyes was observed in a pH range of 2–6 [79,84]. Although the surface area of the AC is known as a key factor for dye adsorption because it provides more spaces for the dye molecules to be attached, it can be perceived from Table 4 that this was not always true for anionic dye adsorption due to ionic interactions, which could have a positive or negative impact, depending on the specific adsorption system. For example, the maximum anionic dye absorption (960 mg/g) was found in AC obtained from pink shower seed pods, although it contained a moderate surface area (283.4 m^2/g) [75], where the strong electrostatic attraction between the functional groups of the AC surface and anionic dye was facilitated with π–π interactions between the free electrons of the aromatic rings of the Congo Red structure, and the delocalized π-electrons on the basal planes of carbon.

4.4. The Effect of Process Conditions

Other than the modification performed on the biomass, if the amount of adsorbent is kept constant, then different process parameters, which include the pH, temperature, dye concentration, and chemical composition of the biomass, can impact the overall adsorption.

The pH of the solution plays a vital role during the adsorption. In numerous studies for anionic dye adsorption, the favorable pH was reported as 2 [27,30,39,43], and in some cases, even 1 [31,35]. This is directly related to the change in the surface charge in lignocellulose that is altered at a lower pH. A lower pH reduces the electronegativity of the surface and can even alter it to a positive surface charge [37], thus possibly attracting anionic dyes more conveniently. This was found to be true in cases of different kinds of anionic dyes such as acid dyes [30,43], reactive dyes [27], and direct dyes [39]. Although this was seen as a common trend, some exceptions were also reported in which the pH showed a minor impact [38], or in which a moderate pH (7.5) was found to be more effective for better adsorption [88].

The bath temperature is also important because a higher temperature can often improve the mobility of dye molecules, and thus, the hindering force from the electronegative adsorbent surface diminishes [89]. As a result, a greater amount of dye can be transported into the adsorbent surface. Added to that, swelling can result in the adsorbent, which can promote intraparticle diffusion inside the structure [89,90], allowing more dyes be adsorbed onto the surface. However, the opposite phenomenon was also frequently reported (exothermic adsorption) [17,37,41], where a higher temperature was unfavorable for dye uptake. This was explained as a possible weakening of the related forces of attraction by a higher temperature [17,40].

The concentration of dye present in a solution largely affects the adsorption rate, and particularly at the initial stage. If the dye concentration is high, then the rate of adsorption will be higher, and vice versa. The content of cellulose in the biomass is also reported as an influencing factor in the adsorption. In a study, three different biomasses (i.e., coconut shells, cauliflower cores, and broccoli stalks) were used for the adsorption of the same anionic dyes, although the chemical compositions of the biomasses were different. The total cellulose, hemicellulose, starch, and pectinic sugar in the cauliflower cores were the highest among the three and resulted in a higher uptake of reactive and acid dyes [89]. Furthermore, the adsorption capacity was particularly significantly affected by the impact of cellulose as the external layer, rather than hemicellulose and lignin [89]. The chemical groups in lignocellulosic biomass (such as –OH, C=O, and C–O) are also good candidates to form hydrogen bonding with anionic dyes [30,43].

5. Adsorption Isotherms

Adsorption isotherms provide dye equilibrium relations that are widely used to assess the appropriateness of an adsorption mechanism. The most commonly used isotherms are the Langmuir and Freundlich models for any dye adsorption, which have also frequently been used to evaluate anionic dye attachment with plant-derived agricultural wastes.

The Langmuir isotherm model can be linearized as:

$$C_e/q_e = C_e/q_{max} + 1/(k_L q_{max}) \tag{1}$$

where q_e is the amount of anionic dye absorbed by the adsorbent at equilibrium; C_e is the amount of dye left in the solution (mg/L) at the same point; q_{max} is the expression of the maximum dye-adsorption capacity (mg/g) by the adsorbent; k_L is the binding affinity (L/mg), which is also known as the Langmuir constant. The Langmuir constant (k_L) can be used to derive the suitability of the adsorption system. For example, the equilibrium parameter (R_L) can be calculated using the value of the k_L by the following equation [91]:

$$R_L = 1/(1 + k_L C_0) \tag{2}$$

where C_0 is the initial dye concentration (mg/L) in the solution before the adsorption begins. The calculated R_L indicates whether the adsorption system was favorable or not. For instance, if $0 < R_L < 1$, then the adsorption is favorable, if $R_L > 1$, then the adsorption is unfavorable, if $R_L = 1$, then the adsorption is linear, and if $R_L = 0$, then the adsorption is irreversible.

The other frequently used isotherm is the Freundlich model, which can be linearized as follows [92]:

$$\ln q_e = \ln k_F - (1/n)\ln C_e \tag{3}$$

where k_F is the adsorption capacity, and n is the adsorption intensity. The n value is also used to measure the deviation from the linearity (or heterogeneity). If $1/n < 1$ (n ranges from 1 to 10), then the adsorption is counted as favorable, and if $1/n > 1$, then the adsorption is considered unfavorable. If 1/n is close to zero, then the adsorbent is more heterogeneous, and vice versa [92]. Further indications were also noted from the value of the n, such as $n = 1$ indicates the linear nature of the adsorption, $n < 1$ indicates chemical adsorption, and $n > 1$ indicates physical adsorption [93].

While the Langmuir isotherm is considered more related to monolayer adsorption, the Freundlich isotherm is considered for multilayer adsorption. The coefficient-of-determination (R^2) values of the Langmuir and Freundlich models listed in Table 5 across different kinds of dyes (acid, reactive, and direct), and after different sorts of modifications (mechanical, chemical, and thermal), show a sporadic result, and often closer values of the R^2 between the Langmuir and Freundlich models. However, as a traditional pattern, often either chemisorption [38,88] or physisorption [7,31] has been claimed as the mechanism of interaction during the adsorption. A combination of multiple mechanisms has also been reported on many occasions. These include different combinations of physical and chemical adsorption [39–41,57], hydrogen bonding [89], ion exchange [34,94], valence forces [58], electrostatic and π–π interactions [75], ion exchange and chemical reaction [84], and so on. These interactions were only predicted and were not always confirmed through chemical analysis. Nevertheless, there are great possibilities for many of these interactions during the adsorption. For instance, the zeta potential (ζ) of chitosan-modified lignocellulose biomass transformed the ζ value from −10.8 mV to +5.1 mV, making it more appropriate for an electrostatic attraction towards anionic dye [7]. Furthermore, because lignocellulose is rich in phenolic groups from lignin, there are chances for π–π interactions between the rings present in the anionic dye structure. A schematic representation of different possible dye–adsorbent interactions during anionic dye separation is shown in Figure 7a. The mechanism of adsorption can also differ based on the concentration of dyes. For example, with lower and higher concentrations of dyes, monolayer and multilayer adsorption were observed, respectively [53], while simultaneous monolayer and multilayer adsorption was also reported [84]. Therefore, often the physical interaction between the dye and adsorbent is likely to be combined with chemical interactions, depending on whether there are enough chemical sites present for the reactions and favorable conditions.

Table 5. Reported R^2 values of Langmuir and Freundlich isotherm models for anionic dye adsorption

Modification Technique	Dye	Adsorbent	R^2		Reference
			Langmuir Model	Freundlich Model	
Mechanical	Congo Red	Jujuba seeds	0.999	0.987–0.989	[34]
Mechanical	Reactive Black 5	Spent mushroom waste	0.997	0.907	[39]
Mechanical	Cibacron Blue	Ash seed	0.9357	0.98	[40]
Mechanical	Cibacron Blue	Bean peel	0.9822	0.9414	[41]
Mechanical	Direct Red 5B	Spent mushroom waste	0.998	0.918	[39]
Mechanical	Direct Black 22	Spent mushroom waste	0.996	0.905	[39]
Mechanical	Acid Blue 25	Waste tea residue	0.9197–0.9309	Not reported	[31]
Mechanical	Congo Red	Bottom ash and deoiled soya	0.79–0.99	0.85–0.97	[88]
Mechanical	Congo Red	Jute processing Waste	0.9916	0.9912	[42]
Chemical	Congo Red	Sawdust	0.9418–0.9867	0.9040–0.9876	[50]
Chemical	Congo Red	Rice husk	0.9939–0.9988	0.9672–0.9861	[52]
Chemical	Congo Red	Palm tree fiber waste	0.996	0.857	[78]
Chemical	Congo Red	Coffee Waste	0.99	0.67	[18]
Chemical	Diamine Green B	Rice husk	0.9209–0.9731	0.8085–0.8988	[52]
Chemical	Reactive Black 5	Palm kernel Shell	0.901	0.854	[53]
Chemical	Reactive Black 5	Coffee Waste	0.96	0.93	[18]
Chemical	Reactive Black 5	Fermentation Wastes	0.994–0.999	Not reported	[56]
Chemical	Acid Blue 25	Cotton gin trash	0.967–0.996	0.875–0.947	[7]
Chemical	Direct Red 89	Aquatic Plant	0.995	0.999	[54]
Chemical	Acid Black 24	Rice husk	0.9887–0.9975	0.9715–0.9823	[52]
Thermal	Eriochrome Black T	Date palm Fronds	0.694–0.953	0.711–0.975	[84]
Thermal	Congo Red	Java Citronella	0.904–0.957	0.994–0.999	[77]
Thermal	Eriochrome Black T	Vegetable Waste	0.853–0.978	0.965–0.996	[28]

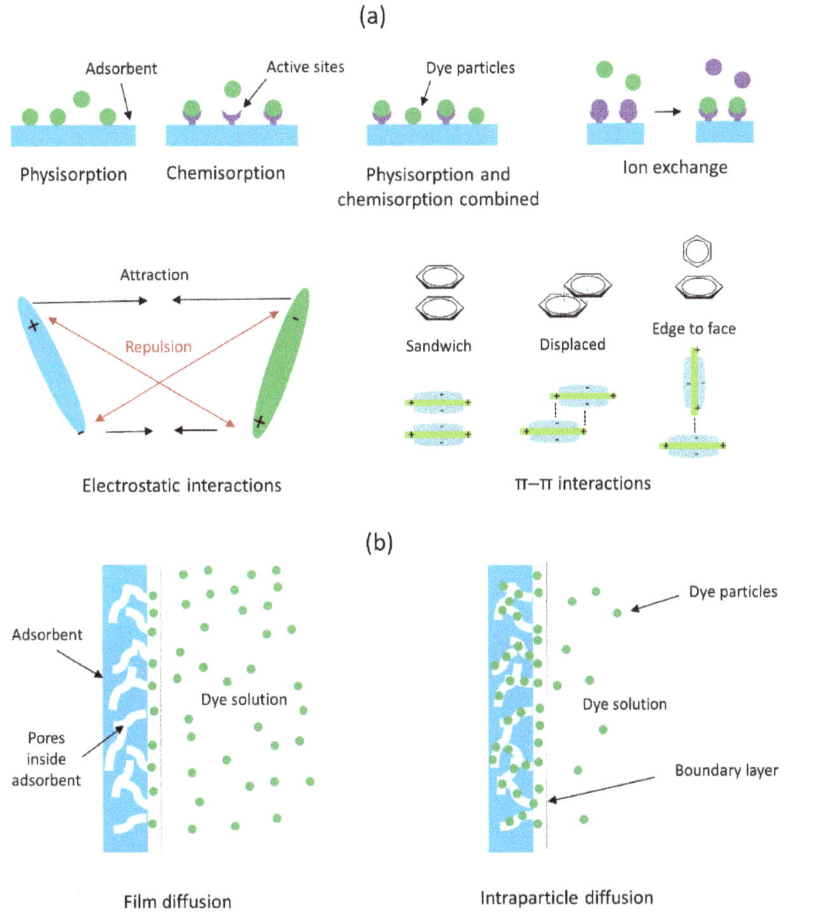

Figure 7. Schematic representations of: (**a**) different interactions between dye particles and lignocellulosic biomass; (**b**) diffusion mechanism of dye particles during an adsorption process by a biomass-derived adsorbent.

Some other less commonly used isotherms for anionic dyes include the Redlich–Peterson model [84], Dubinin–Radushkevich model, Temkin model, and Flory–Huggins model [7].

The Redlich–Peterson model is a three-parameter model, and it is known as a combined form of the Langmuir and Freundlich models. The linear form can be expressed as [84]:

$$\frac{C_e}{q_e} = \frac{1}{K_{RP}} + \frac{a_{RP}}{K_{RP}} C_e^{b^{RP}} \tag{4}$$

where K_{RP} and a_{RP} are the Redlich–Peterson constants, and b_{RP} is an exponent from 0 and 1. The fitting results in a Redlich–Peterson model have occasionally been used for anionic dye adsorption to confirm and represent the adsorption mechanism as a combination of both the Langmuir and Freundlich models [31] (for instance, the simultaneous involvement of monolayer and multilayer adsorption [84]).

The linear form of the Dubinin–Radushkevich isotherm can be expressed as [95]:

$$\ln q_e = \ln q_m - k_{DR}\varepsilon^2 \tag{5}$$

where q_m represents the maximum adsorption capacity, k_{DR} is the isotherm constant, and ε represents the Polanyi potential and is calculated by the following equation:

$$\varepsilon = RT \ln\left(1 + \frac{1}{C_e}\right) \quad (6)$$

where T is the equilibrium temperature, and R represents the ideal gas constant (8.314 $J \cdot mol^{-1} \cdot K^{-1}$) [39]. The isotherm constant (k_{DR}) is used to measure the free energy (E) of adsorption:

$$E = 1/\sqrt{2k_{DR}} \quad (7)$$

If the E is found to be <8 kJ/mol, then this indicates physisorption, and when the E ranges 8–16 kJ/mol, then a chemisorption process is assumed [95]. This isotherm model has sometimes been used to determine between physisorption and chemisorption [39], and to justify the findings from other models [7].

The linear form of the Temkin model can be shown as:

$$q_e = B_1 \ln(k_T C_e) \quad (8)$$

where k_T is the Temkin constant, and B_1 is derived from the below equation:

$$B_1 = \frac{RT}{b_T} \quad (9)$$

where b_T is another Temkin constant related to the heat of adsorption [45]. From a good fitting with the Temkin model for anionic dye adsorption, the adsorption potential is predicted. The value of the Temkin constants can be used to evaluate the decrease and increase in the binding energy at different adsorption temperatures [45].

The linear equation of the Flory–Huggins model can be expressed as [96]:

$$\ln\frac{\theta}{C_0} = \ln k_{FH} + x\ln(1-\theta) \quad (10)$$

where k_{FH} is the Flory–Huggins constant, x represents the remaining dye particles in the adsorption sites, and θ represents the degree of the adsorbent surface coverage by the dye particles and is derived from the following equation:

$$\theta = 1 - \frac{C_e}{C_0} \quad (11)$$

When the value of the k_{FH} is positive, a suitable and spontaneous adsorption system is likely [96]. This model can be used to determine the surface coverage of the dye particles onto the adsorbents, although it is used infrequently for anionic dye adsorption [7].

However, as mentioned above, the use of these models was less frequent for anionic dye adsorption compared with the Langmuir and Freundlich isotherms, and they often showed an inferior fitting.

6. Kinetics and Diffusion

The kinetics of anionic dye adsorption on biomass-derived adsorbents has mainly been investigated by pseudo-first-order and pseudo-second-order kinetic models, and often linear equations have been used to draw the plots and find the fitting of the data [18,40,54,57,75]. The first-order model is also known as the Lagergren equation, which relates to the adsorption rate associated with the number of unoccupied areas of the dye [97]. The linear equation of the pseudo-first-order model is represented as:

$$\ln(q_e - q_t) = \ln q_e - k_1 t \quad (12)$$

where k_1 is the first-order adsorption-rate constant, and q_t is the weight of the dye adsorbed (mg/g) at time (t) [98,99]. When the data better fit with the first-order equation, the adsorption is more likely to be a physisorption process.

The pseudo-second-order model is also parallelly used with the first-order kinetic model to check the adsorption feasibility. There are four linear equations of this model, which can be derived from each other [92,100]:

$$\frac{t}{q_t} = \frac{1}{k_2 q_e^2} + \frac{t}{q_e} \tag{13}$$

$$\frac{1}{q_t} = \frac{1}{q_e} + \frac{1}{k_2 q_e^2}\left(\frac{1}{t}\right) \tag{14}$$

$$q_t = q_e - \frac{1}{k_2 q_e}\left(\frac{q_t}{t}\right) \tag{15}$$

$$\frac{q_t}{t} = k_2 q_e^2 - k_2(q_e)q_t \tag{16}$$

where k_2 is the second-order adsorption-rate constant. In practice, Equation (13) is comprehensively used. However, after plotting the data, this equation often shows a very good fit (e.g., $R^2 > 0.99$), which is related to the presence of the t value in both the x-axis and y-axis [101]. This can be justified by fitting the same data with another linear equation in which t is not present on both sides and can result in a poor fit (Figure 8). When the data better fit with the second-order equation, the adsorption is more related to a chemisorption process. However, fitting with second-order models for anionic dye adsorbents was often not justified in the literature by plotting with more than one equation when linear models were used. Furthermore, the conversion of the data necessary for the linearization of these kinetic models can also lead to an alteration of the error structure and variation in the weight in data points, and it can also introduce errors into the independent variable [102]. However, using a nonlinear model of second-order kinetics does not alter the kinetic parameters, and thus it is considered to be more accurate and is recommended for estimating the kinetic parameters [103,104].

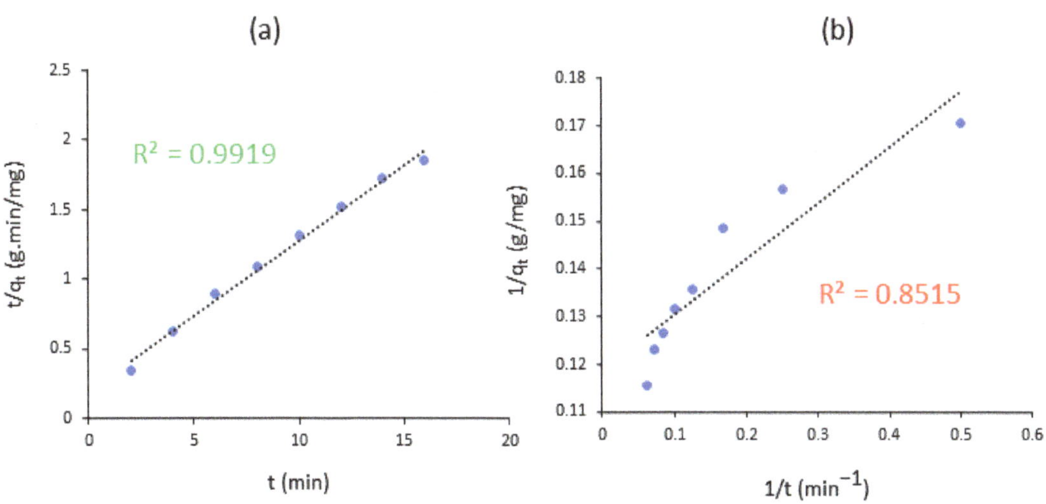

Figure 8. An example of the alteration of the fitting value when graphs were plotted using two different pseudo-second-order equations, showing: (**a**) good fit, and (**b**) poor fit from the same data.

The molecular-diffusion mechanism is another key factor in the adsorption rate that cannot be identified by the pseudo models. If the adsorption is a physisorption phenomenon, then there is more possibility that the diffusion is the rate-controlling step [101]. In an adsorbent/dye-solution adsorption process, the transfer of dye particles into the adsorbent can be controlled by one of the three known mechanisms: film diffusion, intraparticle diffusion, or a combination of both [105]. Film diffusion is the transfer of dye particles to the adsorbent surface, and intraparticle diffusion is the diffusion of dye particles into the adsorbent pores (Figure 7b). It was reported that both film diffusion (also known as external mass transfer) and intraparticle diffusion can play a significant role during anionic dye adsorption [35].

The linear form of the intraparticle-diffusion model can be expressed as [92]:

$$q_t = k_{ID}.t^{1/2} + C \quad (17)$$

where k_{ID} represents the diffusion-rate constant, and C is the thickness of the boundary layer (Figure 7b). If C is considerably higher, then a superior influence of the surface adsorption on the adsorption rate is likely [106]. As per this model, if the fitted line goes through the origin (0, 0), then this is a sign of the sole control of intraparticle diffusion [107]. However, in most cases, more than one step of adsorption is observed from the plotting of the intraparticle-diffusion model. These include the initial film diffusion followed by intraparticle diffusion, and possibly a third and slow adsorbing stage close to an equilibrium.

7. Thermodynamics of Adsorption

Thermodynamic studies of dye adsorption are performed when the adsorption is conducted at different temperatures. From the pseudo-second-order constant (k_2), the activation energy (E_a) of the adsorption process can be determined by the Arrhenius equation:

$$\ln k_2 = \ln A - \frac{E_a}{RT} \quad (18)$$

where A is the Arrhenius factor, E_a is the Arrhenius activation energy (kJ/mol), T represents the absolute temperature (K), and R represents the gas constant [99].

The entropy (ΔS°) and enthalpy (ΔH°) are derived using the Eyring formula, which is expressed as:

$$\ln\left(\frac{k_2}{T}\right) = \ln\left(\frac{K_b}{h}\right) + \frac{\Delta S^\circ}{R} - \frac{\Delta H^\circ}{RT} \quad (19)$$

where K_b is the Boltzman constant, and h is the Planck constant. A positive value of the ΔH° indicates an endothermic reaction (adsorption increases with increasing temperature) [39], while a negative value of the ΔH° indicates an exothermic process (adsorption decreases with increasing temperature) [100]. A positive value of the ΔS° indicates a decrease in the randomness of solid/liquid interfaces, and vice versa [39].

Finally, the Gibbs energy of activation (ΔG°) is calculated from the following formula:

$$\Delta G^\circ = \Delta H^\circ - T\Delta S^\circ \quad (20)$$

A positive value of the ΔG° indicates nonspontaneous reactions, which need an input of energy, while a negative value of the ΔG° indicates spontaneous reactions, which can continue without external energy [108].

Thermodynamic studies have been extensively conducted for anionic dye adsorption and the values of the ΔS°, ΔH°, and ΔG° have been used to determine whether the process is spontaneous/nonspontaneous and exothermic/endothermic, as well as the randomness in the adsorbent/solution interfaces. As shown in Table 6, in most cases, the mechanism was found to be spontaneous (negative ΔG°) for either kind of dye or the modification methods. Although the endothermic process (negative ΔH°) was more observed, the exothermic process (positive ΔH°) was also seen. In most cases, positive ΔS° was also seen, which means more uniformity during the anionic dye adsorption.

Table 6. The trends of the enthalpy (ΔH°), Gibbs energy of activation (ΔG°), and entropy (ΔS°) of anionic dye adsorption onto plant-derived agricultural waste.

Adsorbent	Modification Technique	Dye	ΔH° (kJ/mol)	ΔG° (kJ/mol)	ΔS° (J/mol K)	References
Bean peel	Mechanical	Cibacron Blue	−32.36	−3.97 to −4.89	−92.21	[41]
Ash seed	Mechanical	Cibacron Blue	−28.95	−18.01 to −18.37	−35.36	[40]
Spent mushroom waste	Mechanical	Direct Red 5B	0.99	−0.79 to −1.83	1.63	[39]
Spent mushroom waste	Mechanical	Direct Black 22	0.59	−0.69 to −1.33	1.65	[39]
Spent mushroom waste	Mechanical	Direct Black 71	3.79	−3.32 to −7.47	7.02	[39]
Spent mushroom waste	Mechanical	Reactive Black 5	0.62	−0.15 to −0.77	0.21	[39]
Vegetable waste	Mechanical	Eriochrome Black T	44.51	−2.72 to −5.89	158.51	
Jujuba seeds	Mechanical	Congo Red	12.94	−3.49 to −6.43	57.9	[34]
Sawdust	Chemical	Acid Red G	28.7	−1.07 to −3.14	103.4	[45]
Orange peel	Chemical	Congo Red	19	−1.88 to −3.23	70	[16]
Peanut husk	Chemical	Drimarine Black CL-B	−23.74	−0.22 to −2.9	72	[55]
Coffee waste	Chemical	Reactive Black 5	8.28	−9.66 to −11.08	59.99	[18]
Coffee waste	Chemical	Congo Red	35.05	−3.76 to −8.17	130.59	[18]
Rice husk	Chemical	Diamine Green B	15.06	−3.46 to −4.39	35.11	[52]
Rice husk	Chemical	Acid Black 24	16.32	−2.65 to −3.10	43.52	[52]
Rice husk	Chemical	Congo Red	2.96	−2.98 to −3.37	19.57	[52]
Wheat straw	Chemical	Reactive Red 24	4.53	−29.05 to −32.93	135.1	[12]
Waste coir pith	Chemical	Procion Orange	25.69	−8.43 to −11.15	111.52	[51]
Waste coir pith	Chemical	Acid Brilliant Blue	70.09	−5.65 to −11.37	245.64	[51]
Water hyacinth	Thermal	Congo Red	−189.01 to −918.94	−3001.04 to −12,254.1	16.39 to 53.97	[79]
Coffee husk	Thermal	Indigo Carmine	21.72 to 35.83	−46.71 to −61.05	229.9 to 287.6	[81]
Java citronella	Thermal	Congo Red	−0.007 to −0.029	7.9 to 30.2	−6 to −92	[77]
Pinecone	Thermal	Alizarin Red S	29.13	−2.75 to −4.39	82	[62]
Vegetable waste	Thermal	Eriochrome Black T	−10.08	−2.76 to 3.43	−22.32	[28]

8. Circularity and Sustainability

To prove the adsorbents as sustainable and economically viable, the circularity of the adsorbents has received significant attention. In this regard, the regeneration ability of the adsorbents is a key criterion. If the adsorbed dyes can be successfully desorbed from the adsorbent, then the reuse of both the adsorbent and dye could be possible, and the process becomes more economically feasible [34]. Regeneration studies have been widely considered for anionic dye adsorbents, and the results are promising. These are often performed by altering the pH of the solution. When anionic dyes are physically bound to the adsorbent, an alteration of the pH, often alkaline [33], could be more useful to transfer them into the solution [7]. For example, a desorption experiment using different components (i.e., water, HCl, acetic acid, NaOH, and ethanol) showed a maximum Methyl Orange desorption of 67.85% by a NaOH solution, while the second best was 34.5% by acetic acid [60]. This was also confirmed by different studies in which a 45 min treatment of the adsorbent with 0.1 M NaOH at 20 °C was enough to desorb most of the anionic dyes and reuse the adsorbents [7,109]. In many studies, the regeneration of anionic dyes and the reuse of the adsorbents were proven highly feasible. For example, 45 consecutive cycles from sawdust adsorbent were reported [45], while at least 3 cycles were possible by a cotton-gin-trash bioadsorbent [7] and activated carbon prepared from pine cone [62]. Although high desorption efficiencies, such as 98% [16] or 83.7–90.3% dye desorption have been reported [77], a lower desorption rate, such as 35–43%, was also reported when the dyes were attached with adsorbent through a chemical reaction (chemisorption) [15]. However, more than half of the studies on anionic dye removal have considered desorption studies to justify the effectiveness of the proposed adsorbents, which shows the requirement for sustainability. From an economic evaluation, it was reported that the cost for the preparation and regeneration of adsorbents of biomass waste (USD 42.43/kg) was significantly lower than commercially available activated carbon (USD 111.37/kg) [60]. From both the economic and environmental points of view, the adsorbents prepared from natural plant-derived agricultural wastes seem to be more sustainable, as, on the one hand, they reduce the process cost, and, on the other hand, they diminish the amount of waste from the environment. The use of these adsorbents in a circular manner by successful regeneration is further beneficial. When the adsorbents reach their end of life, the dyes can be desorbed and, as the adsorbents are biodegradable, they can be discarded into the land without any hazard. Overall, the extent of the dye-desorption studies conducted along with the adsorption of anionic dyes indicates the urgency of making these bioadsorbents more sustainable and efficient, validating their more feasible use in the future.

9. Conclusions and Future Scopes

Anionic dyes are widely used in different applications and are one of the major contaminants in dye wastewater. Over the years, lignocellulosic agricultural wastes have been widely used to remove these dyes from wastewater. Adsorbing anionic dyes by these adsorbents is challenging, as they inherently possess some repulsing effect on anionic dyes due to their electronegative surfaces. Nevertheless, given the low cost and abundance of these natural resources, researchers have frequently made them capable of anionic dye adsorption by mechanical, chemical, and thermal modifications. In cases in which only mechanical modifications are performed, adjusting the bath solution to a lower pH can enhance the dye adsorption. Reducing the particle size and tuning the temperature to a favorable condition can also promote dye adsorption. Although thermal modification such as transformation into activated carbon, showed an improvement in the adsorption chemically modifying the surface by inducing new cationic groups was found to be the most effective method of anionic dye removal when using these resources. The chemical modification of these biomasses was most successful when using amine-based and quaternary ammonium compounds. It is worth mentioning that quaternary ammonium compounds often possess environmental hazards, while amine-based compounds can also be hazardous, and thus require careful attention. Even though the use of amine-rich nonhazardous biopolymers, such as chitosan and its derivatives, have been well explored for the removal of anionic dyes, they are not widely considered for the modifications of lignocellulosic biomass (with a few exceptions), which could be a future research area. Furthermore, when the efficiency of a chemically modified biomass adsorbent was reported, often it was not parallelly compared with the chemically unmodified biomass. Thus, the degree of improvement in the dye adsorption was often not identified. However, the extent of regeneration studies shows the keen attention of researchers to not only just proposing an adsorbent to separate anionic dyes, but also to proving its sustainability in terms of cost reduction and effectiveness. The success trend portrays the promising future of using these abundant wastes as the key resources for the essential separation of anionic dyes.

Author Contributions: Conceptualization, A.N.M.A.H.; writing—original draft preparation, A.N.M.A.H., N.S., A.S.M.S. and S.A.S.; writing—review and editing, A.N.M.A.H. and N.S.; visualization, A.N.M.A.H., A.S.M.S. and S.A.S. All authors have read and agreed to the published version of the manuscript.

Funding: This research received no external funding.

Institutional Review Board Statement: Not applicable.

Informed Consent Statement: Not applicable.

Data Availability Statement: Not applicable.

Conflicts of Interest: The authors declare no conflict of interest.

References

1. Haque, A.N.M.A.; Remadevi, R.; Wang, X.; Naebe, M. Sorption properties of fabricated film from cotton gin trash. *Mater. Today Proc.* **2020**, *31*, S221–S226. [CrossRef]
2. Hassan, M.M.; Carr, C.M. A critical review on recent advancements of the removal of reactive dyes from dyehouse effluent by ion-exchange adsorbents. *Chemosphere* **2018**, *209*, 201–219. [CrossRef]
3. Lellis, B.; Fávaro-Polonio, C.Z.; Pamphile, J.A.; Polonio, J.C. Effects of textile dyes on health and the environment and bioremediation potential of living organisms. *Biotechnol. Res. Innov.* **2019**, *3*, 275–290. [CrossRef]
4. Salleh, M.A.M.; Mahmoud, D.K.; Karim, W.A.W.A.; Idris, A. Cationic and anionic dye adsorption by agricultural solid wastes: A comprehensive review. *Desalination* **2011**, *280*, 1–13. [CrossRef]
5. Haque, A.N.M.A.; Remadevi, R.; Rojas, O.J.; Wang, X.; Naebe, M. Kinetics and equilibrium adsorption of methylene blue onto cotton gin trash bioadsorbents. *Cellulose* **2020**, *27*, 6485–6504. [CrossRef]
6. Ray, R.C.; Behera, S.S. Solid state fermentation for production of microbial cellulases. In *Biotechnology of Microbial Enzymes*; Elsevier: Amsterdam, The Netherlands, 2017; pp. 43–79.
7. Haque, A.N.M.A.; Remadevi, R.; Wang, X.; Naebe, M. Adsorption of anionic Acid Blue 25 on chitosan-modified cotton gin trash film. *Cellulose* **2020**, *27*, 9437–9456. [CrossRef]
8. Roa, K.; Oyarce, E.; Boulett, A.; ALSamman, M.; Oyarzún, D.; Pizarro, G.D.C.; Sánchez, J. Lignocellulose-based materials and their application in the removal of dyes from water: A review. *Sustain. Mater. Technol.* **2021**, *29*, e00320. [CrossRef]
9. Mishra, S.; Cheng, L.; Maiti, A. The utilization of agro-biomass/byproducts for effective bio-removal of dyes from dyeing wastewater: A comprehensive review. *J. Environ. Chem. Eng.* **2021**, *9*, 104901. [CrossRef]
10. Mo, J.; Yang, Q.; Zhang, N.; Zhang, W.; Zheng, Y.; Zhang, Z. A review on agro-industrial waste (AIW) derived adsorbents for water and wastewater treatment. *J. Environ. Manag.* **2018**, *227*, 395–405. [CrossRef]
11. Rachna, K.; Agarwal, A.; Singh, N. Rice husk and Sodium hydroxide activated Rice husk for removal of Reactive yellow dye from water. *Mater. Today Proc.* **2019**, *12*, 573–580. [CrossRef]
12. Xu, X.; Gao, B.-Y.; Yue, Q.-Y.; Zhong, Q.-Q. Preparation and utilization of wheat straw bearing amine groups for the sorption of acid and reactive dyes from aqueous solutions. *J. Hazard. Mater.* **2010**, *182*, 1–9. [CrossRef] [PubMed]
13. Doltabadi, M.; Alidadi, H.; Davoudi, M. Comparative study of cationic and anionic dye removal from aqueous solutions using sawdust-based adsorbent. *Environ. Prog. Sustain. Energy* **2016**, *35*, 1078–1090. [CrossRef]
14. Abdelghaffar, F. Biosorption of anionic dye using nanocomposite derived from chitosan and silver Nanoparticles synthesized via cellulosic banana peel bio-waste. *Environ. Technol. Innov.* **2021**, *24*, 101852. [CrossRef]
15. Soldatkina, L.; Zavrichko, M. Equilibrium, kinetic, and thermodynamic studies of anionic dyes adsorption on corn stalks modified by cetylpyridinium bromide. *Colloids Interfaces* **2018**, *3*, 4. [CrossRef]
16. Munagapati, V.S.; Kim, D.-S. Adsorption of anionic azo dye Congo Red from aqueous solution by Cationic Modified Orange Peel Powder. *J. Mol. Liq.* **2016**, *220*, 540–548. [CrossRef]

17. Tahir, M.A.; Bhatti, H.N.; Iqbal, M. Solar Red and Brittle Blue direct dyes adsorption onto Eucalyptus angophoroides bark: Equilibrium, kinetics and thermodynamic studies. *J. Environ. Chem. Eng.* **2016**, *4*, 2431–2439. [CrossRef]
18. Wong, S.; Ghafar, N.A.; Ngadi, N.; Razmi, F.A.; Inuwa, I.M.; Mat, R.; Amin, N.A.S. Effective removal of anionic textile dyes using adsorbent synthesized from coffee waste. *Sci. Rep.* **2020**, *10*, 2928. [CrossRef]
19. Broadbent, A.D. *Basic Principles of Textile Coloration*; Society of Dyers and Colorists Bradford: Bradford, UK, 2001; Volume 132.
20. Khatri, A.; Peerzada, M.H.; Mohsin, M.; White, M. A review on developments in dyeing cotton fabrics with reactive dyes fo reducing effluent pollution. *J. Clean. Prod.* **2015**, *87*, 50–57. [CrossRef]
21. Haque, A.N.M.A.; Islam, M.A. The contribution of different vinyl sulphone-reactive dyes to an effluent. *J. Taibah Univ. Sci.* **2015** *9*, 594–600. [CrossRef]
22. Sekar, N. Acid dyes. In *Handbook of Textile and Industrial Dyeing: Principles, Processes and Types of Dyes*; Clark, M., Ed.; Woodhead Publishing Limited: Cambridge, UK, 2011; pp. 486–514.
23. Chakraborty, J. *Fundamentals and Practices in Colouration of Textiles*; CRC Press: Boca Raton, FL, USA, 2015.
24. Haque, A.N.M.A.; Zhang, Y.; Naebe, M. A review on lignocellulose/poly (vinyl alcohol) composites: Cleaner approaches fo greener materials. *Cellulose* **2021**, *28*, 10741–10764. [CrossRef]
25. Haque, A.N.M.A.; Remadevi, R.; Naebe, M. Lemongrass (Cymbopogon): A review on its structure, properties, applications and recent developments. *Cellulose* **2018**, *25*, 5455–5477. [CrossRef]
26. Sanderson, K. Lignocellulose: A chewy problem. *Nature* **2011**, *474*, S12–S14. [CrossRef] [PubMed]
27. Değermenci, G.D.; Değermenci, N.; Ayvaoğlu, V.; Durmaz, E.; Çakır, D.; Akan, E. Adsorption of reactive dyes on lignocellulosic waste; characterization, equilibrium, kinetic and thermodynamic studies. *J. Clean. Prod.* **2019**, *225*, 1220–1229. [CrossRef]
28. Aziz, E.K.; Abdelmajid, R.; Rachid, L.M.; Mohammadine, E.H. Adsorptive removal of anionic dye from aqueous solutions using powdered and calcined vegetables wastes as low-cost adsorbent. *Arab. J. Basic Appl. Sci.* **2018**, *25*, 93–102. [CrossRef]
29. Rodríguez, A.; García, J.; Ovejero, G.; Mestanza, M. Adsorption of anionic and cationic dyes on activated carbon from aqueou solutions: Equilibrium and kinetics. *J. Hazard. Mater.* **2009**, *172*, 1311–1320. [CrossRef]
30. Stavrinou, A.; Aggelopoulos, C.; Tsakiroglou, C. Exploring the adsorption mechanisms of cationic and anionic dyes onto agricultural waste peels of banana, cucumber and potato: Adsorption kinetics and equilibrium isotherms as a tool. *J. Environ. Chem. Eng.* **2018**, *6*, 6958–6970. [CrossRef]
31. Jain, S.N.; Tamboli, S.R.; Sutar, D.S.; Jadhav, S.R.; Marathe, J.V.; Shaikh, A.A.; Prajapati, A.A. Batch and continuous studies for adsorption of anionic dye onto waste tea residue: Kinetic, equilibrium, breakthrough and reusability studies. *J. Clean. Prod.* **2020** *252*, 119778. [CrossRef]
32. El Messaoudi, N.; Dbik, A.; El Khomri, M.; Sabour, A.; Bentahar, S.; Lacherai, A. Date stones of Phoenix dactylifera and jujube shells of Ziziphus lotus as potential biosorbents for anionic dye removal. *Int. J. Phytoremediat.* **2017**, *19*, 1047–1052. [CrossRef]
33. Namasivayam, C.; Prabha, D.; Kumutha, M. Removal of direct red and acid brilliant blue by adsorption on to banana pith *Bioresour. Technol.* **1998**, *64*, 77–79. [CrossRef]
34. Reddy, M.S.; Sivaramakrishna, L.; Reddy, A.V. The use of an agricultural waste material, Jujuba seeds for the removal of anionic dye (Congo red) from aqueous medium. *J. Hazard. Mater.* **2012**, *203*, 118–127. [CrossRef]
35. Tunç, Ö.; Tanacı, H.; Aksu, Z. Potential use of cotton plant wastes for the removal of Remazol Black B reactive dye. *J. Hazard. Mater.* **2009**, *163*, 187–198. [CrossRef] [PubMed]
36. Parimelazhagan, V.; Jeppu, G.; Rampal, N. Continuous Fixed-Bed Column Studies on Congo Red Dye Adsorption-Desorption Using Free and Immobilized Nelumbo nucifera Leaf Adsorbent. *Polymers* **2021**, *14*, 54. [CrossRef] [PubMed]
37. Ben Arfi, R.; Karoui, S.; Mougin, K.; Ghorbal, A. Adsorptive removal of cationic and anionic dyes from aqueous solution by utilizing almond shell as bioadsorbent. *Euro-Mediterr. J. Environ. Integr.* **2017**, *2*, 20. [CrossRef]
38. Farah, J.Y.; Elgendy, N. Performance, kinetics and equilibrium in biosorption of anionic dye Acid Red 14 by the waste biomass of Saccharomyces cerevisiae as a low-cost biosorbent. *Turk. J. Eng. Environ. Sci.* **2013**, *37*, 146–161.
39. Alhujaily, A.; Yu, H.; Zhang, X.; Ma, F. Adsorptive removal of anionic dyes from aqueous solutions using spent mushroom waste *Appl. Water Sci.* **2020**, *10*, 183. [CrossRef]
40. Grabi, H.; Lemlikchi, W.; Derridj, F.; Lemlikchi, S.; Trari, M. Efficient native biosorbent derived from agricultural waste precurso for anionic dye adsorption in synthetic wastewater. *Biomass Convers. Biorefinery* **2021**, 1–18. [CrossRef]
41. Grabi, H.; Derridj, F.; Lemlikchi, W.; Guénin, E. Studies of the potential of a native natural biosorbent for the elimination of an anionic textile dye Cibacron Blue in aqueous solution. *Sci. Rep.* **2021**, *11*, 9705. [CrossRef]
42. Banerjee, S.; Dastidar, M. Use of jute processing wastes for treatment of wastewater contaminated with dye and other organics *Bioresour. Technol.* **2005**, *96*, 1919–1928. [CrossRef]
43. Mbarki, F.; Kesraoui, A.; Seffen, M.; Ayrault, P. Kinetic, thermodynamic, and adsorption behavior of cationic and anionic dyes onto corn stigmata: Nonlinear and stochastic analyses. *Water Air Soil Pollut.* **2018**, *229*, 95. [CrossRef]
44. Jin, L.; Li, W.; Xu, Q.; Sun, Q. Amino-functionalized nanocrystalline cellulose as an adsorbent for anionic dyes. *Cellulose* **2015**, *22* 2443–2456. [CrossRef]
45. Lyu, W.; Yu, M.; Feng, J.; Yan, W. Highly crystalline polyaniline nanofibers coating with low-cost biomass for easy separation and high efficient removal of anionic dye ARG from aqueous solution. *Appl. Surf. Sci.* **2018**, *458*, 413–424. [CrossRef]
46. Zhou, T.; Lu, W.; Liu, L.; Zhu, H.; Jiao, Y.; Zhang, S.; Han, R. Effective adsorption of light green anionic dye from solution by CPB modified peanut in column mode. *J. Mol. Liq.* **2015**, *211*, 909–914. [CrossRef]

7. Su, Y.; Zhao, B.; Xiao, W.; Han, R. Adsorption behavior of light green anionic dye using cationic surfactant-modified wheat straw in batch and column mode. *Environ. Sci. Pollut. Res.* **2013**, *20*, 5558–5568. [CrossRef] [PubMed]
8. Oei, B.C.; Ibrahim, S.; Wang, S.; Ang, H.M. Surfactant modified barley straw for removal of acid and reactive dyes from aqueous solution. *Bioresour. Technol.* **2009**, *100*, 4292–4295. [CrossRef] [PubMed]
9. Ibrahim, S.; Fatimah, I.; Ang, H.-M.; Wang, S. Adsorption of anionic dyes in aqueous solution using chemically modified barley straw. *Water Sci. Technol.* **2010**, *62*, 1177–1182. [CrossRef] [PubMed]
0. Ansari, R.; Seyghali, B.; Mohammad-Khah, A.; Zanjanchi, M.A. Highly efficient adsorption of anionic dyes from aqueous solutions using sawdust modified by cationic surfactant of cetyltrimethylammonium bromide. *J. Surfactants Deterg.* **2012**, *15*, 557–565. [CrossRef]
1. Namasivayam, C.; Sureshkumar, M. Anionic dye adsorption characteristics of surfactant-modified coir pith, a 'waste'lignocellulosic polymer. *J. Appl. Polym. Sci.* **2006**, *100*, 1538–1546. [CrossRef]
2. Jiang, Z.; Hu, D. Molecular mechanism of anionic dyes adsorption on cationized rice husk cellulose from agricultural wastes. *J. Mol. Liq.* **2019**, *276*, 105–114. [CrossRef]
3. Koay, Y.S.; Ahamad, I.S.; Nourouzi, M.M.; Abdullah, L.C.; Choong, T.S.Y. Development of novel low-cost quaternized adsorbent from palm oil agriculture waste for reactive dye removal. *BioResources* **2014**, *9*, 66–85. [CrossRef]
4. Benabbas, K.; Zabat, N.; Hocini, I. Study of the chemical pretreatment of a nonconventional low-cost biosorbent (Callitriche obtusangula) for removing an anionic dye from aqueous solution. *Euro-Mediterr. J. Environ. Integr.* **2021**, *6*, 54. [CrossRef]
5. Noreen, S.; Bhatti, H.N.; Nausheen, S.; Sadaf, S.; Ashfaq, M. Batch and fixed bed adsorption study for the removal of Drimarine Black CL-B dye from aqueous solution using a lignocellulosic waste: A cost affective adsorbent. *Ind. Crops Prod.* **2013**, *50*, 568–579. [CrossRef]
6. Won, S.W.; Kim, H.-J.; Choi, S.-H.; Chung, B.-W.; Kim, K.-J.; Yun, Y.-S. Performance, kinetics and equilibrium in biosorption of anionic dye Reactive Black 5 by the waste biomass of Corynebacterium glutamicum as a low-cost biosorbent. *Chem. Eng. J.* **2006**, *121*, 37–43. [CrossRef]
7. Saputra, O.A.; Nauqinida, M.; Pujiasih, S.; Kusumaningsih, T.; Pramono, E. Improvement of anionic and cationic dyes removal in aqueous solution by Indonesian agro-waste oil palm empty fruit bunches through silylation approach. *Groundw. Sustain. Dev.* **2021**, *13*, 100570. [CrossRef]
8. Lin, Q.; Wang, K.; Gao, M.; Bai, Y.; Chen, L.; Ma, H. Effectively removal of cationic and anionic dyes by pH-sensitive amphoteric adsorbent derived from agricultural waste-wheat straw. *J. Taiwan Inst. Chem. Eng.* **2017**, *76*, 65–72. [CrossRef]
9. Velinov, N.; Radović Vučić, M.; Petrović, M.; Najdanović, S.; Kostić, M.; Mitrović, J.; Bojić, A. The influence of various solvents' polarity in the synthesis of wood biowaste sorbent: Evaluation of dye sorption. *Biomass Convers. Biorefinery* **2021**, 1–12. [CrossRef]
0. Ahmaruzzaman, M.; Reza, R.A. Decontamination of cationic and anionic dyes in single and binary mode from aqueous phase by mesoporous pulp waste. *Environ. Prog. Sustain. Energy* **2015**, *34*, 724–735. [CrossRef]
1. Demiral, H.; Demiral, I.; Karabacakoğlu, B.; Tümsek, F. Adsorption of textile dye onto activated carbon prepared from industrial waste by ZnCl2 activation. *J. Int. Environ. Appl. Sci.* **2008**, *3*, 381–389.
2. Bhomick, P.C.; Supong, A.; Baruah, M.; Pongener, C.; Sinha, D. Pine Cone biomass as an efficient precursor for the synthesis of activated biocarbon for adsorption of anionic dye from aqueous solution: Isotherm, kinetic, thermodynamic and regeneration studies. *Sustain. Chem. Pharm.* **2018**, *10*, 41–49. [CrossRef]
3. Zhao, B.; Xiao, W.; Shang, Y.; Zhu, H.; Han, R. Adsorption of light green anionic dye using cationic surfactant-modified peanut husk in batch mode. *Arab. J. Chem.* **2017**, *10*, S3595–S3602. [CrossRef]
4. Velinov, N.; Mitrović, J.; Kostić, M.; Radović, M.; Petrović, M.; Bojić, D.; Bojić, A. Wood residue reuse for a synthesis of lignocellulosic biosorbent: Characterization and application for simultaneous removal of copper (II), Reactive Blue 19 and cyprodinil from water. *Wood Sci. Technol.* **2019**, *53*, 619–647. [CrossRef]
5. Ma, H.; Li, J.-B.; Liu, W.-W.; Miao, M.; Cheng, B.-J.; Zhu, S.-W. Novel synthesis of a versatile magnetic adsorbent derived from corncob for dye removal. *Bioresour. Technol.* **2015**, *190*, 13–20. [CrossRef] [PubMed]
6. Markets and Markets, Activated Carbon Market by Type, Application (Liquid Phase (Water Treatment, Foods & Beverages, Pharmaceutical & Medical), Gas Phase (Industrial, Automotive), and Region (APAC, North America, Europe, Middle East, South America)—Global Forecast to 2026. 2022. Available online: https://www.marketsandmarkets.com/Market-Reports/activated-carbon-362.html (accessed on 15 August 2022).
7. Ao, W.; Fu, J.; Mao, X.; Kang, Q.; Ran, C.; Liu, Y.; Zhang, H.; Gao, Z.; Li, J.; Liu, G. Microwave assisted preparation of activated carbon from biomass: A review. *Renew. Sustain. Energy Rev.* **2018**, *92*, 958–979. [CrossRef]
8. Büchel, K.H.; Moretto, H.-H.; Werner, D. *Industrial Inorganic Chemistry*; John Wiley & Sons: Hoboken, NJ, USA, 2008.
9. Bhatnagar, A.; Hogland, W.; Marques, M.; Sillanpää, M. An overview of the modification methods of activated carbon for its water treatment applications. *Chem. Eng. J.* **2013**, *219*, 499–511. [CrossRef]
0. Belyaeva, O.; Krasnova, T.; Semenova, S. Effect of modification of granulated activated carbons with ozone on their properties. *Russ. J. Appl. Chem.* **2011**, *84*, 597–601. [CrossRef]
1. Delamar, M.; Desarmot, G.; Fagebaume, O.; Hitmi, R.; Pinsonc, J.; Savéant, J.-M. Modification of carbon fiber surfaces by electrochemical reduction of aryl diazonium salts: Application to carbon epoxy composites. *Carbon* **1997**, *35*, 801–807. [CrossRef]
2. García, A.B.; Martínez-Alonso, A.; y Leon, C.A.L.; Tascón, J.M. Modification of the surface properties of an activated carbon by oxygen plasma treatment. *Fuel* **1998**, *77*, 613–624. [CrossRef]

73. Otake, Y.; Jenkins, R.G. Characterization of oxygen-containing surface complexes created on a microporous carbon by air an nitric acid treatment. *Carbon* **1993**, *31*, 109–121. [CrossRef]
74. Silva, A.R.; Freire, C.; De Castro, B.; Freitas, M.; Figueiredo, J. Anchoring of a nickel (II) Schiff base complex onto activated carbo mediated by cyanuric chloride. *Microporous Mesoporous Mater.* **2001**, *46*, 211–221. [CrossRef]
75. Theamwong, N.; Intarabumrung, W.; Sangon, S.; Aintharabunya, S.; Ngernyen, Y.; Hunt, A.J.; Supanchaiyamat, N. Activate carbons from waste Cassia bakeriana seed pods as high-performance adsorbents for toxic anionic dye and ciprofloxacin antibioti remediation. *Bioresour. Technol.* **2021**, *341*, 125832. [CrossRef]
76. Nayak, A.; Bhushan, B.; Gupta, V.; Sharma, P. Chemically activated carbon from lignocellulosic wastes for heavy metal wastewate remediation: Effect of activation conditions. *J. Colloid Interface Sci.* **2017**, *493*, 228–240. [CrossRef]
77. Saha, A.; Basak, B.B.; Ponnuchamy, M. Performance of activated carbon derived from Cymbopogon winterianus distillatio waste for scavenging of aqueous toxic anionic dye Congo red: Comparison with commercial activated carbon. *Sep. Sci. Techno* **2020**, *55*, 1970–1983. [CrossRef]
78. Alhogbi, B.G.; Altayeb, S.; Bahaidarah, E.; Zawrah, M.F. Removal of anionic and cationic dyes from wastewater using activate carbon from palm tree fiber waste. *Processes* **2021**, *9*, 416. [CrossRef]
79. Extross, A.; Waknis, A.; Tagad, C.; Gedam, V.; Pathak, P. Adsorption of congo red using carbon from leaves and stem of wate hyacinth: Equilibrium, kinetics, thermodynamic studies. *Int. J. Environ. Sci. Technol.* **2022**, 1–38. [CrossRef]
80. Ravenni, G.; Cafaggi, G.; Sárossy, Z.; Nielsen, K.R.; Ahrenfeldt, J.; Henriksen, U. Waste chars from wood gasification an wastewater sludge pyrolysis compared to commercial activated carbon for the removal of cationic and anionic dyes from aqueou solution. *Bioresour. Technol. Rep.* **2020**, *10*, 100421. [CrossRef]
81. Paredes-Laverde, M.; Salamanca, M.; Diaz-Corrales, J.D.; Flórez, E.; Silva-Agredo, J.; Torres-Palma, R.A. Understanding th removal of an anionic dye in textile wastewaters by adsorption on ZnCl2 activated carbons from rice and coffee husk wastes: combined experimental and theoretical study. *J. Environ. Chem. Eng.* **2021**, *9*, 105685. [CrossRef]
82. Güzel, F.; Sayğılı, H.; Sayğılı, G.A.; Koyuncu, F. New low-cost nanoporous carbonaceous adsorbent developed from caro (Ceratonia siliqua) processing industry waste for the adsorption of anionic textile dye: Characterization, equilibrium and kineti modeling. *J. Mol. Liq.* **2015**, *206*, 244–255. [CrossRef]
83. Malik, P.K. Use of activated carbons prepared from sawdust and rice-husk for adsorption of acid dyes: A case study of Aci Yellow 36. *Dye. Pigment.* **2003**, *56*, 239–249. [CrossRef]
84. Zubair, M.; Mu'azu, N.D.; Jarrah, N.; Blaisi, N.I.; Aziz, H.A.; A Al-Harthi, M. Adsorption behavior and mechanism of methylen blue, crystal violet, eriochrome black T, and methyl orange dyes onto biochar-derived date palm fronds waste produced a different pyrolysis conditions. *Water Air Soil Pollut.* **2020**, *231*, 240. [CrossRef]
85. Periyaraman, P.M.; Karan, S.; Ponnusamy, S.K.; Vaidyanathan, V.; Vasanthakumar, S.; Dhanasekaran, A.; Subramanian, Adsorption of an anionic dye onto native and chemically modified agricultural waste. *Environ. Eng. Manag. J.* **2019**, *18*, 257–27
86. Bouhadjra, K.; Lemlikchi, W.; Ferhati, A.; Mignard, S. Enhancing removal efficiency of anionic dye (Cibacron blue) using wast potato peels powder. *Sci. Rep.* **2021**, *11*, 2090. [CrossRef]
87. Sayğılı, H.; Güzel, F.; Önal, Y. Conversion of grape industrial processing waste to activated carbon sorbent and its performance i cationic and anionic dyes adsorption. *J. Clean. Prod.* **2015**, *93*, 84–93. [CrossRef]
88. Mittal, A.; Mittal, J.; Malviya, A.; Gupta, V. Adsorptive removal of hazardous anionic dye "Congo red" from wastewater usin waste materials and recovery by desorption. *J. Colloid Interface Sci.* **2009**, *340*, 16–26. [CrossRef]
89. Landin-Sandoval, V.; Mendoza-Castillo, D.; Seliem, M.; Mobarak, M.; Villanueva-Mejia, F.; Bonilla-Petriciolet, A.; Navarro-Santo P.; Reynel-Ávila, H. Physicochemical analysis of multilayer adsorption mechanism of anionic dyes on lignocellulosic biomasse via statistical physics and density functional theory. *J. Mol. Liq.* **2021**, *322*, 114511. [CrossRef]
90. Demirbas, E.; Nas, M. Batch kinetic and equilibrium studies of adsorption of Reactive Blue 21 by fly ash and sepiolite. *Desalinatio* **2009**, *243*, 8–21. [CrossRef]
91. Vadivelan, V.; Kumar, K.V. Equilibrium, kinetics, mechanism, and process design for the sorption of methylene blue onto ric husk. *J. Colloid Interface Sci.* **2005**, *286*, 90–100. [CrossRef]
92. Tanzifi, M.; Yaraki, M.T.; Kiadehi, A.D.; Hosseini, S.H.; Olazar, M.; Bharti, A.K.; Agarwal, S.; Gupta, V.K.; Kazemi, A. Adsorptio of Amido Black 10B from aqueous solution using polyaniline/SiO2 nanocomposite: Experimental investigation and artificia neural network modeling. *J. Colloid Interface Sci.* **2018**, *510*, 246–261. [CrossRef] [PubMed]
93. Senthil Kumar, P.; Palaniyappan, M.; Priyadharshini, M.; Vignesh, A.; Thanjiappan, A.; Sebastina Anne Fernando, P.; Tanvi Ahmed, R.; Srinath, R. Adsorption of basic dye onto raw and surface-modified agricultural waste. *Environ. Prog. Sustain. Energ* **2014**, *33*, 87–98. [CrossRef]
94. Iakovleva, E.; Sillanpää, M.; Maydannik, P.; Liu, J.T.; Allen, S.; Albadarin, A.B.; Mangwandi, C. Manufacturing of novel low-cos adsorbent: Co-granulation of limestone and coffee waste. *J. Environ. Manag.* **2017**, *203*, 853–860. [CrossRef] [PubMed]
95. Batool, F.; Akbar, J.; Iqbal, S.; Noreen, S.; Bukhari, S.N.A. Study of isothermal, kinetic, and thermodynamic parameters fo adsorption of cadmium: An overview of linear and nonlinear approach and error analysis. *Bioinorg. Chem. Appl.* **2018**, *201* 3463724. [CrossRef]
96. Dada, A.O.; Latona, D.F.; Ojediran, O.J.; Nath, O.O. Adsorption of Cu (II) onto bamboo supported manganese (BS-Mn) nanocon posite: Effect of operational parameters, kinetic, isotherms, and thermodynamic studies. *J. Appl. Sci. Environ. Manag.* **2016**, *2* 409–422. [CrossRef]

7. Kezerle, A.; Velic, N.; Hasenay, D.; Kovacevic, D. Lignocellulosic materials as dye adsorbents: Adsorption of methylene blue and Congo red on Brewers' spent grain. *Croat. Chem. Acta* **2018**, *91*, 53–65. [CrossRef]
3. Farzana, N.; Uddin, M.Z.; Haque, M.M.; Haque, A.N.M.A. Dyeability, kinetics and physico-chemical aspects of Bombyx mori muslin silk fabric with bi-functional reactive dyes. *J. Nat. Fibers* **2018**, *17*, 986–1000. [CrossRef]
9. Haque, A.N.M.A.; Hussain, M.; Smriti, S.A.; Siddiqa, F.; Farzana, N. Kinetic Study of Curcumin on Modal Fabric. *Tekstilec* **2018**, *61*, 27–32. [CrossRef]
)0. Haque, A.N.M.A.; Hussain, M.; Siddiqa, F.; Haque, M.; Islam, G. Adsorption Kinetics of Curcumin on Cotton Fabric. *Tekstilec* **2018**, *61*, 76–81. [CrossRef]
)1. Hubbe, M.A.; Azizian, S.; Douven, S. Implications of apparent pseudo-second-order adsorption kinetics onto cellulosic materials: A review. *BioResources* **2019**, *14*, 7582–7626. [CrossRef]
02. El-Khaiary, M.I.; Malash, G.F.; Ho, Y.-S. On the use of linearized pseudo-second-order kinetic equations for modeling adsorption systems. *Desalination* **2010**, *257*, 93–101. [CrossRef]
03. Ho, Y.-S. Second-order kinetic model for the sorption of cadmium onto tree fern: A comparison of linear and non-linear methods. *Water Res.* **2006**, *40*, 119–125. [CrossRef]
)4. Ho, Y.-S. Isotherms for the sorption of lead onto peat: Comparison of linear and non-linear methods. *Pol. J. Environ. Stud.* **2006**, *15*, 81–86.
)5. Perez-Ameneiro, M.; Vecino, X.; Cruz, J.; Moldes, A. Wastewater treatment enhancement by applying a lipopeptide biosurfactant to a lignocellulosic biocomposite. *Carbohydr. Polym.* **2015**, *131*, 186–196. [CrossRef]
06. Islam, G.N.; Ke, G.; Haque, A.N.M.A.; Islam, A. Effect of ultrasound on dyeing of wool fabric with acid dye. *Int. J. Ind. Chem.* **2017**, *8*, 425–431. [CrossRef]
07. Roy, A.; Adhikari, B.; Majumder, S. Equilibrium, kinetic, and thermodynamic studies of azo dye adsorption from aqueous solution by chemically modified lignocellulosic jute fiber. *Ind. Eng. Chem. Res.* **2013**, *52*, 6502–6512. [CrossRef]
08. Haque, A.N.M.A.; Farzana, N.; Smriti, S.; Islam, M. Kinetics and thermodynamics of silk dyeing with turmeric extract. *AATCC J. Res.* **2018**, *5*, 8–14. [CrossRef]
09. Ibrahim, S.M.; Badawy, A.A.; Essawy, H.A. Improvement of dyes removal from aqueous solution by Nanosized cobalt ferrite treated with humic acid during coprecipitation. *J. Nanostructure Chem.* **2019**, *9*, 281–298. [CrossRef]

sustainability

MDPI

Article

Sustainable Dyeing Process for Nylon 6 Fabrics by Rhubarb Flower Using Different Bio-Mordants

Fatemeh Shahmoradi Ghaheh [1], Aminoddin Haji [2,*] and Elaheh Daneshvar [3]

1 Department of Textile Engineering, Faculty of Environmental Science, Urmia University of Technology, Urmia 5756151818, Iran; f.shahmoradi@uut.ac.ir
2 Department of Textile Engineering, Yazd University, Yazd 8915818411, Iran
3 Colour & Imaging Lab, School of Computing Sciences, University of East Anglia, Norwich NR4 7TJ, UK; e.daneshvar@uea.ac.uk
* Correspondence: ahaji@yazd.ac.ir

Abstract: The purpose of this study is to propose a fully sustainable dyeing process for nylon 6. In order to achieve this goal, Rhubarb flower parts were used to produce a brown hue on nylon 6 fabric. The effects of dyeing parameters such as dyeing time, temperature, dyebath pH, M:L, salt addition, dispersing agent, and dye concentration on color strength were investigated. Using 100%owf dye in an acidic medium at boil and the material to liquor ratio of 1:30 for 75 min was determined to be the optimal condition for dyeing nylon 6 with rhubarb flower. In order to achieve acceptable color fastness, four natural mordants were applied, including walnut husks, pistachio hulls, pine cones and green coffee. Colorimetric measurements revealed that mordanting did not affect the hue of the color compared to the non-mordant sample. In addition, diverse natural mordants produced the same color (i.e., brown) with varying color strengths, of which 10%owf walnut husk generated the strongest color. Bio-mordanted samples were also found to have excellent color fastness, thereby providing an effective substitute for metal mordants.

Keywords: natural dyeing; polyamide; eco-friendly; tannin; color strength; fastness

check for updates

Citation: Shahmoradi Ghaheh, F.; Haji, A.; Daneshvar, E. Sustainable Dyeing Process for Nylon 6 Fabrics by Rhubarb Flower Using Different Bio-Mordants. *Sustainability* **2023**, *15*, 9232. https://doi.org/10.3390/su15129232

Academic Editor: Dario Donno

Received: 4 May 2023
Revised: 4 June 2023
Accepted: 5 June 2023
Published: 7 June 2023

1. Introduction

In today's world, industry and scholars are emphasizing sustainability, and the textile industry is striving to adopt this approach. Dyeing is an essential process in the textile industry that utilizes numerous synthetic dyes. However, chemicals and synthetic dyes which make up the majority of those used in the textile industry, result in the emission of vast quantities of highly polluted wastewater. Therefore, there is a current research focus on developing eco-friendly substitutes for synthetic dyes. Various reports have been published on the coloration and functionalization of different textile fibers using natural dyes. However, most studies have been focused on natural fibers such as wool, silk, and cotton [1–3], while the number of studies on natural dyeing of synthetic fibers is comparably low. Recently, several attempts have been made to apply natural dyes on synthetic fibers such as polyester [4–6].

Nylon fibers have been a game-changer in the textile and other industries since their invention in the 1930s. These synthetic fibers are known for their durability, strength, and versatility, making them ideal for a wide range of applications. In the textile industry, nylon fibers are used to create fabrics that are lightweight, breathable, and resistant to wear and tear. They are also used in the production of carpets, upholstery, and other household items. Nylon fibers have also found their way into other industries such as automotive manufacturing, where they are used to make seat belts and airbags. Nylon fibers are commonly dyed using synthetic dyes due to their ability to produce bright and vibrant colors. However, the use of synthetic dyes can have negative impacts on the environment and human health. Natural dyes, on the other hand, are derived from plant or animal sources and are considered to be more sustainable and eco-friendly. They also have the potential to produce unique and subtle shades that cannot be achieved with synthetic dyes. The use of natural dyes for dyeing of nylon fibers is important in promoting sustainable textile practices and reducing the environmental impact of textile production [7,8]. In this regard, there are some publications on utilizing natural dyes for polyamides such as madder [9,10], berberine (*Berberis vulgaris*) [11], weld and pomegranate peel [12], Henna leaves [13], onion outer shell [14], curcumin and saffron [15], mangrove bark [16], leaves of the almond tree [17], and betel nut (*Areca catechu*) [18]. However, the common point among all these natural dyes is that they were utilized with various metal mordants resulting in wastewater pollution. Nevertheless, Esfand (*P. harmala*) was employed as a natural dye in nylon dyeing and numerous bio-mordants were suggested as alternatives to the metal mordants [19]. Furthermore, in our previous study, we used the dragon's blood resin together with different bio-mordants to dye nylon 6 fabric. Our previous findings revealed that artemisia displayed the best efficiency of bio-mordanting in nylon 6 dyeing with dragon's blood resin [20]. Therefore, to mitigate the hazardous consequences of the dyeing processes, researchers are now replacing metal mordants with bio-mordants.

The introduction of new natural sources for dyes has the potential to expand the range of colors available and introduce new specific features. Rhubarb (*Rheum ribes* L.) is a member of the polygonaceae family, native to Asia, particularly Iran, Pakistan, India, and China. Rhubarb contains different tannoids such as Catechin, Glucogalline, Tetrarine, and Gallic acid, among others. Previous studies have exclusively used rhubarb (i.e., root part) as a natural colorant for natural fibers such as wool [21–23] and cotton [24]. However, there has been no research on applying rhubarb flowers on synthetic fibers. Rhubarb flowers are produced in large clusters at the end of the main stem and have a red hue due to various anthraquinone derivatives present in them. Among the various anthraquinone derivatives identified in the rhubarb flower, chrysophanol and emodin are the most abundant coloring compounds (Figure 1) [25–27]. These derivatives have great potential for textile dyeing. This paper explores using the flower component of rhubarb to achieve a deep brown hue on nylon 6 fabric. In addition, the importance of new bio-mordants to achieve a completely eco-friendly process for dyeing synthetic fibers is investigated. The objective of this study is to introduce a fully eco-friendly dyeing process for nylon 6 fabric using plant dye and mordants, exhibiting similar hues and comparable fastness properties with the metal mordants.

Figure 1. *Cont.*

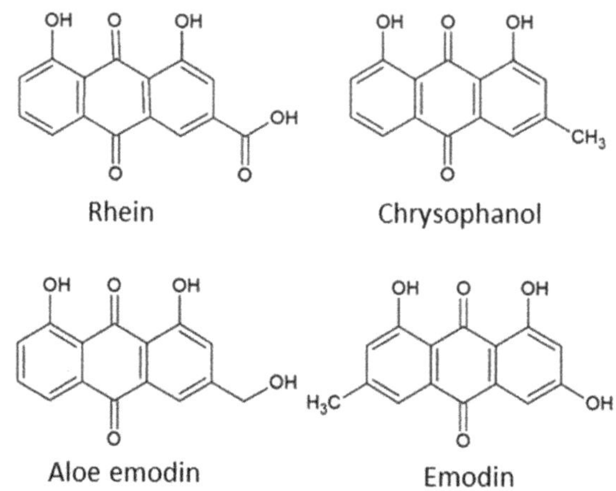

Figure 1. Rhubarb flower and its important coloring compounds.

2. Experimental Section

2.1. Materials

Naz Harir Co., Mashhad, Iran, provided a nylon 6 warp knitted raw fabric (60 g/m^2 Dried *Rheum ribes* L. flowers, collected in spring 2022, were purchased from a local store i Tehran, Iran. Four different bio-mordants which contain high amounts of polyphenols suc as tannin, ellagic acid, and chlorogenic acid were utilized. Among them, pistachio hul walnut husk, and pinecone (collected in autumn 2022) were procured from a local stor in Tehran, Iran and green coffee was an import from Vietnam. All plant materials wer stored in their original form in a cool and dry place, away from sunlight, and were usec for dyeing in less than 6 months from the date of collecting. Furthermore, common meta salts including iron(II) sulphate ($FeSO_4$), aluminium potassium sulphate ($KAl(SO_4)_2$), zin sulfate ($ZnSO_4$), copper sulphate ($CuSO_4$) and sodium sulphate (Na_2SO_4) were supplie by Merck (Germany). To adjust the pH of the dyebath, acetic acid and sodium hydroxid were purchased from Merck (Darmstadt, Germany). Nonionic detergent (Welly wash) an dispersing agent (Heptamol SHT) were provided by Kohan Taj Kimia, Iran.

2.2. Dye and Bio-Mordant Extraction

Initially, the rhubarb flowers were dried and then pulverized into a fine powder usin a mechanical grinder. The powder was then sifted through a 40-mesh nylon screen. Nex 10 g of the rhubarb flower powder was extracted by using 200 mL of distilled water i a Soxhlet extractor for an hour. The resulting solution was filtered through a Whatma filter No. 42, and the filtrate was used as the stock solution. The same process was used t extract the bio-mordants.

2.3. Pre-Mordanting

Initially, the nylon 6 fabrics underwent scouring at 60 °C that employed distille water and a nonionic detergent (1 g/L) with a material to liquor ratio of 40:1 for 30 mi Subsequently, the fabrics were rinsed with tap water and left to air dry. Bio-mordant bath were prepared by adding varying amounts of mordant extracts (10, 20, and 50%owf) i water at the material to liquor ratio (M:L) of 1:40. The wetted fabrics were then immerse in the mordanting solutions and the temperature was gradually raised to the boiling poin at a rate of 2 °C/min. Mordanting was continued at the boil for an hour before rinsing th mordanted fabrics with tap water and drying them at room temperature.

To compare the effectiveness of bio-mordants with that of metal mordants, severa common metal salts were employed. Metal mordant solutions using the aforementione metal salts were prepared by adding 1 %owf in water at a liquor ratio of 1:30, and metal mordanting was performed similarly to the bio-mordanting procedure.

2.4. Dyeing Process

Mordanted fabrics were immersed in the dyeing solution and the dyeing was performed using exhaustion method under different conditions. In this research, dyeing variables including temperature, time, pH, M:L, dye concentration, salt addition, and dispersing agent were optimized through experimental tests. Therefore, the nylon 6 fabric was dyed at 40 °C, 60 °C, 80 °C, and 100 °C to evaluate the optimum dyeing temperature. Dyeing time was optimized at intervals of 15, 30, 45, 60, 75, and 90 min. Dye concentration was varying at different percentages including 10, 20, 30, 40, 50, and 100 %owf (on the weight of fabric). The pH of dyebath was varying at 3, 5, 7, 9, and 11, while different levels of M:L, i.e., 1:10, 1:20, 1:30, 1:40, and 1:50 were applied. In addition, different concentrations of dispersing agent were used (0, 0.5, and 1 g/L) to determine their effect on dye exhaustion. Different amounts of sodium sulfate (1, and 2.5 g/L) were applied to determine the effect of salt addition on the dyeing levelness. The temperature of the dyebath was raised from 25 °C to the desired temperature during with the rate of 2 °C/min. In addition, to adjust the pH of the dye bath, a few drops of acetic acid (1% *v/v*) or sodium carbonate (1% *w/v*) were used. A schematic presentation of the dyeing procedure is shown in Figure 2 [20].

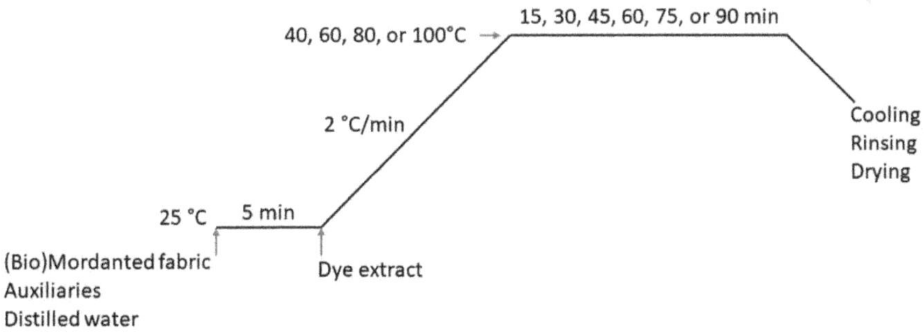

Figure 2. Schematic presentation of the dyeing procedure.

2.5. Color Measurement

The reflectance spectra of dyed samples were measured using a Color-Eye 7000 A spectrophotometer (X-rite, Grand Rapids, MI, USA) in the visible light range from 400 to 750 nm with 10 nm steps. Then, the color strength (K/S value) of the samples was calculated using the Kubelka–Munk equation according to Equation (1),

$$\frac{K}{S} = \frac{(1-R)^2}{2R}, \quad (1)$$

where R is the reflectance value at each wavelength, K is the absorbance coefficient, and S is the scattering coefficient. In addition, the CIEL*a*b* coordinates of the samples under CIE standard illuminant D65 and CIE 1964 standard observer were calculated.

2.6. Color Fastness

The wash fastness of the dyed samples was measured according to the ISO 105-C06 (A1S):2010 method [28] using a Launderometer. The light fastness was assessed according to the ISO 105-B02:2013 method [29].

3. Results and Discussion

This section first discusses the optimization of the variables of dyeing of nylon 6 fabric using Rhubarb flower without applying any mordant. Then, the efficacy of the bio-mordants as eco-friendly alternatives for metal mordants is described. Lastly, the remarkable color fastness of the samples dyed in the presence of bio-mordants is presented.

3.1. Dyeing Parameters Optimization

Figure 3 shows the effect of dye bath pH in the visible range of wavelengths. The effect of dye bath pH was investigated to achieve maximum color strength of the dyed samples. The color strength decreased as the pH increased from 3 up to 11. In fact, maximum color strength was achieved when dyeing was performed at pH 3. It has been shown that the point of zero charge for nylon 6 fiber is around pH 4–5 [30]. Above this pH, the surface charge of the fibers is negative, which repels the anionic dye molecules, while at pH values less than 5, the -NH and NH_2 groups of polyamide chain gain positive charge, which attracts the anionic dye molecules causing an increase in dye exhaustion.

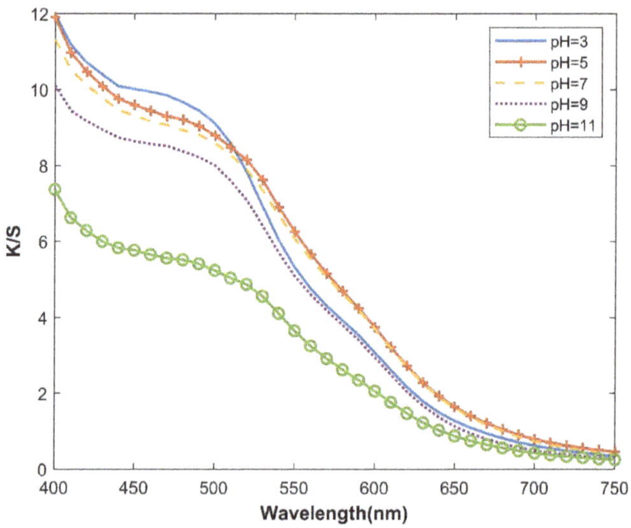

Figure 3. The effect of dye bath pH on color strength (dyeing with 50%owf dye at M:L = 1:40, and 100 °C for 1 h, without salt and dispersing agent).

The influence of dyeing temperature on color strength (K/S) was investigated for 60 min at temperatures ranging from 40 °C to 100 °C, and the results are shown in Figure 4. The color strength clearly increased as the dyeing temperature increased, and thus the maximum color strength was obtained at 100 °C. As the temperature rises, the polymeric structure of nylon 6 opens up [24] and more dye molecules becomes separated from the aggregates [31], making dye adsorption easier and yielding a greater K/S value.

According to Figure 5, the results of changing dyeing time show that the nylon 6 fabric's color strength increased until the dye exhaustion reached equilibrium at 75 min. It is well known that extending the dyeing time may increase the opportunity for dye molecules to be placed in the dyeing sites of polyamide fibers [32]. Therefore, the optimal time for dyeing nylon 6 fabrics was discovered to be 75 min since after increasing the dyeing time to 90 min, the K/S value decreased. Decreasing K/S values over a longer time may be attributed to the shift in the equilibrium leading the dye molecules to migrate from the fabric into the dye bath [33].

As for dye concentration, according to Figure 6, as the dye concentration increased, the K/S values gradually increased until at 100 %owf an extreme shift occurred, and the maximum color strength was yielded. This may reveal that the saturation point of nylon 6 has not been reached even when using 100 %owf. In fact, the defect of natural dyes is that the gaining of fiber saturation occurs at high dye concentrations [10].

Figure 7 shows the effect of the M:L ratio on the K/S values for the wavelength of 400 nm to 750 nm with 10 nm intervals. If we consider K/S curves as a metric for determining the optimal parameters, increase in M:L causes an increase in K/S until the M:L = 1:30 has the highest K/S curve. However, a further increase in M:L achieves lower K/S due to the dilution of the dye solution phase.

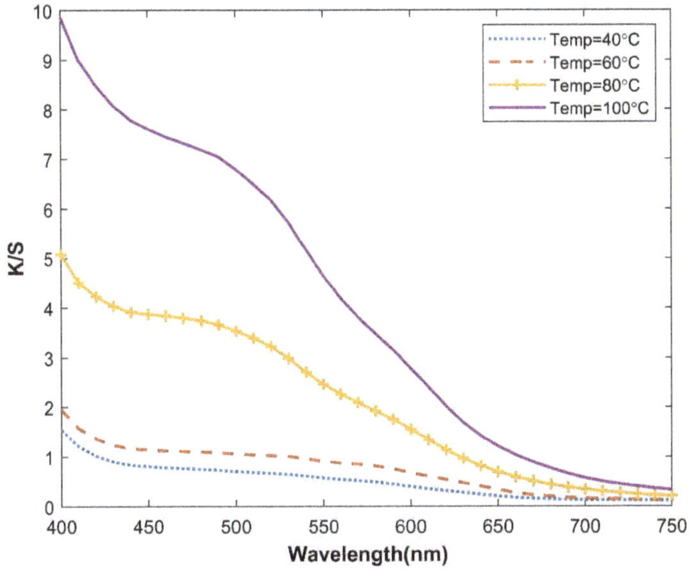

Figure 4. The effect of dyeing temperature on the color strength of dyed samples (dyeing with 50%owf dye at M:L = 1:40 and pH = 3 for 1 h without salt and dispersing agent).

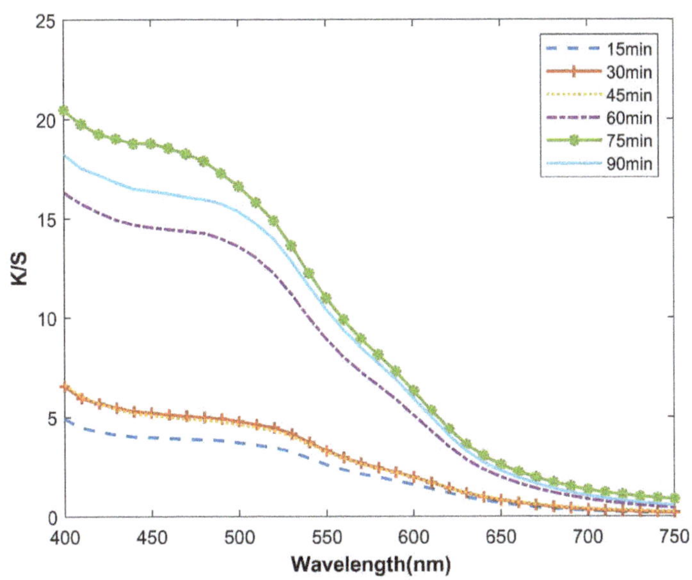

Figure 5. The effect of dyeing time on the color strength of dyed samples (dyeing using 50%owf dye at M:L = 1:40, 100 °C, and pH = 3, without salt and dispersing agent).

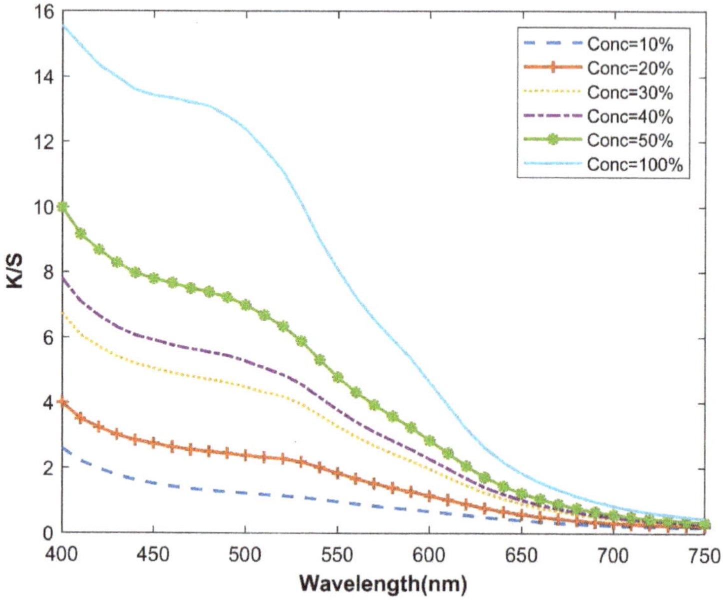

Figure 6. The effect of dye concentration on color strength (dyeing at M:L = 1:40, 100 °C, and pH = 3 for 75 min, without salt and dispersing agent).

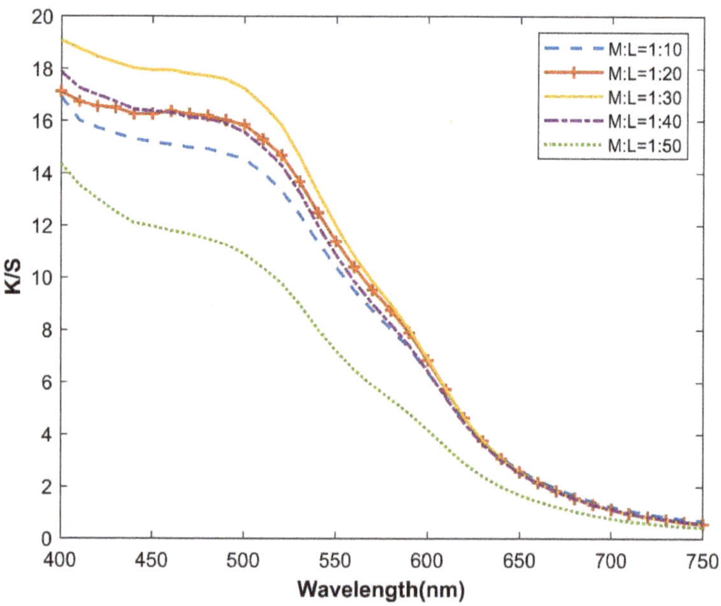

Figure 7. The effect of M:L ratio on the color strength of dyed samples (dyeing using 100%owf dye a 100 °C, and pH = 3 for 75 min, without salt and dispersing agent).

It is well known that salt addition in dyeing is necessary in the case of using high affinity dyes as it retards the dye migration, thus obtaining better level dyeing. Sodium sulfate is the common leveling agent in dyeing synthetic fibers, and thus we investigated the effect of salt in the dyeing procedure of nylon 6 with a natural dye. According to Figure 8, it is clearly indicated that in the absence of salt, the color strength is higher than it is when using a 1 and 2.5 g/L salt, in which when the salt concentration increases, the color strength decreases slightly. Sodium sulfate releases Na^+ and SO_4^{2-} as free ions in the dyebath. The interaction of sulfate anion with cationic sites on the nylon 6 fiber may cause their neutralization and consequently reduce dye ionic attraction between the fibers and anionic dye molecules. In addition, charge neutralization of the dye anions by the electrolyte cations (Na^+) affects the dye substantivity and inhibits the interaction with the nylon 6 fiber cations. Increase in salt concentration therefore inhibits dye exhaustion and consequently decreases the color strength of the dyed nylon 6 fabric [34,35].

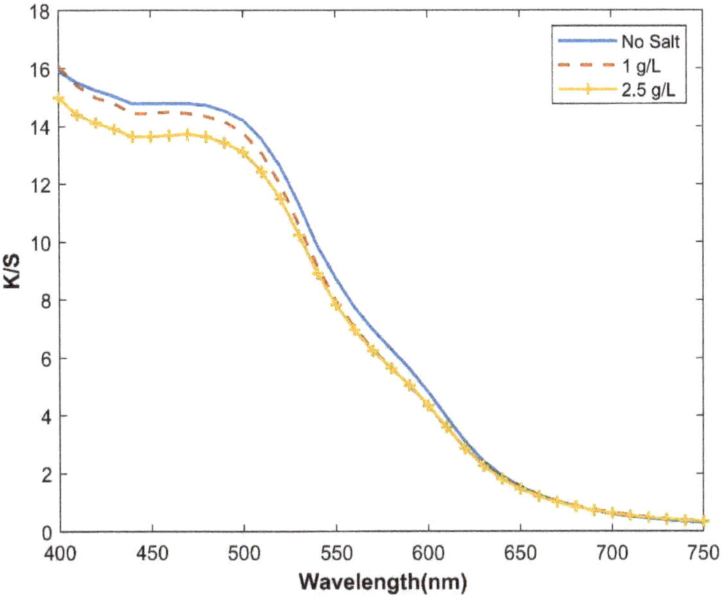

Figure 8. The effect of salt addition on color strength of dyed samples (dyeing using 100%owf dye at M:L = 1:30, 100 °C, and pH = 3 for 75 min, without dispersing agent).

Applying a dispersing agent in natural dyeing is important as the dye molecules may be aggregated. Therefore, the effect of dispersing agent addition on the color strength of dyed samples is shown in Figure 9. Expectedly, using a 0.5 g/L dispersing agent increased the color strength, and a further increase had a negative effect on the dyeing performance, probably due to the increase in the water solubility of dyes by formation of micelles with a hydrophilic outer shell and keeping the dye molecules in the solution led to these molecules avoiding the penetration into nylon 6 as a hydrophobic fiber [36]. Therefore, using a 0.5 g/L of dispersing agent may result in slight improvements in the dyeing yield.

3.2. Statistical Analysis

To analyze whether dyeing parameters have a significant effect on color strength, one sample *t*-test analysis ($\alpha = 0.05$) was employed based on the maximum values of color strength (λ_{max}). The t-statistic, degree of freedom, and *p*-value of dyeing parameters are listed in Table 1. It is clear that except for dyeing temperature, other parameters are significant in the case of color strength. Therefore, to achieve the desired shade, controlling dye bath pH, concentration, M:L, and dyeing time should be considered carefully. Meanwhile, dyeing temperature is the least significant parameter in dyeing nylon 6 with Rhubarb dye.

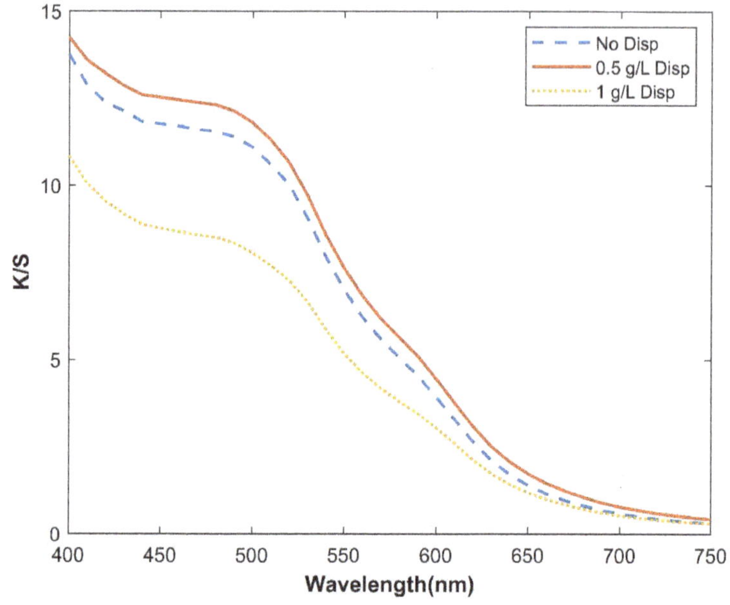

Figure 9. The effect of dispersing agent on color strength of dyed samples (dyeing using 100%ow dye at M:L = 1:30, 100 °C, and pH = 3 for 75 min, without salt).

Table 1. One-sample *t*-test of dyeing parameters.

Dyeing Parameter	T Statistic	Degree of Freedom	*p*-Value (Significance)
Dye bath pH	11.6270	2	3.1×10^{-4}
Dyeing Temperature	2.2589	3	0.1090
Dye Concentration	3.8582	5	0.0119
M:L ratio	19.1920	4	4.3×10^{-5}
Dyeing Time	4.1610	5	0.0088

3.3. *Effect of Mordanting on Color Strength*

3.3.1. Effect of Metal Mordants

Most natural dyes have a poor affinity for synthetic fibers. Therefore, applying mordants is necessary to increase the dye affinity. Figure 10 shows the effect of using different metal mordants on color strength. It is clear that Iron enhances the color strength while Copper, Zinc, and Aluminium decrease the color strength. The formation of a complex between the metal ions and dye molecules affects the L*, a*, and b* color coordinates of the dyed samples. Previous studies have shown that some mordants such as aluminium can lead to a decrease in color strength and cause an increase in the lightness and increase the a* and b* values. As can be seen in Table 2, the same results have been obtained when applying Cu, Zn, and Al mordants on nylon 6. This means a hypochromic shift caused by the formation of the complex between the dye molecules and Cu, Zn, and Al ions. Similar results have been reported elsewhere [37–39].

Finally, Iron achieves the highest color strength among the other metals used. It can be attributed to the chelation of the dye molecules with iron to generate a complex with hyperchromic shift in the spectrum as schematically illustrated in Scheme 1 [40,41]. The -COOH group of Rhein can be ionized and form an ionic bond with the positively charged amine end groups of nylon 6 fibers under acidic condition.

Table 2. Color difference in mordanted sample compared to unmordanted sample (dyeing was performed using 100%owf dye at M:L = 1:30, 100 °C, and pH = 3 for 75 min, without salt and dispersing agent).

Mordants	L*	a*	b*	K/S (λ_{max})	ΔE
Unmordanted	29.2	16.17	14.32	15.7	Control sample
Zn	29.75	17.35	15.37	15.3	1.7
Cu	29.45	17.38	16.18	16.87	2.2
Fe	28.03	15.15	13.92	16.87	1.6
Al	31.04	18.07	17.1	15.3	3.8
Pistachio hull	29.67	17.52	15.66	16.07	1.9
Walnut husk	26.64	15.1	13.52	18.77	2.8
Green coffee	31.47	16.89	15.76	13.94	2.7
Pinecone	27.19	16.86	14.55	18.62	2.1

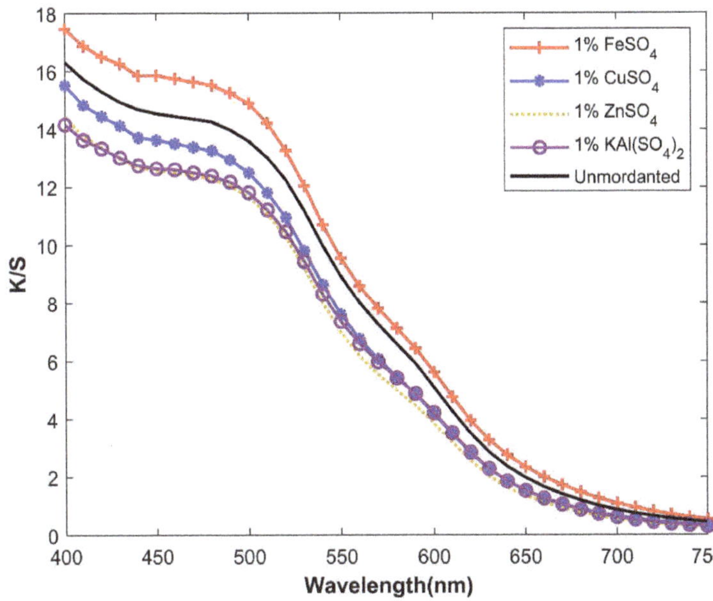

Figure 10. Color strength of dyed samples with pre-mordanting method using different metal mordants (dyeing was performed using 100%owf dye at M:L = 1:30, 100 °C, and pH = 3 for 75 min, without salt and dispersing agent).

3.3.2. Effect of Bio-Mordants

To investigate the effect of using different concentrations of bio-mordants on the color strength, the K/S plots of samples dyed after being mordanted with different bio-mordants including walnut husk, pistachio hull, pinecone, and green coffee are shown in Figure 11. It can be seen that the λ_{max} of all mordanted samples as well as unmordanted sample occurs at 400 nm and the shapes of the K/S curves are similar.

Except for green coffee, all used natural mordants enhanced the color strength of the dyed samples. However, pistachio hull has the least effect on enhancing color strength among the other used mordants. Walnut husk and pinecone reveal considerable performance in increasing color strength. In fact, all concentrations of walnut husk increased the color strength, of which 10%owf had the highest efficiency.

All the mentioned bio-mordants in this research contain several hydroxyl functional groups. For example, the green walnut husk contains two hydrolyzable tannins, i.e., ellagic acid and tannic acid. These tannins contain a lot of -OH groups that have the potential for hydrogen bonding with the dye molecules as well as the nylon 6 fiber [42]. Therefore the -OH groups of these bio-mordants link with the amide group of nylon 6 and also form hydrogen bonding with the -OH groups of Emodin and Chrysophanol of Rhubarb dye (see Scheme 2). The mentioned mechanism leads to increasing the color strength of bio-mordanted samples compared with the unmordanted sample [20,43].

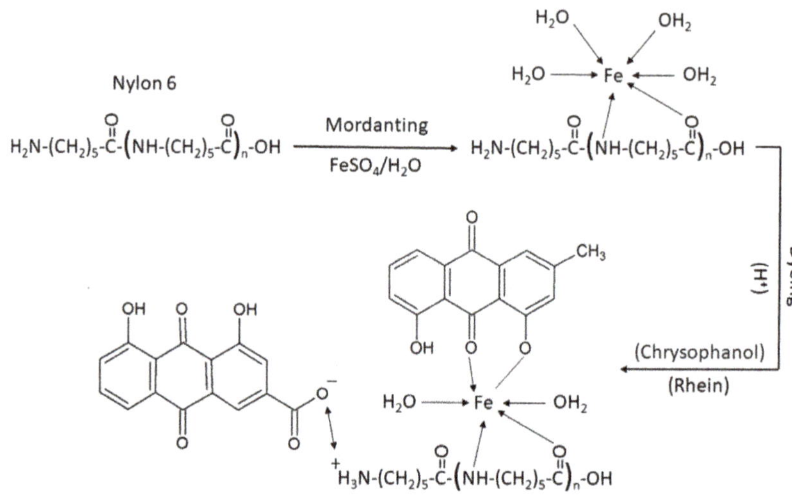

Scheme 1. Schematic presentation of the interactions between metal mordant, nylon 6 fiber, and dye molecules.

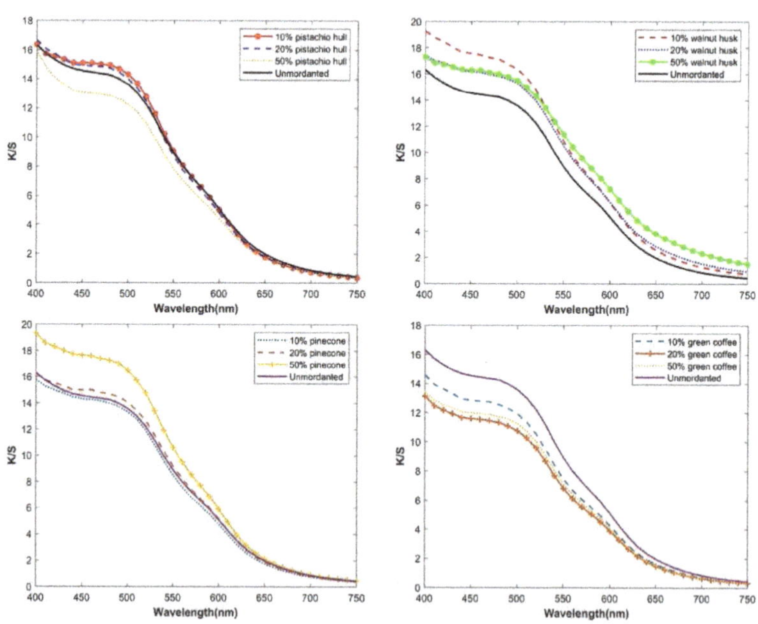

Figure 11. The effect of various concentrations of different bio-mordants on color strength of dyed samples (dyeing was performed using 100%owf dye at M:L = 1:30, 100 °C, and pH = 3 for 75 min without salt and dispersing agent).

3.4. Color Coordinates of Mordanted Samples

Figure 12 indicates the color chromaticity of mordanted samples to investigate whether mordanting process changes the hue of dyeing. Except for Aluminum, all mordants including metal and natural types are close to the unmordanted sample in the chromaticity diagram indicating mordanting did not change the hue of rhubarb applied on nylon 6. Furthermore, the color coordinates of the pistachio and Cu are so close together, and thus the pistachio could be an excellent substitution for copper metal. It is clear that walnut could be a good alternative for iron metal as the same color is generated. To clarify this result, Table 2 shows the value of the CIELAB color difference between mordanted and unmordanted samples. It could be seen that the color differences are fair in the field of natural dyeing as the maximum color difference is 3.8 in the case of aluminum. Meanwhile, other color differences are close to 2, which proves the pre-mordanting process did not change the color of rhubarb on nylon 6 fabric significantly.

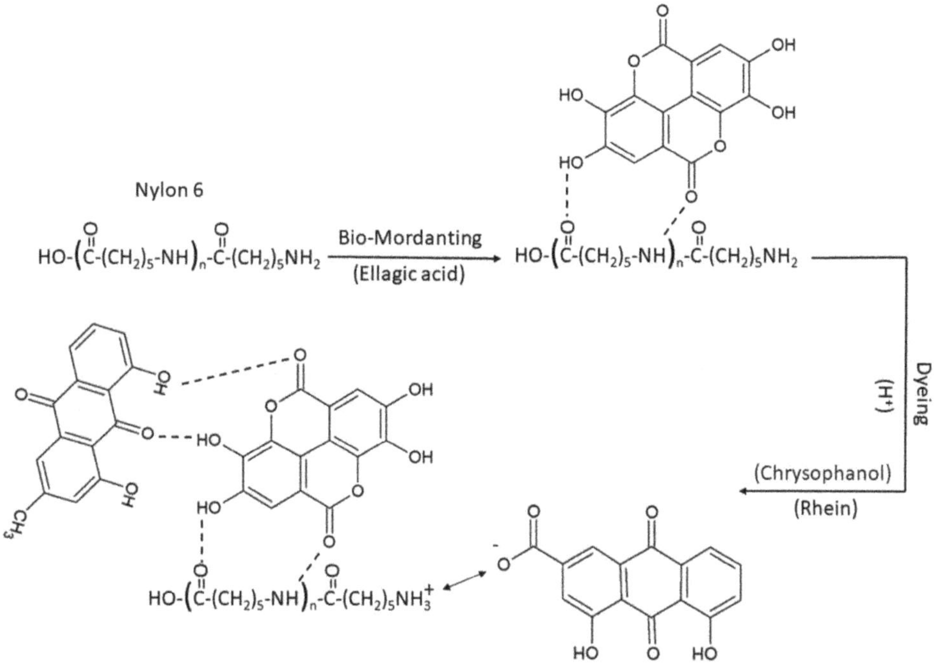

Scheme 2. Schematic presentation of the interactions between bio-mordant, nylon 6 fibers and dye molecules.

3.5. Evaluation of Color Fastness

There are multiple factors that affect the fastness properties such as the chemical structure of dye and mordant, the physical property of fiber, the interaction between dye–mordant–fiber, and the mordanting method that should be considered during the dyeing procedure. Therefore, selecting appropriate mordants based on the dye structure has high importance to achieve high efficiency in color strength and fastness. Therefore, we used proper bio-mordants as well as metal mordants to compare the color fastness of different types of mordants.

We investigated the effect of four common metallic mordants on fastness to washing and light using the bottom mordanting process. The results are shown in Table 3. The washing fastness of unmordanted nylon 6 fabric is evaluated as fair (i.e., 3), which reveals that the mordanting procedure is a necessary step for this natural dye. It is clear that all of the metal and natural mordants alter the wash fastness score from fair to excellent, whicl is due to the strong bonding between dye, metal or bio-mordant, and fiber as indicatec schematically in Schemes 1 and 2. Therefore, it is concluded that the used bio-mordant could be excellent substitutions for metal mordants since the same color fastness is providec and, in addition, they do not lead to environmental hazards.

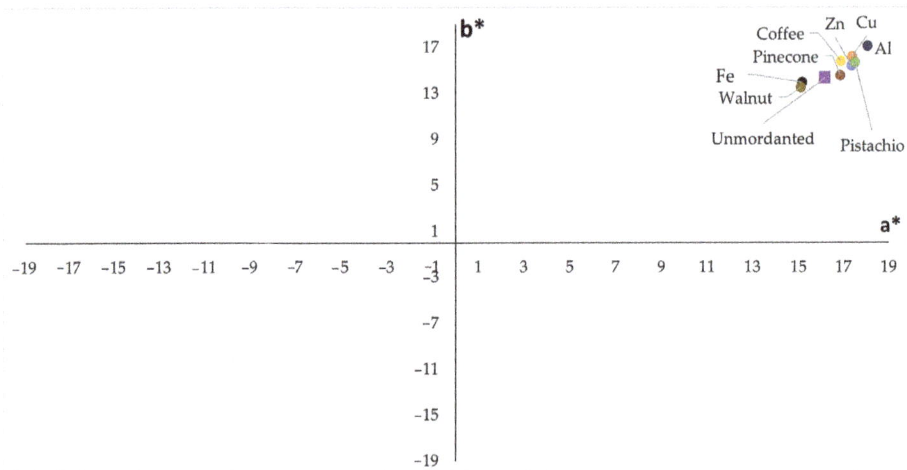

Figure 12. Chromaticity coordinates of different bio-mordants used.

Table 3. Colorfastness of nylon 6 fabric dyed with and without different mordants.

Mordant Type	Wash Fastness			Light Fastness
	Color Change	Stain on Nylon 6	Stain on Cotton	
Unmordanted	3	4	4	2
Zn	5	5	5	8
Cu	5	5	5	8
Fe	5	5	5	8
Al	5	5	5	8
Pistachio hull	5	5	5	8
Walnut husk	5	5	5	8
Green coffee	5	5	5	8
Pinecone	5	5	5	8

Like washing fastness, all mordants including metal and natural mordants enhance the light fastness in which an excellent rating (i.e., 8 score) was achieved. Regarding ligh fastness, there was an amazing result in which light exposure increased the color strengtl of the samples compared to the unexposed samples.

4. Conclusions

In this paper, the feasibility of dyeing the nylon 6 fabric with the *Rhubarb* flower as a natural dye was investigated. The optimal conditions for dyeing nylon 6 with *Rhubarb* were found to be the 100 %owf dye at boiling temperature in an acidic environment for 75 mir Various metallic and bio-mordants were employed to improve the dyeability and fastness properties. The results showed that almost all mordants (except Al, Zn, and green coffee increased the color strength of the dyed samples. However, the 10%owf walnut husk has the best efficiency in case of increasing the color strength. In addition, the color coordinates were not affected significantly by mordanting, especially in the case of bio-mordanting Mordanting enhanced the fastness against washing and light significantly. The results o this study confirm that the mentioned bio-mordants can be used as excellent substitutes of metal mordants and natural dyeing can be performed in a more environmentally friendly way to avoid wastewater pollution and hazards.

Author Contributions: Conceptualization, A.H. and F.S.G.; methodology, A.H. and F.S.G.; software, A.H., F.S.G. and E.D.; validation, A.H.; formal analysis, A.H., F.S.G. and E.D.; investigation, F.S.G.; resources, F.S.G.; data curation, A.H.; writing—original draft preparation, F.S.G. and E.D.; writing—review and editing, A.H.; visualization, E.D.; supervision, A.H.; project administration, A.H. All authors have read and agreed to the published version of the manuscript.

Funding: This research received no external funding.

Data Availability Statement: Not applicable.

Conflicts of Interest: The authors declare no conflict of interest.

References

1. Buyukakinci, Y.B.; Guzel, E.T.; Karadag, R. Organic cotton fabric dyed with dyer's oak and barberry dye by microwave irradiation and conventional methods. *Ind. Text.* **2021**, *72*, 30–38. [CrossRef]
2. Kiran, S.; Hassan, A.; Adeel, S.; Qayyum, M.A.; Yousaf, M.S.; Abdullah, M.; Habib, N. Green dyeing of microwave treated silk using coconut coir based tannin natural dye. *Ind. Text.* **2020**, *71*, 227–234. [CrossRef]
3. Haji, A.; Qavamnia, S.S.; Nasiriboroumand, M. The use of D-optimal design in optimization of wool dyeing with Juglans regia bark. *Ind. Text.* **2018**, *69*, 104–110. [CrossRef]
4. Vadood, M.; Haji, A. A hybrid artificial intelligence model to predict the color coordinates of polyester fabric dyed with madder natural dye. *Expert Syst. Appl.* **2022**, *193*, 116514. [CrossRef]
5. Haji, A.; Vadood, M. Environmentally Benign Dyeing of Polyester Fabric with Madder: Modelling by Artificial Neural Network and Fuzzy Logic Optimized by Genetic Algorithm. *Fiber. Polym.* **2021**, *22*, 3351–3357. [CrossRef]
6. Liman, M.L.R.; Islam, M.T.; Hossain, M.M.; Sarker, P. Sustainable Dyeing Mechanism of Polyester with Natural Dye Extracted from Watermelon and Their UV Protective Characteristics. *Fiber. Polym.* **2020**, *21*, 2301–2313. [CrossRef]
7. Lara, L.; Cabral, I.; Cunha, J. Ecological Approaches to Textile Dyeing: A Review. *Sustainability* **2022**, *14*, 8353. [CrossRef]
8. Militký, J.; Venkataraman, M.; Mishra, R. The chemistry, manufacture, and tensile behavior of polyamide fibers. In *Handbook of Properties of Textile and Technical Fibres*, 2nd ed.; Bunsell, A.R., Ed.; Woodhead Publishing: Cambridge, UK, 2018; pp. 367–419. [CrossRef]
9. Gupta, D.; Gulrajani, M.; Kumari, S. Light fastness of naturally occurring anthraquinone dyes on nylon. *Color. Technol.* **2004**, *120*, 205–212. [CrossRef]
10. Sadeghi-Kiakhani, M. Eco-friendly dyeing of wool and nylon using madder as a natural dye: Kinetic and adsorption isotherm studies. *Int. J. Environ. Sci. Technol.* **2015**, *12*, 2363–2370. [CrossRef]
11. Haji, A.; Mousavi Shoushtari, A.; Mirafshar, M. Natural dyeing and antibacterial activity of atmospheric-plasma-treated nylon 6 fabric. *Color. Technol.* **2014**, *130*, 37–42. [CrossRef]
12. Shams-Nateri, A.; Hajipour, A.; Dehnavi, E.; Ekrami, E. Colorimetric Study on Polyamides Dyeing With Weld and Pomegranate Peel Natural Dyes. *Cloth. Text. Res. J.* **2014**, *32*, 124–135. [CrossRef]
13. Mirjalilia, M.; Karimib, L.; Paydara, H.; Chizarifarda, G. Effect of henna natural dye on antibacterial properties of dyed nylon fabric with various mordants. *Iran. J. Org. Chem* **2014**, *6*, 1389–1395.
14. Miah, M.R.; Hossain, M.; Dipto, A.; Telegin, F.; Quan, H. Eco-dyeing of Nylon Fabric Using Natural Dyes Extracted from Onion Outer Shells: Assessment of the Effect of Different Mordant on Color and Fastness Properties. *Int. J. Sci. Eng. Res.* **2016**, *7*, 1030–1042.
15. Elnagar, K.; Abou Elmaaty, T.; Raouf, S. Dyeing of polyester and polyamide synthetic fabrics with natural dyes using ecofriendly technique. *J. Text.* **2014**, *2014*, 363079. [CrossRef]
16. Vuthiganond, N.; Nakpathom, M.; Somboon, B.; Narumol, N.; Rungruangkitkrai, N.; Mongkholrattanasit, R. Dyeing properties of nylon fabrics with mangrove bark extract via mordanting methods. *Mater. Sci. Forum* **2016**, *857*, 495–498.
17. Vadwala, Y.; Kola, N. Dyeing on nylon fabric with natural dye extracted from waste leaves of *Terminalia catappa* locally known as tropical almond tree. *Int. J. Home Sci.* **2017**, *3*, 175–181.
18. Kabir, S.M.M.; Dhar, A.K.; Bhattacharjee, M. The use of natural *Areca catechu* dyes for silk and nylon and its halochromic effect. *J. Text. Inst.* **2020**, *111*, 882–889. [CrossRef]
19. Adeel, S.; Hasan, M.u.; Batool, F.; Ozomay, M.; Hosseinnezhad, M.; Amin, N.; Hussaan, M. Eco-friendly bio-dyeing of bio-treated nylon fabric using Esfand (*P. harmala*) based yellow natural colorant. *J. Eng. Fibers Fabr.* **2022**, *17*, 15589250221091265. [CrossRef]
20. Haji, A.; Shahmoradi Ghaheh, F.; Mohammadi, L. Dyeing of polyamide 6 fabric with new bio-colorant and bio-mordants. *Environ. Sci. Pollut. Res.* **2022**, *30*, 37981–37996. [CrossRef]
21. Khan, S.A.; Ahmad, A.; Khan, M.I.; Yusuf, M.; Shahid, M.; Manzoor, N.; Mohammad, F. Antimicrobial activity of wool yarn dyed with *Rheum emodi* L. (Indian Rhubarb). *Dye. Pigment.* **2012**, *95*, 206–214. [CrossRef]

22. Khan, S.A.; Shahid-ul-Islam; Shahid, M.; Khan, M.I.; Yusuf, M.; Rather, L.J.; Khan, M.A.; Mohammad, F. Mixed metal mordan dyeing of wool using root extract of *Rheum emodi* (Indian Rhubarb/Dolu). *J. Nat. Fibers* **2015**, *12*, 243–255. [CrossRef]
23. Ali, N.; Elkatibe, E.; Elmohamedy, R.; Nassar, S.; Elshemy, N. Dyeing properties of wool fibers dyed with rhubarb as natural dy via ultrasonic and conventional methods. *Egypt. J. Chem.* **2019**, *62*, 119–130.
24. Ahmed, N.; Nassar, S.; El-Shishtawy, R.M. Novel green coloration of cotton fabric. Part I: Bio-mordanting and dyeing characteris tics of cotton fabrics with madder, Alkanet, Rhubarb and Curcumin Natural Dyes. *Egypt. J. Chem.* **2020**, *63*, 1605–1617.
25. Singh, N.P.; Gupta, A.P.; Sinha, A.K.; Ahuja, P.S. High-performance thin layer chromatography method for quantitative determi nation of four major anthraquinone derivatives in *Rheum emodi*. *J. Chromatogr. A* **2005**, *1077*, 202–206. [CrossRef]
26. Ansari, B.; Mehrizi, M.K.; Haji, A. Dyeing of Oxygen Plasma Treated Wool Fibers with *Rhuem ribes* L. Flowers. *J. Color Sci. Techno.* **2015**, *9*, 135–143.
27. Xiang, H.; Zuo, J.; Guo, F.; Dong, D. What we already know about rhubarb: A comprehensive review. *Chin. Med.* **2020**, *15*, 88 [CrossRef]
28. *ISO 105-C06:2010*; Textiles—Tests for Colour Fastness—Part C06: Colour Fastness to Domestic and Commercial Laundering. ISC Geneva, Switzerland, 2010.
29. *ISO 105-B02:2013*; Textiles—Tests for Colour Fastness—Part B02: Colour Fastness to Artificial Light: Xenon Arc Fading Lamp Tes ISO: Geneva, Switzerland, 2013.
30. Roy, S.; Ghosh, S.; Bhowmick, N.; Roychoudhury, P. Study the effect of denier and fiber cut length on zeta potential of nylon an polyester fibers for sustainable dyeing process. *J. Environ. Res. Dev.* **2016**, *11*, 392.
31. Mirnezhad, S.; Safapour, S.; Sadeghi-Kiakhani, M. Dual-mode adsorption of cochineal natural dye on wool fibers: Kinetic equilibrium, and thermodynamic studies. *Fibers Polym.* **2017**, *18*, 1134–1145. [CrossRef]
32. Sadeghi-Kiakhani, M.; Tehrani-Bagha, A.R.; Gharanjig, K.; Hashemi, E. Use of pomegranate peels and walnut green husks a the green antimicrobial agents to reduce the consumption of inorganic nanoparticles on wool yarns. *J. Clean. Prod.* **2019**, *231* 1463–1473. [CrossRef]
33. Tayade, P.B.; Adivarekar, R.V. Dyeing of cotton fabric with *Cuminum cyminum* L. as a natural dye and its comparison with synthetic dye. *J. Text. Inst.* **2013**, *104*, 1080–1088. [CrossRef]
34. Nnorom, O.O.; Anyanwu, P.I.; Oguzie, C.K.; Okonkwo, S.D. Inhibitive Effect of NaCl and Citric Acid on the Colour Yield of Acic Dye on Nylon Fabric. *J. Text. Sci. Technol.* **2019**, *6*, 49–58. [CrossRef]
35. Yang, Y. Effect of Salts on Physical Interactions in Wool Dyeing with Acid Dyes. *Text. Res. J.* **1998**, *68*, 615–620. [CrossRef]
36. Chakraborty, J.N. 18—Dyeing with disperse dye. In *Fundamentals and Practices in Colouration of Textiles;* Chakraborty, J.N., Ed Woodhead Publishing: New Delhi, India, 2010; pp. 192–201. [CrossRef]
37. Haji, A.; Rahimi, M. RSM Optimization of Wool Dyeing with *Berberis thunbergii* DC Leaves as a New Source of Natural Dye *J. Nat. Fibers* **2022**, *19*, 2785–2798. [CrossRef]
38. Shabbir, M.; Islam, S.U.; Bukhari, M.N.; Rather, L.J.; Khan, M.A.; Mohammad, F. Application of *Terminalia chebula* natural dye o wool fiber—Evaluation of color and fastness properties. *Text. Cloth. Sustain.* **2016**, *2*, 1. [CrossRef]
39. Shabbir, M.; Rather, L.J.; Shahid-ul, I.; Bukhari, M.N.; Shahid, M.; Khan, M.A.; Mohammad, F. An eco-friendly dyeing of wooler yarn by *Terminalia chebula* extract with evaluations of kinetic and adsorption characteristics. *J. Adv. Res.* **2016**, *7*, 473–482 [CrossRef]
40. Räisänen, R.; Nousiainen, P.; Hynninen, P.H. Emodin and Dermocybin Natural Anthraquinones as Mordant Dyes for Wool an Polyamide. *Text. Res. J.* **2001**, *71*, 1016–1022. [CrossRef]
41. Bukhari, M.N.; Shahid-ul-Islam; Shabbir, M.; Rather, L.J.; Shahid, M.; Singh, U.; Khan, M.A.; Mohammad, F. Dyeing studies an fastness properties of brown naphtoquinone colorant extracted from *Juglans regia* L. on natural protein fiber using different meta salt mordants. *Text. Cloth. Sustain.* **2017**, *3*, 3. [CrossRef]
42. Jahanban-Esfahlan, A.; Ostadrahimi, A.; Tabibiazar, M.; Amarowicz, R. A comprehensive review on the chemical constituent and functional uses of walnut (*Juglans* spp.) husk. *Int. J. Mol. Sci.* **2019**, *20*, 3920. [CrossRef] [PubMed]
43. Hosseinnezhad, M.; Gharanjig, K.; Jafari, R.; Imani, H.; Razani, N. Cleaner colorant extraction and environmentally wool dyein using oak as eco-friendly mordant. *Environ. Sci. Pollut. Res.* **2021**, *28*, 7249–7260. [CrossRef]

sustainability

MDPI

Article

Sustainable and Environmental Dyeing with MAUT Method Comparative Selection of the Dyeing Recipe

Meral Özomay

Department of Textile Engineering, Marmara University, 34722 Istanbul, Turkey; meral.akkaya@marmara.edu.tr or meralozomay@gmail.com

Abstract: The textile industry is one of the most complex sectors, in terms of the materials and chemical processes used from petroleum and the environmental degradation during its production and disposal. It is therefore a sector looking for new possibilities and for more sustainable materials and applications. One option is to use natural dyes, as they are considered biodegradable, do not pollute the environment, and have potential use for many sectors, including the fashion industry. In this study, Alanya silk was dyed by a natural dyeing method with crocus sativus, *Helichrysum arenarium,* and *Glycyrrhiza glabra* L., plants that grow in and around the Alanya region. *Quercus aegilops* L. grown in the region was preferred as mordant, a natural binder, and is one of the plants with the highest tannin content, and it was used with a more environmentally friendly and sustainable approach to increase the binding in natural dyeing instead of chemical mordants. The aim is to provide an environmental and scientific contribution to the dyeing producers in this region. According to the MAUT (Multi-Attribute Utility Theory) method, the best dyes in terms of fastness and color efficiency were determined as the dyes made with the *Glycyrrhiza glabra* L. plant.

Keywords: MAUT; natural mordant; Alanya silk; natural dyeing; sustainable dyeing

check for updates

Citation: Özomay, M. Sustainable and Environmental Dyeing with MAUT Method Comparative Selection of the Dyeing Recipe. *Sustainability* **2023**, *15*, 2738. https://doi.org/10.3390/su15032738

Academic Editor: Abu Naser Md Ahsanul Haque

Received: 27 December 2022
Revised: 21 January 2023
Accepted: 31 January 2023
Published: 2 February 2023

1. Introduction

Textile materials are dyed using various dyestuffs [1,2] and methods [3–5] to add aesthetic and commercial value to the product. Until the discovery and commercialization of synthetic dyestuffs, these processes were carried out only with natural dye extracts [6]. With the introduction of synthetic dyes into our lives, we moved away from natural dyes. This was due to our rapid access to synthetic dyes. With the increasing use of synthetic dyes, a large amount of water is used in the textile sector and disposal of this water polluted with the chemicals used [7] also mixes into the environment and causes significant pollution [8].

'Synthetic dyes' are part of the major chemical pollutants category in terms of textile waste [9]. 'Synthetic dyes' are the primary source of waste from the textile industry [10] and one of the causes of environmental pollution [11]. In recent years, although the use of some heavy metals such as [12] iron, titanium oxide, and derivatives [13] was restricted due to their environmental hazards, synthetic dyes continue to cause severe damage, particularly to groundwater [3] and also to the environment [14]. Synthetic dyes and chemicals not only pollute the environment during the production of textile materials, but also adversely affect human health during their use [15], and even discarded synthetic dyed textiles after their use also cause environmental pollution [5]. In this context, natural dyes have gradually started to gain importance in the market because they are biodegradable and do not cause health risks; therefore, can be easily used without as many environmental concerns [16,17]. Due to all these adverse effects of synthetic dyes, the world today is entering significantly into a trend of returning to nature and natural processes [18], and the level of awareness about environmentalist approaches is increasing. Restricting the use of chemicals [19] and treatment of wastewater [20], reducing the use of plastics [21] and recycling waste [22], and increasing the use of environmentally friendly and biodegradable natural materials [23] are just some of the steps taken to protect the environment.

Natural dye plants [24] are generally annual or biennial plants that do not have adverse effects on human and environmental health [25], and do not cause environmental pollution during production, during use, and when they become waste after use [18,26]. Natural dyes are used not only in the coloring of textiles but also in many other fields [27] such as pharmacy [28], cosmetics [29], food [30], and packaging [31]. The cost of dyeing with natural dyes is almost the same as dyeing with synthetic dyes [32]. In addition, the light, washing, and rubbing fastness [33] of the obtained products are better than synthetic dyes and inks.

Natural dyestuffs are generally classified into four main categories [34]: plant, fungal, animal, and mineral-sourced natural dyestuffs. Dyestuffs obtained from flowers [35], leaves [36], bark/peel [37,38], roots [39], seeds [40], and fruits [41] of plants are called galenic natural dyestuffs.

The mordanting operation is generally performed in order to color the textile surfaces obtained from natural fibers. Mordanting is a process that increases the affinity between the textile surface and the dyestuff [42]. It increases the retention of the dyestuff to the fibers and the absorption ability of the fibers. It also provides brighter and more permanent color. In natural dyeing, substances such as potassium aluminum sulfate, soda, sodium sulfate, sodium carbonate, copper sulfate, aluminum sulfate [43], and tannin [44] are generally utilized.

MAUT "Multi-Attribute Utility Theory"

Multiple Criteria Decision Making (MCDM) is shortly defined as selecting the right alternative by solving multiple and conflicting criteria [45,46]. Today, there are more than 70 [47] MCDM. One of these methods is the MAUT Method [48]. MAUT is also defined as Multi-attribute Value Theory (MAVT) basically [49]. The method is based on ranking a set of alternatives and selecting the best one among them. The MAUT method is used to find the most suitable alternative based on both qualitative and quantitative criteria. This method aims to find the most beneficial alternative by making subjective data calculable. Although the MAUT method seems to have a disadvantage in that it requires a lot of sensitive input data, it is also quite advantageous in finding a solution by considering many uncertainties [50].

The recently used multi-criteria decision-making method, utility theory (MAUT), was started to be applied by Ralph L. Keeney in 1974 in his paper "Multiplicative Utility Functions" [51]. After Keeney, James S. Dyer et al. in 1992 published "Multiple Criteria Decision Making, Multiattribute Utility Theory": In their writing named "The Next Ten Years", they found the different research topics and developments exciting for management science [52]. In 1999, Tim Bedford and Roger Cooke in their paper, "A new generic model for applying MAUT" [53], presented the Theory and Methodology of a new generic model for applying MAUT. Subsequently, many researchers applied the MAUT method to ranking and selection prioritization studies [54,55]. Studies on natural dyeing generally give CieLab values, light, washing, and friction fastness with color efficiency values. The relationship between these values is interpreted depending on variables such as the amount of dyestuff or mordant. In this study, the results of light and washing fastness and color efficiency of fabrics dyed according to the amount of mordant and different dyestuffs are interpreted; however, in this interpretation, a statistical analysis was used based on the MAUT method from MCDM methods. At the same time, the MAUT Method is used to rank the recipes in terms of light and washing fastness. The MAUT Method has not been utilized for such a rating in the literature.

2. Materials and Methods

2.1. Material

In this research, Alanya silk woven in a plain weave was used with 100% silk fiber. Fabric weight is 136 g/m^2, weft density is 11.4 cm, and warp density is 16.3 cm. *Quercus aegilops* L. was used as mordant, and crocus sativus, *Helichrysum arenarium*, and *Glycyrrhiza glabra* L. extracts were used as natural dyes. The selected dyestuff plants are plants that grow naturally in many regions of Alanya and can be easily reached by the people of Alanya. Acorn, which is selected as a mordant plant, also grows naturally in many regions of Alanya. It is selected as a mordant material because it is a plant with a high proportion of tannin. Tannin has been used as a binder in natural dyeing in the past.

2.2. Method

By applying the MAUT method to the data obtained as a result of light and washing fastnesses, the most suitable dye plant and mordant amount were determined according to the binder *Quercus aegilops* L. mordant. According to the MAUT Method, the test results of the fabrics dyed with *Quercus aegilops* L. mordant were determined as the criteria of the problem, and the options belonging to mordant rates were determined as alternatives. For the application, tests were carried out using 5 different concentrations of *Quercus aegilops* L. mordant (3%, 5%, 10%, 15%, 20%) and 3 plant extracts for a total of 15 different dyes and the appropriate recipe alternative from different concentrations of *Quercus aegilops* L. mordant, which provides the binding of the color in the dyeing of Alanya silk, was selected according to the MAUT method.

Sericin Removal: The silk fabric was pretreated and dried for 60′ to remove sericin in a bath prepared with 1 g/L soap and 2 g/L sodium carbonate with 98 °C and 1/50 Flotte ratio [32]. Color coordinates, whiteness, yellowness, brightness and K/S values of untreated and pretreated silk fabrics are given in Table 1.

Table 1. Whiteness values of untreated and treated silks.

Parameter	Untreated Silk	Pretreated Silk
L	80.16	91.122
a	0.798	−0.150
b	9.912	7.722
Whiteness	53.645	67.158
Yellowness	20.42	10.8
Brightness	50.769	69.972
K/S (360 λ)	3.87	6.76

Preparing extract: *Quercus aegilops* L. as a natural mordant, and crocus sativus, *Helichrysum arenarium,* and *Glycyrrhiza glabra* L. as the dyestuffs of the study, were boiled separately in distilled water for 1 h each, then filtered and cooled [35].
Mordanting: *Quercus aegilops* L. mordant was prepared at 3%, 5%, 10%, 15%, 15%, and 20% concentration according to the material weight. When the Flotte with a Flotte ratio of 1/100 reached the boiling temperature, the fabric was added, and the process was carried out by stirring for 60′. After 24 h keeping it in this bath, it was dried without washing [34].
Dyeing: The dyeing was carried out using 100% of the plant, so mL of the extracts corresponding to 1 g of plant per 1 g of sample was added to the dye bath by calculating the ratio and proportion method. The Flotte was prepared as 1/100 at 30 °C and it was brought to boiling temperature in 20 min, then the previously mordanted and dried sample was placed in the dye bath. Dyeing was performed by boiling at 100 °C for 60 min. The samples were kept in the dye bath for 24 h, washed 3 times, and dried [26].

Effect of dye temperature and dye concentration on exhaustion are given in Figure 1:

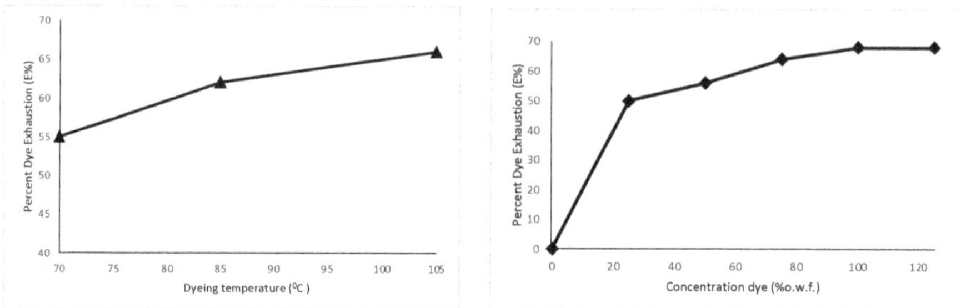

Figure 1. Effect of dye temperature and dye concentration on exhaustion.

Color Measurement: Color measurements of the dyed samples were performed with a Datacolor Spectra Flash 600 plus reflectance spectrophotometer using a Datamaster computer program according to the CIE L*a*b system. Color measurements were made using a D65 light source with a 10° observer [56]. Equation (1) was used to calculate color values according to the CIELab system [57].

$$\Delta E^* = [(\Delta L^*)^2 + (\Delta a^*)^2 + (\Delta b^*)^2]^{1/2} \quad (1$$

In the above formula, L* is lightness–darkness, a* is redness–greenness, and b* is yellowness–blueness values.

K/S Values: The color of the dyed samples was evaluated with color strength (K/S calculated using the Kubelka–Munk equation (Equation (2)). R is the reflectance value of the fiber at the wavelength at maximum absorption, K is the absorption coefficient, and S is the scattering coefficient [34].

$$K/S = (1 - R)^2/2R \quad (2$$

Fastnesses: The washing fastnesses of the test samples was carried out in a washing fastness test machine (Gyrowash Washer Tester) according to ISO 105-C06 [58] standard and the evaluation was carried out with reflectance spectrophotometer according to ISO A05 [59] standard. Light fastnesses were applied in a light fastness tester (Solarbox 1500E according to TS 1008 EN ISO 105-B02 standard [60] and the total color difference (ΔE* between the untested sample and the test samples was evaluated in a spectrophotometer
MAUT Method solution stages: In this study, MAUT (Multi-Attribute Utility Theory) defined as "Multi-Attribute Utility Theory", is used in multi-criteria decision-making methods that focus on ranking and selecting from a set of alternatives.

The main purpose of the MAUT method is that in every decision problem, there is a real-valued utility function U defined over the set of feasible alternatives, and the decision maker chooses the largest feasible one [61]. The steps given below are followed when applying the MAUT method:

Step 1: The set of alternatives subject to the decision problem (an) and the attributes/criteria (x_m) that will help in selecting the alternatives should be determined.
Step 2: Weight values (w_i) are assigned or calculated, which ensures that the criteria are evaluated correctly, and priorities are set. The sum of all w_i values must equal to 1 see Equation (3).

$$\sum_{i=1}^{m} w_i = 1 \quad (3$$

Step 3: Data values for the criteria are entered or their values are calculated. In this calculation, while quantitative values obtained from experimental results are for quantitative criteria, assigning specified values is carried out for qualitative criteria (x_m).

Step 4: The calculated values are placed in the decision matrix, and the normalization process is started. In the normalization process, the best and worst values for each criterion are first determined; the best value is assigned a value of 1, the worst value is assigned a value of 0, and the following Equation (4) is used to calculate the other values within this range:

$$u_i(x_i)\frac{x - x_i^-}{x_i^+ - x_i^-} \quad (4)$$

The terms used in this equation are shown below.

x_i^+: Best value for criteria
x_i^-: Worst value for criteria
X: Current benefit value in the calculated row

Step 5: Immediately after the normalization process, utility values are determined. The utility function Equation (5) is as follows [62]:

$$U(x) = \sum_{i=1}^{m} u_i(x_i) * w_i \quad (5)$$

$U(x)$: Utility value of the alternative
$u_i(x_i)$: Normalized utility values for each criterion and each alternative
w_i: Weight values

3. Result and Discussion

Codes for dye plant and mordant concentration are given in Table 2:

Table 2. Codes for dye plant and mordant concentration.

Code	Mordant Amount	Plant
A-1	%3	*Crocus Sativus*
A-2	%5	*Crocus Sativus*
A-3	%10	*Crocus Sativus*
A-4	%15	*Crocus Sativus*
A-5	%20	*Crocus Sativus*
B-1	%3	*Helichrysum arenarium*
B-2	%5	*Helichrysum arenarium*
B-3	%10	*Helichrysum arenarium*
B-4	%15	*Helichrysum arenarium*
B-5	%20	*Helichrysum arenarium*
C-1	%3	*Glycyrrhiza glabra* L.
C-2	%5	*Glycyrrhiza glabra* L.
C-3	%10	*Glycyrrhiza glabra* L.
C-4	%15	*Glycyrrhiza glabra* L.
C-5	%20	*Glycyrrhiza glabra* L.

The codes used for the information of total color differences (ΔE*), K/S values, washing (color change and staining), and light fastness of the samples dyed with crocus sativus, *Helichrysum arenarium* and *Glycyrrhiza glabra* L. extracts evaluated by spectrophotometer are given below.

ΔE: Color Difference Compared to Dyeing without Mordant
ΔE Y: Washing Fastness (Color Change)
ΔE 1: Washing Fastness (Staining) Wool
ΔE 2: Washing Fastness (Staining) Polyacrylonitrile
ΔE 3: Washing Fastness (Staining) Polyester
ΔE 4: Washing Fastness (Staining) Polyamide
ΔE 5: Washing Fastness (Staining) Cotton
ΔE 6: Washing Fastness (Staining) Acetate

ΔE I: Light Fastness

K/S: Color Yield (Kubelka–Munk) (Absorption coefficients/Scattering coefficients)

L, a, b, C, and h values obtained by dyeing Alanya silk with crocus sativus, *Helichrysum arenarium*, and *Glycyrrhiza glabra* L. plants using *Quercus aegilops* L. mordant are given altogether in Table 3.

Table 3. CıeLab Values of Dyeings.

Code	Mordant Substance	L	a	b	C	h	ΔE	Color
Crocu Sativus	Without Mordant	85.07	−1.04	18.63	18.66	93.21	-	
A-1	%3- *Quercus aegilops* L.	82.62	1.04	21.11	21.14	87.18	4.064	
A-2	%5- *Quercus aegilops* L.	82.78	0.94	20.52	20.54	87.38	3.573	
A-3	%10- *Quercus aegilops* L.	74.81	2.66	24.92	25.06	83.91	12.592	
A-4	%15- *Quercus aegilops* L.	72.49	3.18	21.82	22.05	81.70	13.644	
A-5	%20- *Quercus aegilops* L.	70.37	3.83	27.27	27.53	82.01	17.733	
Helichrysum arenarium	Without Mordant	66.70	3.23	20.69	20.95	81.12	-	
B-1	%3- *Quercus aegilops* L.	67.56	3.32	22.19	22.44	81.49	1.728	
B-2	%5- *Quercus aegilops* L.	70.04	2.89	21.17	21.37	82.23	3.395	
B-3	%10- *Quercus aegilops* L.	64.99	4.07	21.65	22.02	79.36	2.125	
B-4	%15- *Quercus aegilops* L.	67.09	4.06	23.01	23.36	79.99	2.489	
B-5	%20- *Quercus aegilops* L.	63.37	4.09	21.04	21.43	79.00	3.451	
Glycyrrhiza glabra L.	Without Mordant	73.69	4.03	22.00	22.36	79.61	-	
C-1	%3- *Quercus aegilops* L.	67.93	5.08	24.12	24.65	78.11	6.226	
C-2	%5- *Quercus aegilops* L.	74.11	3.74	22.80	23.11	80.63	0.954	
C-3	%10- *Quercus aegilops* L.	71.78	3.95	22.24	22.59	79.93	1.925	
C-4	%15- *Quercus aegilops* L.	68.53	4.24	22.76	23.15	79.45	5.218	
C-5	%20- *Quercus aegilops* L.	65.09	5.12	23.82	24.36	77.87	8.860	

Within the framework of the data obtained in Table 3, it was determined that the color became darker, the red nuance of the color increased, and the brightness decreased, in general, as the amount of mordant increased in the dyeings with crocus sativus, *Helichrysum arenarium*, and *Glycyrrhiza glabra* L. It was determined that the color of dyeing without mordant was lighter and brighter than dyeing with mordant.

The ΔE differences and color yield results obtained after washing and light fastnesses by dyeing Alanya silk with crocus sativus, *Helichrysum arenarium*, and *Glycyrrhiza glabra* L. plants using *Quercus aegilops* L. mordant are given altogether in Table 4.

Table 4. Color difference and color yield experimental results.

Code	ΔE	ΔE Y	ΔE 1	ΔE 2	ΔE 3	ΔE 4	ΔE 5	ΔE 6	ΔE I	K/S
A-1	4.064	6.064	1.352	0.815	0.159	0.643	5.554	4.047	7.929	1.11
A-2	3.573	6.821	2.184	0.765	0.11	0.599	5.435	3.246	6.403	1.13
A-3	12.592	10.651	1.022	0.861	0.387	0.535	6.27	3.056	7.915	2.11
A-4	13.644	8.729	0.797	0.934	0.422	0.617	5.691	2.476	7.488	2.26
A-5	17.733	11.423	1.637	1.233	0.368	0.958	4.606	3.005	8.586	2.75
B-1	1.728	7.407	1.79	0.639	0.271	0.679	0.431	4.016	5.764	6.22
B-2	3.395	4.914	1.746	0.356	0.846	0.67	0.379	4.998	5.632	5.1
B-3	2.125	10.615	1.325	0.39	0.856	0.628	1.526	3.261	6.039	7.3
B-4	2.489	3.884	2.171	0.682	1.101	0.981	1.097	4.089	6.211	6.85
B-5	3.451	6.452	2.225	0.583	0.978	0.978	1.313	3.943	5.429	7.76
C-1	6.226	7.202	1.6	0.513	0.669	2.053	1.688	3.235	7.009	7.14
C-2	0.954	5.674	1.146	0.363	0.71	0.892	1.669	3.46	5.856	4.9
C-3	1.925	3.762	0.846	0.322	0.942	0.98	1.774	3.186	6.573	4.64
C-4	5.218	4.327	1.023	0.372	0.612	0.803	1.527	3.282	5.189	5.31
C-5	8.86	5.41	1.246	0.319	0.731	1.009	1.392	3.942	9.36	6.6

The data given in Table 5 show the decrease and increase in ΔE differences depending on the change in the amount of mordant.

Table 5. Effect of mordant amount on ΔE differences.

CODE	ΔE Y	ΔE 1	ΔE 2	ΔE 3	ΔE 4	ΔE 5	ΔE 6	ΔE I	K/S
A-1 A-2 A-3 A-4 A-5	↓	↓	↓	↓	↑↓	↑	↑	↓	↓
B-1 B-2 B-3 B-4 B-5	↑	↓	↑	↓	↓	↓	↑	↑	↓
C-1 C-2 C-3 C-4 C-5	↑↓	↑↓	↑	↓	↑↓	↑	↓	↑↓	↑↓

While the biggest color difference compared to mordant-free dyeing was obtained from dyeing with the crocus sativus plant, it was determined that the colors made with *Helichrysum arenarium* were the closest to mordant-free dyeing. The sample with the highest color change after washing was obtained from dyeing with the highest concentration of the crocus sativus plant and *Quercus aegilops* L. mordant. The lowest color difference after washing is generally seen in the dyeings made with *Glycyrrhiza glabra* L. There was no significant difference between the mordants in terms of staining on Acetate, Cotton, Polyamide, Polyester, Polyacrylonitrile, and Wool, and staining was realized at rates too low to be visible to the eye. When we look at the color difference values obtained as a result of light fastness, the highest color differences were obtained from the dyes made with the crocus sativus plant, and the lowest color differences were obtained from the dyes made with the *Helichrysum arenarium* plant. When we examined the dyeings in terms of colo yield, it was concluded that there was no significant difference between the dyeings mad with *Helichrysum arenarium* and *Glycyrrhiza glabra* L., but the color yields of the dyeing made with crocus sativus were low.

According to the data obtained in Tables 6 and 7, it was determined that the dye made with crocus sativus and *Glycyrrhiza glabra* L. were darker than the dyeing withou mordant. In general, it was determined that the red and yellow color nuances increased a low rates in the dyeings. In addition, it was determined that all dyeings were more vivi than the dyeing without mordant.

Table 6. Spectrophotometric color measurement values of the samples.

Mordant	ΔL	Δa	Δb	ΔC	ΔH	ΔE
A-1	−2.453	2.083	2.483	2.479	−2.087	4.064
A-2	−2.289	1.985	1.894	1.886	−1.992	3.573
A-3	10.259	3.705	6.293	6.405	−3.507	12.592
A-4	−12.576	4.225	3.187	3.388	−4.065	13.644
A-5	14.703	4.870	8.637	8.874	−4.421	17.733
B-1	0.860	0.085	1.496	1.492	0.142	1.728
B-2	3.344	−0.344	0.478	0.423	0.409	3.395
B-3	−1.707	0.834	0.952	1.080	0.660	2.125
B-4	0.394	0.826	2.314	2.418	−0.434	2.489
B-5	−3.326	0.854	0.343	0.486	−0.782	3.451
C-1	−5.760	1.044	2.120	2.282	−0.614	6.226
C-2	0.421	−0.290	0.805	0.744	0.424	0.954
C-3	−1.907	−0.083	0.243	0.225	0.124	1.925
C-4	−5.158	0.208	0.761	0.786	−0.066	5.218
C-5	−8.602	1.087	1.823	2.000	−0.709	8.860

Table 7. Color measurement variation of samples.

Mordant	ΔL*	Δa*	Δb*	ΔC*
A-1	darker	redder	yellower	more vivid
A-2	darker	redder	yellower	more vivid
A-3	lighter	redder	yellower	more vivid
A-4	darker	redder	yellower	more vivid
A-5	lighter	redder	yellower	more vivid
B-1	lighter	redder	yellower	more vivid
B-2	lighter	greener	yellower	more vivid
B-3	darker	redder	yellower	more vivid
B-4	lighter	redder	yellower	more vivid
B-5	darker	redder	yellower	more vivid
C-1	darker	redder	yellower	more vivid
C-2	lighter	greener	yellower	more vivid
C-3	darker	greener	yellower	more vivid
C-4	darker	redder	yellower	more vivid
C-5	darker	redder	yellower	more vivid

Application of the MAUT Method in the Light of the Data Obtained

Step 1: First of all, the alternatives (a_n) subject to the decision problem and th attributes/criteria (xm) that will help in selecting the alternatives were determined. In thi study, the criteria were determined by using standard tests (light and washing fastness tes results and color yield (K/S)) and the alternatives were dyeing recipes using five differen concentrations (3%, 5%, 10%, 15%, 15%, and 20%) of *Quercus aegilops* L. mordant wit crocus sativus, *Helichrysum arenarium*, and *Glycyrrhiza glabra* L. plants, which also grow i Alanya region.

Thus, while rows were created for a total of 15 different alternatives with three dyestuffs × 5 mordant concentrations, the matrix for the recipe was designed using nine different criteria columns and the matrix was constructed for the start and solution.

Step 2: The values of the criteria are calculated. In this calculation, quantitative values obtained from the experimental results and read from the device are used for quantitative criteria, and their values are written in the cell places corresponding to the rows and columns in the matrix (x_m). The data set of the best and worst values of the criteria is determined under the criteria and written as rows. The values are given in Table 8 below.

Table 8. Using experimental results as data input.

			Min	Min	Min	Min	Min	Min	Min	Min	Max
	Code	Mord.	ΔE Y	ΔE 1	ΔE 2	ΔE 3	ΔE 4	ΔE 5	ΔE 6	ΔE I	K/S
Crocus Sativus	A-1	3%	6.064	1.352	0.815	0.159	0.643	5554	4.047	7.929	1.11
	A-2	5%	6.821	2.184	0.765	0.11	0.599	5.435	3.246	6.403	1.13
	A-3	10%	10.651	1.022	0.861	0.387	0.535	6.27	3.056	7.915	2.11
	A-4	15%	8.729	0.797	0.934	0.422	0.617	5.691	2.476	7.488	2.26
	A-5	20%	11.423	1.637	1.233	0.368	0.958	4.606	3.005	8.586	2.75
Helichrysum arenarium	B-1	3%	7.407	1.79	0.639	0.271	0.679	0.431	4.016	5.764	6.22
	B-2	5%	4.914	1.746	0.356	0.846	0.67	0.379	4.998	5.632	5.1
	B-3	10%	10.615	1.325	0.39	0.856	0.628	1.526	3.261	6.039	7.3
	B-4	15%	3.884	2.171	0.682	1.101	0.981	1.097	4.089	6.211	6.85
	B-5	20%	6.452	2.225	0.583	0.978	0.978	1.313	3.943	5.429	7.76
Glycyrrhiza glabra L.	C-1	3%	7.202	1.6	0.513	0.669	2.053	1.688	3.235	7.009	7.14
	C-2	5%	5.674	1.146	0.363	0.71	0.892	1.669	3.46	5.856	4.9
	C-3	10%	3.762	0.846	0.322	0.942	0.98	1.774	3.186	6.573	4.64
	C-4	15%	4.327	1.023	0.372	0.612	0.803	1.527	3.282	5.189	5.31
	C-5	20%	5.41	1.246	0.319	0.731	1.009	1.392	3.942	9.36	6.6
		x^-	11.423	2.225	1.233	1.101	2.053	6.27	4.998	9.36	1.11
		x^-	3.762	0.797	0.319	0.11	0.535	0.379	2.476	5.189	7.76

Step 3: The calculated values are placed in the decision matrix and the normalization process is started. In the normalization process, the best and worst values for each criterion are first determined, the best value is assigned a value of one, the worst value is assigned a value of zero, and Table 9 below is created to calculate the other values within this range:

Table 9. Table of normalized values.

ui(xi)	ΔE Y	ΔE 1	ΔE 2	ΔE 3	ΔE 4	ΔE 5	ΔE 6	ΔE I	K/S
A-1	0.700	0.611	0.457	0.951	0.929	0.122	0.377	0.343	0.000
A-2	0.601	0.029	0.512	1.000	0.958	0.142	0.695	0.709	0.003
A-3	0.101	0.842	0.407	0.720	1.000	0.000	0.770	0.346	0.150
A-4	0.352	1.000	0.327	0.685	0.946	0.098	1.000	0.449	0.173
A-5	0.000	0.412	0.000	0.740	0.721	0.282	0.790	0.186	0.247
B-1	0.524	0.305	0.650	0.838	0.905	0.991	0.389	0.862	0.768
B-2	0.850	0.335	0.960	0.257	0.911	1.000	0.000	0.894	0.600
B-3	0.105	0.630	0.922	0.247	0.939	0.805	0.689	0.796	0.931
B-4	0.984	0.038	0.603	0.000	0.706	0.878	0.360	0.755	0.863
B-5	0.649	0.000	0.711	0.124	0.708	0.841	0.418	0.942	1.000
C-1	0.551	0.438	0.788	0.436	0.000	0.778	0.699	0.564	0.907
C-2	0.750	0.756	0.952	0.395	0.765	0.781	0.610	0.840	0.570
C-3	1.000	0.966	0.997	0.160	0.707	0.763	0.718	0.668	0.531
C-4	0.926	0.842	0.942	0.493	0.823	0.805	0.680	1.000	0.632
C-5	0.785	0.686	1.000	0.373	0.688	0.828	0.419	0.000	0.826

Step 4: Weight values (w_i) are assigned, which ensure that the criteria are evaluated correctly and priorities are set. The sum of all w_i values must equal to one. At this stage, by accepting the weights of the criteria as equal (1/9 = 0.111), the operations were performed on the values given in Table 10.

Table 10. Equal weighting of criteria.

Weight	ΔE Y	ΔE 1	ΔE 2	ΔE 3	ΔE 4	ΔE 5	ΔE 6	ΔE I	K/S	
w_i	0.1111	0.1111	0.1111	0.1111	0.1111	0.1111	0.1111	0.1111	0.1111	=1

At this stage, the MAUT method will be ranked by giving equal values to the criterion weights w_i. The weighted normalization values of the data are given in Table 11.

Table 11. Normalization values weighted according to equal weight.

W_i=	0.1111	0.1111	0.1111	0.1111	0.1111	0.1111	0.1111	0.1111	0.1111
*wi*ui(xi)*	**ΔE Y**	**ΔE 1**	**ΔE 2**	**ΔE 3**	**ΔE 4**	**ΔE 5**	**ΔE 6**	**ΔE I**	**K/S**
A-1	0.078	0.068	0.051	0.106	0.103	0.014	0.042	0.038	0.000
A-2	0.067	0.003	0.057	0.111	0.106	0.016	0.077	0.079	0.000
A-3	0.011	0.094	0.045	0.080	0.111	0.000	0.086	0.038	0.017
A-4	0.039	0.111	0.036	0.076	0.105	0.011	0.111	0.050	0.019
A-5	0.000	0.046	0.000	0.082	0.080	0.031	0.088	0.021	0.027
B-1	0.058	0.034	0.072	0.093	0.101	0.110	0.043	0.096	0.085
B-2	0.094	0.037	0.107	0.029	0.101	0.111	0.000	0.099	0.067
B-3	0.012	0.070	0.102	0.027	0.104	0.089	0.077	0.088	0.103
B-4	0.109	0.004	0.067	0.000	0.078	0.098	0.040	0.084	0.096
B-5	0.072	0.000	0.079	0.014	0.079	0.093	0.046	0.105	0.111
C-1	0.061	0.049	0.088	0.048	0.000	0.086	0.078	0.063	0.101
C-2	0.083	0.084	0.106	0.044	0.085	0.087	0.068	0.093	0.063
C-3	0.111	0.107	0.111	0.018	0.079	0.085	0.080	0.074	0.059
C-4	0.103	0.094	0.105	0.055	0.091	0.089	0.076	0.111	0.070
C-5	0.087	0.076	0.111	0.041	0.076	0.092	0.047	0.000	0.092

According to the results of the calculations made up to this stage, the decision stage is reached. If the weights are taken equal (1/9 = 0.11), our recommendation for the best dyeing recipe is C-4 dyeing with *Glycyrrhiza glabra* L. at 15% *Quercus aegilops* L. concentration. It is seen that the top 3 best dyeings with *Glycyrrhiza glabra* L. are in this group. The results are given in Figure 2 below.

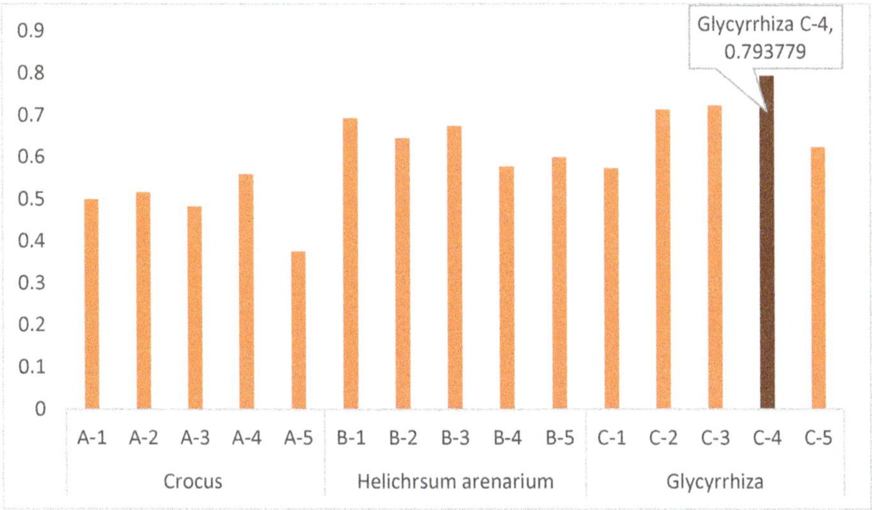

Figure 2. Determining the best dyeing.

4. Conclusions

MCDM methods are taking their place in research with a new method every day. Controversial solutions are presented as to which method is reliable, which method is suitable for use, which method to choose, or which experiment is reproducible.

In this study, Alanya silk fabric was mordanted in five different concentrations (3%, 5%, 10%, 15%, and 20%) with *Quercus aegilops* L., which is a natural mordant, and dyed separately with crocus sativus, *Helichrysum arenarium,* and *Glycyrrhiza glabra* L. plants. According to the results of the MAUT method applied to the fastness results obtained from the dyes, the plant that gave the best three results was *Glycyrrhiza glabra* L. It was observed that the dyeings with the crocus sativus plant had the five worst dyes in terms of fastness properties.

According to the results of the obtained method, the recommended dyeing recipe in terms of fastness and color yield in the dyeing with the crocus sativus, *Helichrysum arenarium,* and *Glycyrrhiza glabra* L., where color tones close to each other are obtained, is given in Table 12.

Table 12. Recommended Coloring Recipe.

	Coloring Recipe	
Mordanting	Mordant Type	*Quercus aegilops* L.
	Mordant (%)	15
	Temperature (°C)	100
	Duration (min)	60
	Waiting	24 h
Dyeing	Dye Type	*Glycyrrhiza glabra* L.
	Dye Plant (%)	100
	Temperature (°C)	100
	Duration (min)	60
	Color	Light Brown
	Waiting	24 h

Funding: This research received no external funding.

Institutional Review Board Statement: Not applicable.

Informed Consent Statement: Not applicable.

Data Availability Statement: Not applicable.

Acknowledgments: This study was carried out in the "Textile and Manuscripts Conservation, an Restoration Central Laboratory" established by the project numbered TR10/18/YMP/0137 supporte by İSTKA (Istanbul Development Agency). I would like to thank Lecturer Ferhat GÜNGÖR from M.Ü Faculty of Technology, Department of Mechanical Engineering, who contributed to the calculation of this study and examined the results.

Conflicts of Interest: The authors declare no conflict of interest.

References

1. Salman, M.; Adeel, S.; Batool, F.; Usama, M.; Hassan, A.; Ozomay, M.; Hosseinnezhad, M. Sustainable Application of Flavonoic Based Natural Colorants. In *Textile Dyes and Pigments: A Green Chemistry Approach*; Wiley-Scrivener: Hoboken, NJ, USA, 2022 pp. 295–321.
2. Batool, F.; Adeel, S.; Azeem, M.; Iqbal, N. Natural dye yielding potential and compounds of selected vegetable residues belongin to Brassicaceae: An approach towards sustainability. *Pak. J. Bot.* **2022**, *54*, 329–336. [CrossRef] [PubMed]
3. Repon, R.; Tauhidul Islam, M.; Al Mamun, A.; Abdur Rashid, M. Comparative study on natural and reactive dye for cotto coloration. *J. Appl. Res. Technol.* **2018**, *16*, 160–169. [CrossRef]
4. Adeel, S.; Rehman, F.-U.; Amin, A.; Amin, N.; Batool, F.; Hassan, A.; Ozomay, M. Sustainable exploration of coffee extracts (*Coffe arabica* L.) for dyeing of microwave-treated bio-mordanted cotton fabric. *Pigment. Resin Technol.* **2022**. [CrossRef]
5. Lara, L.; Cabral, I.; Cunha, J. Ecological Approaches to Textile Dyeing: A Review. *Sustainability* **2022**, *14*, 8353. [CrossRef]
6. Rodiah, M.; Hafizah, S.N.; Asiah, H.N.; Nurhafizah, I.; Norakma, M.; Norazlina, I. Extraction of natural dye from the mesocar and exocarp of *Cocos nucifera*, textile dyeing, and colour fastness properties. *Mater. Today Proc.* **2022**, *48*, 790–795. [CrossRef]
7. Srivastava, A.; Rani, R.M.; Patle, D.S.; Kumar, S. Emerging bioremediation technologies for the treatment of textile wastewate containing synthetic dyes: A comprehensive review. *J. Chem. Technol. Biotechnol.* **2022**, *97*, 26–41. [CrossRef]
8. Tkalec, M.; Sutlovic, A.; Glogar, M.I. Ecological, Economic and Social Aspects of Textile Dyes. In *Economic and Social Development Book of Proceedings*; Cakovec, 6-7 May 2022; Varazdin Development and Entrepreneurship Agency: Varazdin, Croatia, 2022 pp. 69–79.
9. Kishor, R.; Purchase, D.; Saratale, G.D.; Saratale, R.G.; Ferreira, L.F.R.; Bilal, M.; Chandra, R.; Bharagava, R.N. Ecotoxicologica and health concerns of persistent coloring pollutants of textile industry wastewater and treatment approaches for environmenta safety. *J. Environ. Chem. Eng.* **2021**, *9*, 105012. [CrossRef]
10. Manzoor, J.; Sharma, M. Impact of textile dyes on human health and environment. In *Impact of Textile Dyes on Public Health and th Environment*; IGI Global: Hershey, PA, USA, 2020; pp. 162–169.
11. Dartan, G.; Cevik, M.; Aksu, M.; Can, Z.; Keskins, Y.; Yurdun, T.; Süsleyici, B. Investigation the Effects of Treatment Plants o Heavy Metal Levels and Mutagenicity of Wastewaters. *Fresenius Environ. Bull.* **2022**, *31*, 470–477.
12. Holkar, C.R.; Jadhav, A.J.; Pinjari, D.V.; Mahamuni, N.M.; Pandit, A.B. A critical review on textile wastewater treatments: Possibl approaches. *J. Environ. Manag.* **2016**, *182*, 351–366. [CrossRef]
13. Tognella, M.M.P.; Falqueto, A.R.; Espinoza, H.D.C.F.; Gontijo, I.; Gontijo, A.B.P.L.; Fernandes, A.A.; Albino, J. Mangroves as trap for environmental damage to metals: The case study of the Fundão Dam. *Sci. Total Environ.* **2022**, *806*, 150452. [CrossRef]
14. Ardila-Leal, L.D.; Poutou-Piñales, R.A.; Pedroza-Rodríguez, A.M.; Quevedo-Hidalgo, B.E. A brief history of colour, the enviror mental impact of synthetic dyes and removal by using laccases. *Molecules* **2021**, *26*, 3813. [CrossRef] [PubMed]
15. Chowdhary, P.; Bharagava, R.N.; Mishra, S.; Khan, N. Role of industries in water scarcity and its adverse effects on environmen and human health. In *Environmental Concerns and Sustainable Development*; Springer: Singapore, 2020; pp. 235–256.
16. Adeel, S.; Anjum, F.; Zuber, M.; Hussaan, M.; Amin, N.; Ozomay, M. Sustainable Extraction of Colourant from Harmal Seed (*Peganum harmala*) for Dyeing of Bio-Mordanted Wool Fabric. *Sustainability* **2022**, *14*, 12226. [CrossRef]
17. Samanta, A.K.; Konar, A. *Dyeing of Textiles with Natural Dyes*; IntechOpen: London, UK, 2011; pp. 30–56.
18. Adeel, S.; Hasan, M.U.; Batool, F.; Ozomay, M.; Hosseinnezhad, M.; Amin, N.; Hussaan, M. Eco-friendly bio-dyeing of bio-treatec nylon fabric using Esfand (*P. harmala*) based yellow natural colorant. *J. Eng. Fibers Fabr.* **2022**, *17*, 15589250221091265. [CrossRef
19. Patra, A.K.; Pariti, S.R.K. Restricted substances for textiles. *Text. Prog.* **2022**, *54*, 1–101. [CrossRef]
20. Deng, D.; Lamssali, M.; Aryal, N.; Ofori-Boadu, A.; Jha, M.K.; Samuel, R.E. Textiles wastewater treatment technology: A reviev *Water Environ. Res.* **2020**, *92*, 1805–1810. [CrossRef]
21. Patti, A.; Acierno, D. Towards the Sustainability of the Plastic Industry through Biopolymers: Properties and Potential Application to the Textiles World. *Polymers* **2022**, *14*, 692. [CrossRef]
22. Filho, W.L.; Ellams, D.; Han, S.; Tyler, D.; Boiten, V.J.; Paço, A.; Moora, H.; Balogun, A.-L. A review of the socio-economi advantages of textile recycling. *J. Clean. Prod.* **2019**, *218*, 10–20. [CrossRef]

3. Wróblewska-Krepsztul, J.; Rydzkowski, T.; Borowski, G.; Szczypiński, M.; Klepka, T.; Thakur, V.K. Recent progress in biodegradable polymers and nanocomposite-based packaging materials for sustainable environment. *Int. J. Polym. Anal. Charact.* **2018**, *23*, 383–395. [CrossRef]
4. Liman, L.R.; Islam, M.T.; Repon, R.; Hossain, M.; Sarker, P. Comparative dyeing behavior and UV protective characteristics of cotton fabric treated with polyphenols enriched banana and watermelon biowaste. *Sustain. Chem. Pharm.* **2021**, *21*, 100417. [CrossRef]
5. Hayat, T.; Adeel, S.; Rehman, F.U.; Batool, F.; Amin, N.; Ahmad, T.; Ozomay, M. Waste black tea leaves (*Camelia sinensis*) as a sustainable source of tannin natural colorant for bio-treated silk dyeing. *Environ. Sci. Pollut. Res.* **2022**, *29*, 24035–24048. [CrossRef]
6. Özomay, M.; Güngör, F.; Özomay, Z. Determination of optimum dyeing recipe with different amount of mordants in handmade cotton fabrics woven with olive leaves by grey relational analysis method. *J. Text. Inst.* **2022**, *113*, 1048–1056. [CrossRef]
7. Gupta, V.K.; Singh, G.D.; Singh, S.; Kaul, A. *Medicinal Plants: Phytochemistry, Pharmacology and Therapeutics*; Daya Publishing House: New Delhi, India, 2010; Volume 1.
8. González-Minero, F.J.; Bravo-Díaz, L. The use of plants in skin-care products, cosmetics and fragrances: Past and present. *Cosmetics* **2018**, *5*, 50. [CrossRef]
9. Arnason, T.; Hebda, R.J.; Johns, T. Use of Plants for Food and Medicine by Native Peoples of Eastern Canada. *Can. J. Bot.* **1981**, *59*, 2189–2325. [CrossRef]
0. Ural, E.; Özomay, Z.; Özdemir, L. Determination of the effect of palm oil ink on print quality. *Mus Alparslan Univ. J. Sci.* **2018**, *6*, 533–537.
1. Gungor, F.; Akkaya, M. Choosing the appropriate mordant via multi-criteria decision making methods in natural dyeing with tea extract. *Sci. Rev.* **2016**, *2*, 14–21.
2. Özomay, M. Kekik ile Doğal Boyama Yapılan Hatay İpeğinin Haslık Sonuçlarının Çok Kriterli Karar Verme Yöntemiyle En Uygun Seçeneğinin Belirlenmesi. *Avrupa Bilim Ve Teknol. Derg.* **2021**, *31*, 531–538.
3. Rasool, W.; Adeel, S.; Batool, F.; Ahmad, S.A.; Mumtaz, S.; Hussaan, M. Environmental friendly silk and cotton dyeing using natural colorant of Bougainvillea (*Bougainvillea glabra*) flowers: The sustainable approach towards textile industry. *Environ. Sci. Pollut. Res.* **2022**, 1–9. [CrossRef]
4. Özomay, M. *Türkiye'de Yöresel Dokunan bez örneklerinin Doğal Boyarmaddeler ile gri Ilişkisel Analiz Yöntemi Kullanılarak Boyama özelliklerinin Belirlenmesi*; Marmara Üniversitesi Fen Bilimleri Enstitüsü: Marmara, Turkey, 2016.
5. Özomay, M.; Akalın, M. Optimization of fastness properties with gray relational analysis method in dyeing of hemp fabric with natural and classic mordant. *J. Nat. Fibers* **2022**, *19*, 2914–2928. [CrossRef]
6. Adeel, S.; Anjum, M.N.; Ahmad, M.N.; Rehman, F.; Saif, M.J.; Azeem, M.; Amin, N. Eco-friendly dyeing of cotton fabric with waste tea leaves-based tannin natural dye. *Glob. Nest J.* **2021**, *23*, 365–369.
7. Adeel, S.; Habib, N.; Batool, F.; Amin, N.; Ahmad, T.; Arif, S.; Hussaan, M. Environmental friendly exploration of cinnamon bark (*Cinnamomum verum*) based yellow natural dye for green coloration of bio-mordanted wool fabric. *Environ. Prog. Sustain. Energy* **2022**, *41*, e13794. [CrossRef]
8. Naveed, R.; Bhatti, I.A.; Adeel, S.; Ashar, A.; Sohail, I.; Khan, M.U.H.; Masood, N.; Iqbal, M.; Nazir, A. Microwave assisted extraction and dyeing of cotton fabric with mixed natural dye from pomegranate rind (*Punica granatum* L.) and turmeric rhizome (*Curcuma longa* L.). *J. Nat. Fibers* **2020**, *19*, 248–255. [CrossRef]
9. Özomay, M.; Özomay, Z.; Güney, H. Boraks Ile Mordanlanmıs Ipeklı Kumasların Renk ve Haslık Özelliklerinin Araştırılması. In Proceedings of the V INES Science Technolohy and Innovation Congress (STI 2019), London, UK, 26 July 2019.
0. Akkaya, M.; Eyupoglu, S. The examination of the effects of *Quercus aegilops* extract used as natural mordant on colourfulness features of natural dyeing. In Proceedings of the 2016 IEEE NW Russia Young Researchers in Electrical and Electronic Engineering, St. Petersburg, Russia, 2–3 February 2016; pp. 108–111. [CrossRef]
1. Batool, F.; Adeel, S.; Iqbal, N.; Azeem, M.; Hussaan, M. Sustainable natural coloring potential of bitter gourd (*Momordica charantia* L.) residues for cotton dyeing: Innovative approach towards textile industry. *Environ. Sci. Pollut. Res.* **2022**, *29*, 34974–34983. [CrossRef] [PubMed]
2. Islam, M.T.; Repon, M.R.; Liman, M.L.R.; Hossain, M.M.; Al Mamun, M.A. Functional modification of cellulose by chitosan and gamma radiation for higher grafting of UV protective natural chromophores. *Radiat. Phys. Chem.* **2021**, *183*, 109426. [CrossRef]
3. Botteri, L.; Miljković, A.; Glogar, M.I. Influence of Cotton Pre-Treatment on Dyeing with Onion and Pomegranate Peel Extracts. *Molecules* **2022**, *27*, 4547. [CrossRef]
4. Islam, M.T.; Repon, M.R.; Liman, M.L.R.; Hossain, M.M.; Al Mamun, M.A. Plant tannin and chitosan-templated cellulose for improved absorption of UV protective natural chromophores. *Sustain. Chem. Pharm.* **2021**, *21*, 100452. [CrossRef]
5. Güngör, F.; Akkaya, M. Bayburt Yünü İçin Meyankökü İle Boyama Reçetesinin Çok Ölçütlü Karar Verme Yöntem-leriyle Karşılaştırmalı Seçimi. In Proceedings of the Uluslararası İstanbul Tekstil Konferansı—Anadolu'ya Dokunan Bezler, Istanbul, Turkey, 21–23 March 2016; pp. 713–721.
6. Aruldoss, M.; Lakshmi, T.M.; Venkatesan, V.P. A survey on multi criteria decision making methods and its applications. *Am. J. Inf. Syst.* **2013**, *1*, 31–43.
7. Özomay, M.; Massadikova, G. Hazırgiyim Üretiminde Çok Kriterli Karar Verme. In *Üretim ve Hizmet Sektörlerinde Tematik Yak-laşımlar: Performans ve Verimlilik Önerileri*; Gazi Kitabevi: Ankara, Türkiye, 2022; pp. 137–158.

48. Güngör, F.; Akkaya, M. Multi Moora Yöntemi ile Eflani Bezi nin Uygun Mordan Konsantrasyonu İçin Bir Uygulama. In Proceedings of the Uluslararası İstanbul Tekstil Konferansı—Anadolu'ya Dokunan Bezler, Istanbul, Turkey, 21–23 March 2016; pp. 692–699.
49. Pergher, I.; Almeida, A.T.D. Determining production and inventory parameters: An integrated simulation and mavt approach with tradeoff elicitation. *Pesqui. Oper.* **2018**, *38*, 87–97. [CrossRef]
50. Velasquez, M.; Hester, P.T. An analysis of multi-criteria decision making methods. *Int. J. Oper. Res.* **2013**, *10*, 56–66.
51. Keeney, R.L. Multiplicative Utility Functions. *Oper. Res.* **1974**, *22*, 22–34. [CrossRef]
52. Dyer, J.S.; Fishburn, P.C.; Steuer, R.E.; Wallenius, J.; Zionts, S. Multiple Criteria Decision Making, Multiattribute Utility Theory: The Next Ten Years. *Manag. Sci.* **1992**, *38*, 645–654. [CrossRef]
53. Bedford, T.; Cooke, R. A new generic model for applying MAUT. *Eur. J. Oper. Res.* **1999**, *118*, 589–604. [CrossRef]
54. Taufik, I.; Alam, C.N.; Mustofa, Z.; Rusdiana, A.; Uriawan, W. Implementation of Multi-Attribute Utility Theory (MAUT) method for selecting diplomats. *IOP Conf. Ser. Mater. Sci. Eng.* **2021**, *1098*, 032055. [CrossRef]
55. Cegan, J.C.; Filion, A.M.; Keisler, J.M.; Linkov, I. Trends and applications of multi-criteria decision analysis in environmental sciences: Literature review. *Environ. Syst. Decis.* **2017**, *37*, 123–133. [CrossRef]
56. Özomay, M.; Özomay, Z. The effect of temperature and time variables on printing quality in sublimation transfer printing on nylon and polyester fabric. *Avrupa Bilim Ve Teknol. Derg.* **2021**, *23*, 882–891. [CrossRef]
57. Brudzyńska, P.; Sionkowska, A.; Grisel, M. Leather Dyeing by Plant-Derived Colorants in the Presence of Natural Additives. *Materials* **2022**, *15*, 3326. [CrossRef] [PubMed]
58. *ISO 105-C06*; Test for Colour Fastness of Textiles-Colour Fastness to Washing. International Standard Organization: Geneva, Switzerland, 2021.
59. *ISO A05*; Textiles-Tests of Colour Fastness-Part A05: Instrumental Assessment of Change in Colour for Determination of Grey Scale Rating. International Standard Organization: Geneva, Switzerland, 1996.
60. *TS 1008 EN ISO 105-B02*; Textiles-Tests for Colour Fastness-Part B02: Colour Fastness to Artificial Light: Xenon Arc Fading Test. International Standard Organization: Geneva, Switzerland, 2013.
61. Alp, İ.; Öztel, A.; Köse, M.S. Entropi Tabanlı Maut Yöntemi İle Kurumsal Sürdürülebilirlik Performansı Ölçümü: Bir Vaka Çalışması. *Ekon. Ve Sos. Araştırmalar Derg.* **2015**, *11*, 2.
62. Güngör, F. Maut Yöntemi İle 'Ayancık Keteni' nin Şap Mordanına Göre Uygun Boyama Reçetesinin Belirlenmesi. In Proceedings of the Uluslararası İstanbul Tekstil Konferansı—Anadolu'ya Dokunan Bezler, Istanbul, Turkey, 21–23 March 2016; pp. 678–792.

sustainability

MDPI

rticle

\ Sustainable Approach to Study on Antimicrobial and Mosquito Repellency Properties of Silk Fabric Dyed with Neem *Azadirachta indica)* Leaves Extractions

Jusrat Jahan [1,2,*] and Sharfun Nahar Arju [1]

[1] Department of Wet Process Engineering, Bangladesh University of Textiles, Tejgaon, Dhaka 1208, Bangladesh
[2] Department of Textile Engineering, Northern University Bangladesh, Dhaka 1230, Bangladesh
* Correspondence: nusrat.jahandte@nub.ac.bd; Tel.: +88-01677091204

Abstract: The present research work was conducted on developing sustainable production of mulberry filament silk fabric dyed with different extracted dye solutions based on extraction ratios of 1:4, 1:6, 1:8, and 1:10 from neem *(Azadirachta indica)* leaves. The research work focused on evaluating the antimicrobial and mosquito repellent properties of dyed silk fabric. In the experiment, the samples were dyed using the exhaust method at different dye bath concentrations i.e., 15 g/L, 20 g/L, and 25 g/L in the presence and absence of mordant at 80 °C maintaining 1:60 dyeing liquor ratio. The absorbance of the extracted dye solutions was determined with a UV/VIS spectrophotometer, which detected the highest absorbance of 7.73 at the peak 490 nm of λmax of 1:4 extracted dye solution. Fourier transform infrared (FTIR) spectroscopy was used to investigate the chemical structure of dyed fabrics; however, no chemical changes or bond formation occurred; instead, dye particles were deposited on the fabric layer, indicating the presence of bioactive components. Allergy test was also performed to confirm allergic reactions of neem extract on human skin. The antimicrobial activity of extracted dye solutions and dyed samples was estimated against Gram-positive (*Staphylococcus aureus*) and Gram-negative (*Escherichia coli*) bacteria using the agar diffusion method and mosquito repellency of fabrics were examined by the cage method. The results emphasized that dyed fabric with the highest concentrated dye solution, 1:4 extraction, and highest dye bath concentration, 25 g/L along with mango bark mordant solution, possesses the highest antimicrobial activity in terms of an inhibition zone of 0.67 mm against Gram-negative bacteria and 0.53 mm against Gram-positive bacteria obtained after incubation, and the highest mosquito repellent of 75% due to the absorption of active bio constituents. The experimental results also determined that the dyed fabric with 1:4 extracted dye solution exhibited good antimicrobial (inhibition zone, 0.65 mm against *E. coli* and 0.52 mm against *S. aureus*) and mosquito repellent property (66.67%). The experimental study also revealed that Potassium dichromate mordant reduced the antimicrobial (inhibition zone, 0.05 mm against *E. coli* and no inhibition against *S. aureus*) and mosquito repellent action (33.33%). In conclusion, the data revealed that the increase in the extraction ratio of dye solution and dye bath concentration has no impact on the silk fibroin; it only impacts what is deposited on the fabric layer that improves its antimicrobial and mosquito repellency. The current research showed that neem leaves were found to have a beneficial effect in controlling microorganisms and mosquitoes through a sustainable approach.

Keywords: silk filament; neem extraction; spectrophotometer; FTIR; antimicrobial; mosquito repellency; sustainable

itation: Jahan, N.; Arju, S.N. A ıstainable Approach to Study on ntimicrobial and Mosquito epellency Properties of Silk Fabric yed with Neem *(Azadirachta indica)* eaves Extractions. *Sustainability*)22, *14*, 15071. https://doi.org/ .3390/su142215071

cademic Editor: Abu Naser Md hsanul Haque

eceived: 13 October 2022
ccepted: 9 November 2022
ıblished: 14 November 2022

1. Introduction

Silk is a fibrous fibroin polymer that reveals a breathtaking selection of fabric from antiquity to the early twenty-first century [1]. Traditionally, silk has been used from royal wardrobes to silk bedding with different brand-new commercial and industrial applications.

One modern application of silk is suture materials and medical dressing incorporated into the manufacture of holograms and drug delivery systems [2–4]. Silk is hypoallergenic and thus a suitable choice of fabric for those with asthma, allergies, or sensitive skin, and making it effective in medical textiles. Silk can help people ease their sensitivities, as it naturally wards off some of the world's most common allergens, such as dust, mites, and mold. Silk is a naturally biodegradable and recyclable material [5]. Synthetic materials cannot be broken down by bacteria alone over time. Consequently, clothing made out of synthetic materials is generally not recyclable or biodegradable [6]. Instead of several commercial species of silkworm larvae, *Bombyx mori* is widely used for silk farming. Today silk production is being maintained according to both traditional and newer methods [7 A total of 192,692.45 metric tonnes of silk were produced worldwide in 2016, representing an increase of 1.26 times over 2012. Silk production in China and India accounted for 97.94 percent of all silk production worldwide in 2016. The value of silk exports in 2016 was USD 2149 Million, which represents a reduction of 31.49 percent from 2012. Among the top silk exporters are China (53.9%), Italy (13.4%), India (4.2%), Romania (4.2%), and France (3.7%), which together total 79.2% of the world's silk exports [8]. Neem, *Azadiracht indica,* is regarded as a promising tree species [9], considered a safe medicinal plant that has been used traditionally for centuries [10,11]. Throughout the world, especially in the Indian subcontinent, the neem tree is referred to as a village pharmacy because it cures diseases and is a rich source of antioxidants [12]. Neem displays various medicinal properties such as antioxidant, anti-inflammatory, anticancer, antiviral, antibacterial, antifungal, antiulcer hepatoprotective, antipyretic, wound healing activities, etc. [13–15].

Today, with the increase in awareness of the importance of sustainability, the demand for naturally sourced fibers and dyes in the textile sector has been receiving great attention due to biodegradability and health protection properties, such as microbial and insect resistance [16–18]. Silk fiber has a great appeal in clothing for its comfort, hand feel durability, and aesthetic look, dyed either with synthetic or natural dyes. Silk fiber itself is microbial and insect resistant by nature, while adding natural dyes will enhance the applicability in the medical textile industry. In addition, the neem plant has medicinal value in the Indian subcontinent and tropical countries as a source of therapeutic agents Previous studies using the neem plant showed that it contains active substances [19] with multiple medicinal properties. Additionally, it has antibacterial properties [20,21] and might be used for controlling airborne bacterial contamination in residential premises In recent years, dyeing of textile materials with natural dyes is gaining popularity as synthetic dyes are persistent, bio-accumulative, and non-biodegradable. Various studies have been performed in the field of natural dyeing on textiles materials [22–24]. A study on enzymatic natural dyeing of cotton and silk fabrics without metal mordants was performed by Padma S. Vankar, et. al. (2007) [25]. A great deal of research has been performed to incorporate the medicinal properties of neem extract as natural dye on textile materials especially cotton and cotton blended fabric. A recent work was found on antimicrobial activity of silk fabric treated with natural dye extract from neem leaves by Abd El Aty et. al (2018) [26]. They used aqueous neem extract as a natural dye to develop antimicrobial properties of silk fabric. They found that the highest neem dye bath concentration (240 g/L exhibited good inhibitory effects among different dye bath concentrations (40, 80, 120 160, 200, and 240 g/L), while using one extracted dye solution from neem leaves. The neem-dyed silk fabrics offered an excellent antibacterial activity against Gram-positive bacteria and Gram-negative bacteria but no inhibitory activities against yeast and fungi were detected. The limitation of the research was that there was no evidence of the influence of mordant and the efficacy of antimicrobial properties after several wash cycles. Another limitation was that they did not show other medicinal properties of neem leaves, especially anti-insect properties, which might show the mosquito repellency of dyed fabric. In view of the fact that neem leaves contain azadiradirachtin, which is effective against in-house flies and is responsible for providing mosquito-repellent properties to neem-dyed fabric the present study will demonstrate both antibacterial and mosquito-repellency properties using mulberry filament silk fabric dyed with the extracted dye solutions from neem leaves. This study will also show the influence of synthetic and natural mordant on antimicrobial and mosquito-repellent properties.

Therefore, in tropical countries, disease from mosquitoes is one of the greatest problems faced by humans every day. Mosquitoes cause more human suffering than any other organism. In general, mosquitoes are more active around dawn and dusk, although they may also seek hosts during the day. It is possible to avoid being bitten using attractants to attract mosquitoes to other areas, using a repellent, and avoiding actions that may decrease the effectiveness of the repellent. One of the active principle ingredients of natural insecticide is liminoid, such as nimocinolide and isonimocinolide, that is contained in the leaves of the neem plants and is responsible for the bitter test of neem leaves [27]. The neem plant also contains azadiradirachtin, which is effective against in-house flies and when a mosquito consumes this active compound, it destroys its reproductive system. It has been established that neem oil and its leaves extract at an appropriate amount have a repellency power against mosquitoes [28]. Many researchers researched adding a mosquito-repellent finish to textiles, especially to cotton fabric using chemical repellents and natural extraction such as mint, basil, neem plant, etc. [29–32]. Dhara Shukla, et.al made an experimental work on effective formulation of mosquito repellent from plant sources [33]. A. Gupta and Dr. A Singh did research on the development of mosquito repellent finished cotton fabric using eco-friendly mint and achieved moderate repellency on textile materials [34]. However, there is no research that considered silk fabric for imparting mosquito repellent property.

This study of antimicrobial and mosquito repellency properties of dyed silk fabric with different extracted dye solutions from neem leaves is unique from other researches as no study considered mulberry filament silk fiber as a dyeing material with neem extracted solutions of different absorbance using synthetic and natural mordant as well as evaluating dyed samples' properties in terms of mosquito repellency and microbial resistance in a single study. A few studies demonstrated natural dyes that are used for textiles, in particular cotton and cotton blended fabric, to evaluate the antimicrobial activity, and found good inhibitory results with coating technology, with a synthetic antibiotic, resin or cross linker involved in achieving antimicrobial properties [35,36]. Furthermore, protein-based fiber silk has better absorbance and wettability [37] than cellulosic fibers and has better dyeing and functional properties [38]. This is the reason, the current study is using mulberry filament silk fabric as textile materials and neem leaves extract as a natural dye to demonstrate the antimicrobial and mosquito repellent properties of the dyed fabrics. A new line of protective medical equipment and dressing materials for wound healing has been opened in the medical sector through this experimental work. Sustainable dyed silk fabric with neem extracts will be a reliable substitute for antimicrobial and mosquito-repellent textile products.

2. Experimental Procedure

2.1. Materials

2.1.1. Fabric

A well-degummed pure mulberry filament silk (7 filament ply per yarn) fabric (Figure 1) was used in the experiment. Degumming of silk involves the subsequent removal of sericin attached to the cleavage of peptide bonds of silk fibroin either by hydrolytic or enzymatic methods [39]. Silk fabric was sourced from Shopura Silk Mill Ltd., Rajshahi, Bangladesh. The ends per inch (EPI), Picks per inch (PPI), Grams per square meter (GSM), and pH of silk fabric was 110, 85, 37, and 7.6, respectively. The CIE whiteness and tint (D65 illuminator, 10 degree observer) of degummed fabric were 42.53% and −2.04.

2.1.2. Dye and Mordant

Neem (*Azadirachta indica*) leaves were used as a natural dye source for aqueous extraction that was collected from Dhaka, Bangladesh. Synthetic mordant, Potassium dihromate (Brand: MERCK, CAS No. 7778-50-9, Origin: India) was purchased from the local market in Dhaka of Bangladesh and the chemical content of K2Cr2O7 (Assay) was 99% and others were chloride, sulfate, and sodium. A natural mordant, Mango bark solution was extracted from the bark of the mango (*Mangifera indica*) plant. A pictorial view of the leaves of neem plant, bark of mango plant and Potassium dichromate is represented in Figure 2.

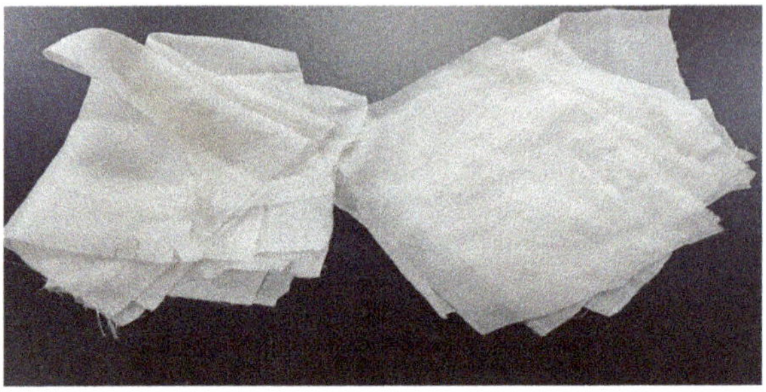

Figure 1. Mulberry filament silk fabric.

(**a**) (**b**) (**c**)

Figure 2. Pictorial view of natural dye source and mordant; (**a**) Dried Neem leaves, (**b**) dried and powdered Mango bark, and (**c**) Potassium dichromate.

2.2. Methods

Experimental works used aqueous dye extraction from the leaves of the neem plant and exhaust dyeing of mulberry filament silk fabric to develop properties of antimicrobial and mosquito repellent. To examine the properties of silk fabrics, different characterization methods were performed. A schematic diagram of experimental work is shown in Figure 3. Different characterizing devices were used for experimental procedures. A Lambda 365 UV/Vis spectrophotometer (Brand: Perkin Elmer, Model: 365K7062102, Software version: UV Express-Version 4.1.1 origin: Germany) was used for measuring the absorbance of extracted solutions. A lab IR smart dyeing machine (Brand: Rapid, Origin: Turkey), digital precision balance (Brand: Kern, Model: EW-220-3NM, Capacity 0.001—220 gm, Origin: Germany), and digital pH meter (Brand: Mettler Toledo 30019029 Model S220 seven compact, Origin: Switzerland,) were used for dyeing experiment. A spectrophotometer (Brand: Data Color, Model: Data color 850, Software: DCI 2:1:1, Origin Switzerland) was used for analyzing the color strength of dyed samples. Additionally used different color fastness devices such as Crockmaster (Brand: James H. Heal, Origin: United Kingdom), Wascator, or Electrolux (Brand: James H. Heal, Model: 290, Origin: United Kingdom), Perspirometer (Brand: James H. Heal, Origin: United Kingdom), and Xenon Test Chamber (Brand: Q-SUN, Model B02, Origin: USA). Fourier Transform Infrared (FTIR Spectroscopy (Brand: Perkin Elmer, Model: Spectrum 2, Origin: Germany) was used to identify bioactive materials.

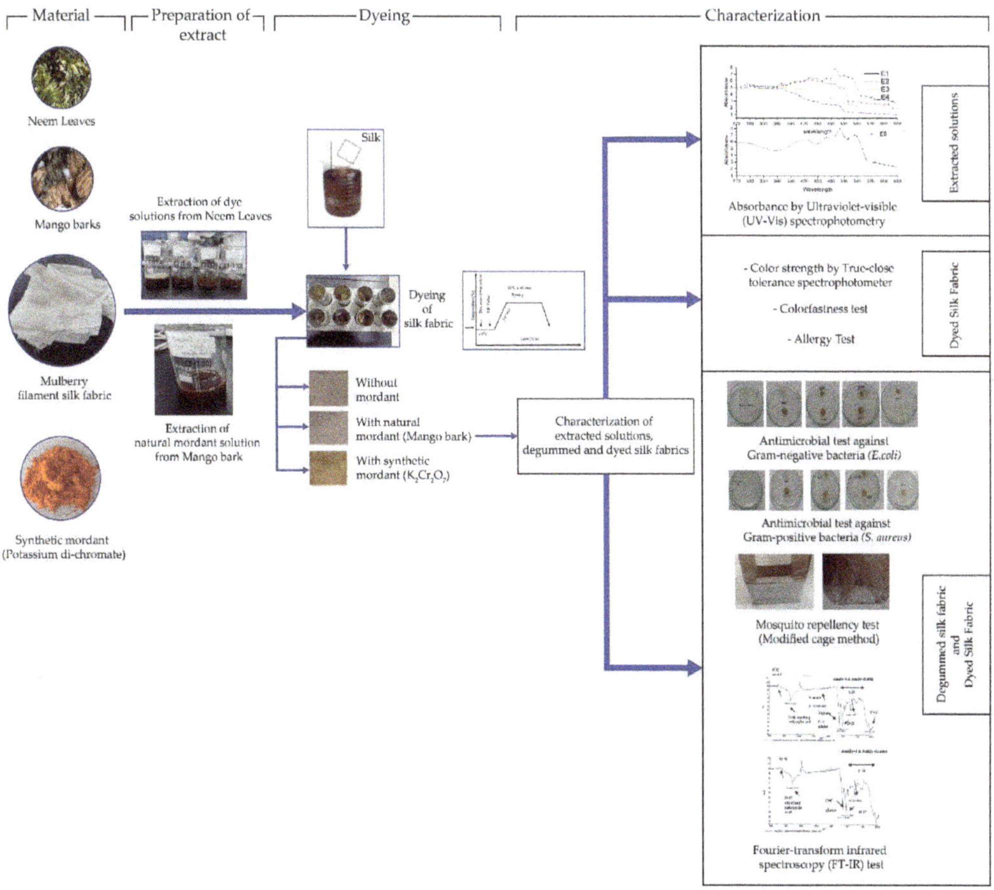

Figure 3. A schematic diagram of experimental design.

2.2.1. Extraction

The aqueous extraction method was employed for extracting coloring matter from the leaves of the neem plant. To begin with, neem leaves were collected, then washed, cleaned, and dried in sunlight for 24 h to remove moisture from the leaves, and then weighed. Again, the mango bark of the mango plant was collected, cleaned, dehydrated for 10 days in proper sunlight, and powdered. The extracted dye solutions were prepared with different material liquor ratios of 1:4, 1:6, 1:8, and 1:10 whereas dried materials were 30 g each. To facilitate extractions, an electric burner was used to maintain 100 °C temperature by covering the lid for 100 min and simmering for 8 h to cool the extraction and filter the dye solutions. Similarly, mango bark mordant solution was also prepared by aqueous extraction maintaining a 1:30 liquor ratio of 30 g of powdered material (Figure 4). Calculated the pH of the extracted solutions to facilitate the dyeing process as shown in Figure 5. The pH meter was calibrated with pH 4, pH 7, and pH 10 buffer solutions before measuring pH. Every 8-h shift, the pH meter was calibrated in the laboratory. For acidic pH 4, a phosphate buffer was used and for alkaline pH, a 10 ammonium buffer solution was used. For pH 7, a sodium phosphate dibasic dihydrate buffer solution was used. Due to differences in pH and liquor ratios, the color of the extracted solutions varied from dark brown to light brown, represented in Figure 6.

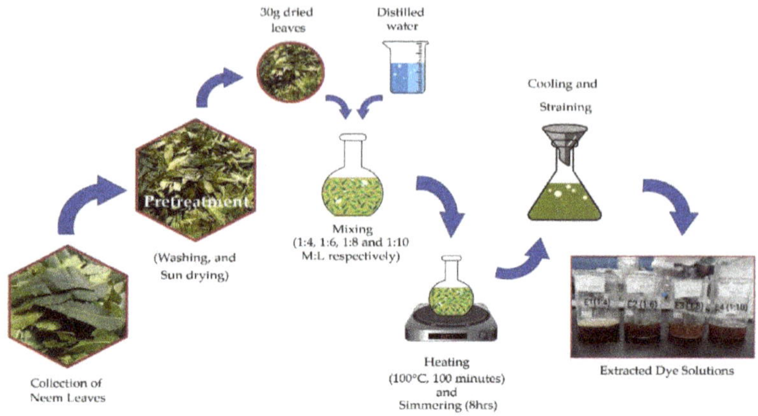

(**a**)

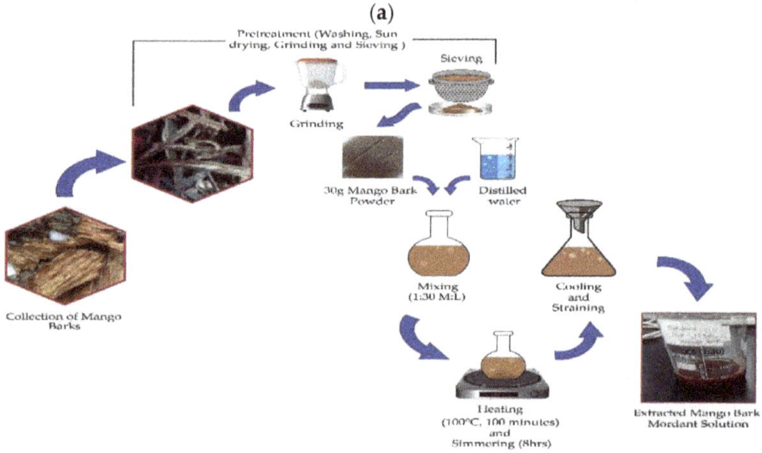

(**b**)

Figure 4. A schematic diagram of aqueous extraction: (**a**) dye solutions; (**b**) mango bark mordant solution.

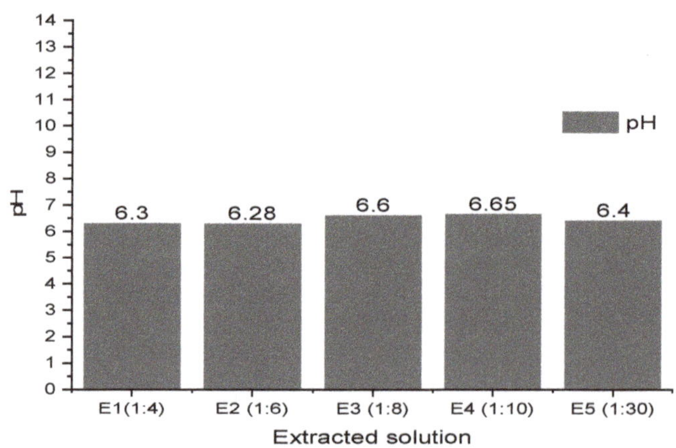

Figure 5. pH of extracted solutions.

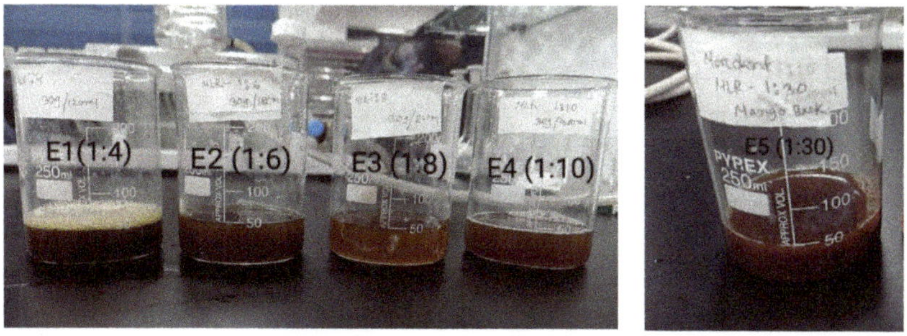

(a) (b)

Figure 6. Extracted solutions, (**a**) dye solutions (**b**) mango bark mordant solution.

2.2.2. Dyeing

Samples of silk fabric were dyed with the extractions individually and also in the presence of mordant (5% owf) through a simultaneous mordanting process in the CHT laboratory in Bangladesh. Samples were dyed using different dye bath concentrations such as 15 g/L, 20 g/L, and 25 g/L with a 1:60 dyeing ratio [40,41] for 60 min at 80 °C for each extraction of 1:4, 1:6, 1:8, and 1:10. The samples and dye bath were prepared and loaded into the dyeing machine at 37 °C with a gradient of 2 °C per minute. After dyeing, samples were rinsed with cold water and squeezed and dewatered. Then samples were dried for 1 min at 100 °C and ironed. A Figure S1 of pictorial view of different dyed samples is given in the supplementary materials.

2.2.3. Sample Labeling

Sample labels were defined according to material composition, type of mordant, dye extraction ratio, and bath concentration during dyeing.

S K/M ExBy

where S stands for silk fiber, K/M stands for the type of mordant (K stands for Potassium dichromate and M stands for extracted Mango bark solution), Ex stands for the dye extraction ratio from neem leaves (E1 for 1:4 extraction, E2 for 1:6 extraction, E3 for 1:8 extraction and E4 for 1:10 extraction), and By stands for bath concentration of dyeing (B1 for 25 g/L dye bath, B2 for 20 g/L dye bath and B3 for 15 g/L dye bath). Examples of a few sample labels for understanding: SE1B1 stands for silk fabric dyed with 1:4 extraction of 25 g/L dye bath concentration, SKE1B1 stands for silk fabric dyed with Potassium dichromate mordant and 1:4 extracted dye solution with 25 g/L dye bath concentration. SME1B1 stands for silk fabric dyed with Mango bark mordant solution and 1:4 extracted dye solution with 25 g/L dye bath concentration. A Table S1 of sample's identification with sample description is given in supplementary materials.

2.3. *Characterization*

2.3.1. Ultraviolet-Visible (UV-Vis) Spectrophotometry

Extracted solutions were characterized by assessing absorbance through Lambda 365 UV/VIS spectrophotometer that works on Beer-lambert law (qualitative analysis) [42] whereas absorbance was measured against wavelength of the 200 nm to 800 nm which determine maximum absorbance or optical strength at a particular wavelength of extracted solutions. Additionally, the antimicrobial activity of extracted solutions was assessed following by AATCC TM 147 standard. The antimicrobial was examined against Gram-positive (*Staphylococcus aureus*) and Gram-negative bacteria (*Escherichia coli*) in terms of bacteria inhibition zone in millimeter (qualitative analysis).

2.3.2. True-Close Tolerance Spectrophotometer

The K/S value was assessed through Datacolor 850 spectrophotometer by CMC color space and tolerance system to determine color strength of different dyed samples, which works on the Kubelka-Munk formula [43] that shown in Equation (1):

$$\frac{K}{S} = \frac{(1 - R)^2}{2R} \tag{1}$$

2.3.3. Colorfastness

Different color fastness (C/F) test methods: C/F to rubbing (ISO 105-X12), C/F to wash (ISO 105 C06), C/F to water (ISO 105 E01), C/F to perspiration (ISO 105 E04), and C/F to light (ISO 105 B02) were done to assess fastness properties of dyed samples. Dyed samples of SE1B1, SE2B1, SE3B1, SE4B1, SKE1B1, and SME1B1 were examined to determine colorfastness results.

2.3.4. Antimicrobial Analysis

The antimicrobial property was assessed by AATCC TM 147 standard [44]. Degummed silk fabric, silk samples dyed with the extracted dye solutions (SE1B1, SE1B2, SE2B1, SE2B2, SE3B1, SE3B2, SE4B1, and SE4B2), silk samples dyed with Potassium dichromate and extracted solution from mango bark individually with 1:4 extracted dye solution (SKE1B1, SKE1B2, SME1B1 and SME1B2) were considered to examine of antimicrobial property. The antimicrobial analysis was examined against Gram-positive (*Staphylococcus aureus*) and Gram-negative (*Escherichia coli*) bacteria that expressed in terms of the diameter of zone of inhibition in millimeter.

2.3.5. Mosquito Repellency Analysis

The modified cage method according to the WHO standard [45] was considered to examine mosquito repellency property and many researchers followed the modified cage method for obtaining accurate results [46–48]. A cage with dimensions of 30 cm × 30 cm × 30 cm (WHO cage standard) was constructed from paper board and plastic materials, as shown in Figure 7. The cage was constructed of paper board, and its top was covered with transparent plastic film. Similarly, a 3 cm square opening was made for mosquitoes and samples to enter the cage. To begin with, we used samples that were 4 cm × 3 cm in size, but the results were not satisfactory. Consider a larger sample size (10 × 5 cm) for testing. The samples were placed in various locations in the cage in order to observe the mosquitoes sitting behavior and the cage was shaken two times each at a three-minute interval and three trials were conducted for each test. Samples SE1B1, SE1B2, SE2B1, SE2B2, SE3B1, SE3B2, SE4B1, SE4B2, SKE1B1, SKE1B2, SME1B1, and SME1B2 were considered to examine mosquito repellent properties. Untreated degummed silk fabric was used as a control for evaluating the mosquito repellency percentage of treated samples. The percentage of mosquito repellency was calculated using Equation (2):

$$MR\% = \frac{U - T}{U} \times 100 \tag{2}$$

where U corresponds to the number of mosquitoes on untreated samples or control samples and T represents the number of mosquitoes on treated samples. This is the regularly used formula as stated by different researchers in their research [49].

2.3.6. Allergy Test

An allergy test was made to assess the allergic reaction of the dyed sample. To assess the allergic reaction of the dyed sample, we considered three age groups. For assessing itching or rashes, we tested on a small area of the selected person's skin by wrapping the dyed sample (SE1B1) and then waited for 8 h to observe. If they have a rash, blistering, and swelling in the area or any redness on the skin, then it is considered to be an allergic reaction.

(a)

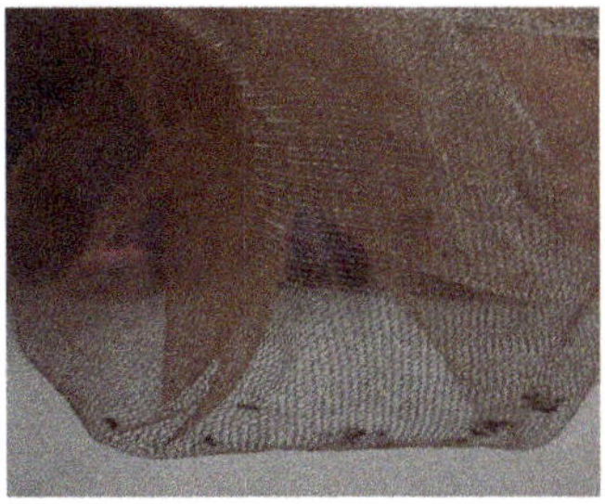

(b)

Figure 7. Preparation for cage method, (**a**) cage (**b**) collected mosquitoes.

2.3.7. Fourier Transform Infrared Spectroscopy (FT-IR)

FT-IR Spectroscopy was used to identify the characteristic structure of untreated and dyed silk fabric. Dyed Samples of SE1B1 SE2B1, SE3B1, SE4B1, SKE1B1, and SME1B1 were tested to observe chemical changes during dyeing with the extracted solutions of neem leaves. The Scanning area was in the range of 4000–400 cm^{-1} and data were plotted into the spectrum.

3. Results and Discussions

3.1. Absorbance of Extracted Solutions

Absorbance of extracted solutions were measured against wavelength of 200–700 nm to determined maximum absorbance at a particular wavelength. After analyzing results, highest absorbance of the extracted solutions was found at 300 nm, 370 nm, 430 nm, 490 nm, 510 nm, and 620 nm of the spectrum (Table 1). In the absorbance spectrum, E1 dye solution detected the highest absorbance, 7.73 at the peak of λmax of 490 nm. E2 detected the highest absorbance of 6.21 at the peak of λmax of 430 nm, E3 detected highest absorbance peak of 5.05 at the λmax of 370 nm and E4 detected highest absorbance peak of 5.21 at the λmax of 300 nm represented in Figure 8. Recent research by Abd El Aty et. al. (2018) showed three peaks for absorbance of 3.765, 3.402, and 3.569 at 320 nm, 370 nm, whereas 1:10 material liquor extraction form neem leaves was considered [26]. Optical wavelengths for the green hue in the spectrum range from 300 nm to 520 nm, while optical wavelengths for the green-yellow hue are from 380 to 510 nm. Based on the results of the test, it was determined that E1 extracted solution had the highest absorbance value at 490 nm in comparison to other extracted dye solutions, and that is why dye solution appears green-yellow color when viewed in that region of the spectrum. The maximum absorbance value of E1 extracted solution suggests that this solution would have the highest concentration of dye among the other extracted solutions [50]. In addition, it had less aqueous content during the extraction process. The absorbance spectrum of the extracted mango bark mordant solution, E5 detected three peaks of λmax: the first peak at λmax, 430 nm of absorbance 6.22; the second peak at λmax, 500 nm of absorbance 7.95; the third peak at λmax, 540 nm of absorbance 6.84 displayed in Table 2.

Table 1. The absorbance of extracted solutions.

Extraction	Wavelength λmax (nm)					
	300 nm	370 nm	430 nm	490 nm	510 nm	620 nm
E1	5.49	5.61	6.44	7.73	6.78	2.96
E2	5.37	5.29	6.21	5.51	4.97	2.01
E3	4.89	5.05	4.89	3.23	2.66	1.97
E4	5.21	4.86	3.19	2.48	1.58	1.06

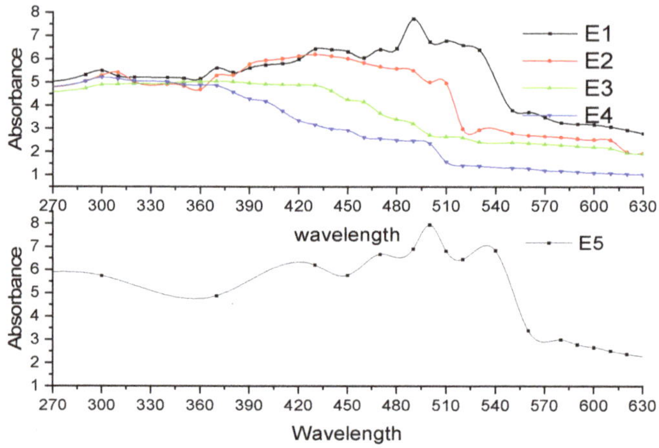

Figure 8. Spectral curve of extracted solutions.

Table 2. The absorbance of mango bark mordant solution.

Extraction	Wavelength, λmax (nm)	Absorbance
E5	430	6.22
	500	7.95
	540	6.84

3.2. Color Strength

The color strength of an object is related to the reflectance value. The darker shade o samples has a high reflectance value and high K/S value. Test results revealed that the colo strength value of dyed samples mordanted with Potassium dichromate, SKE1B1 exhibitec the highest color strength value of 13.8. The dyed sample (SE1B1) with 1:4 extractec dye solution showed the highest color strength value of 13.2 among other dyed sample with other extractions. Test results (Table 3) also revealed that the color strength valu decreased in the case of dyed samples mordanted with mango bark mordant solution. Ir tests, samples dyed with most concentrated dye bath with different extracted solution and mordant exhibited deeper shades and increased K/S values. Thus, in the case o natural dyes, shade development could be improved if concentrated dye baths are used. A Table S2 of color co-ordinate value of dyed samples by spectrophotometer is given in th supplementary materials.

Table 3. Color strength value (K/S) of dyed samples.

Sample	K/S	Sample	K/S	Sample	K/S
SE1B1	13.2	SKE1B1	13.8	SME1B1	11.7
SE1B2	12.4	SKE1B2	12.7	SME1B2	10.6
SE1B3	11.8	SKE1B3	11.6	SME1B3	9.6
SE2B1	12.1	SKE2B1	12.4	SME2B1	10.5
SE2B2	11.7	SKE2B2	11.3	SME2B2	9.7
SE2B3	11.22	SKE2B3	11.1	SME2B3	9.2
SE3B1	11.8	SKE3B1	12.2	SME3B1	8.4
SE3B2	11.41	SKE3B2	11.4	SME3B2	8.2
SE3B3	9.7	SKE3B3	10.1	SME3B3	8.1
SE4B1	11.14	SKE4B1	10.2	SME4B1	8.6
SE4B2	11.12	SKE4B2	9.87	SME4B2	8.1
SE4B3	10.2	SKE4B3	9.85	SME4B3	7.4

3.3. Colorfastness

All color fastness tests were done in terms of color change and color staining except rubbing fastness and light fastness. The color change under dry rubbing was excellent for all dyed samples compared to wet rubbing. Wet rubbing ratings were only 3 for dyed silk with Potassium dichromate mordant (Table 4). Interestingly, all dyed samples had better water fastness ratings of 4 to 4/5 (Table 5) than Potassium dichromate mordanted sample (SKE1B1) since the Potassium dichromate mordant enhanced the color of neem extractions and produced darker and brighter shades. Additionally, dyed samples demonstrated very good colorfastness to wash greyscale rating (4 to 4/5) displayed in Table 6. In the case of colorfastness to perspiration, all dyed samples had better colorfastness to perspiration in terms of color change (4 to 4/5). In terms of color staining, acid perspiration of dyed samples had better results, as demonstrated in Table 7, than alkali perspiration. Compared to the alkali perspiration (Table 8), all dyed samples had better greyscale ratings of 4 and 4/5 on the greyscale, whereas dyed sample SKE1B1 had poor alkali perspiration based on color staining on the multifiber fabric-acetate, cotton, and polyester had 3/4 and all the other samples had better ratings i.e., 4 and 4/5. The light fastness results of all dyed samples was 4 out of 8 on the wool blue scale, except for dyed silk with Potassium dichromate mordant (3/4 blue scale rating) showed in Table 9. Overall, dyed samples were more colorfast to wash based on the comparison of all fastness test results.

Table 4. Greyscale rating of color fastness to rubbing.

Samples	Dry Rubbing	Wet Rubbing
SE1B1	4/5	4
SE2B1	4/5	4
SE3B1	4/5	4
SE4B1	4/5	4
SKE1B1	3/4	3
SME1B1	4/5	4

Table 5. Color change and color staining rating of color fastness to water.

Sample	Color Change	Color Staining					
		Acetate	Cotton	Nylon	Polyester	Acrylic	Wool
SE1B1	4	4	4	4	4	4	4
SE2B1	4	4	4	4	4	4	4
SE3B1	4/5	4/5	4/5	4	4/5	4	4
SE4B1	4	4	4	4/5	4	4	4
SKE1B1	3/4	3/4	3/4	3/4	3/4	3/4	3/4
SME1B1	4	4	4	4	4/5	4	4

Table 6. Color change and color staining rating of color fastness to wash.

Sample	Color Change	Color Staining					
		Acetate	Cotton	Nylon	Polyester	Acrylic	Wool
SE1B1	4/5	4/5	4	4.5	4/5	4	4/5
SE2B1	4	4	4	4	4.5	4	4
SE3B1	4	4	4	4.5	4	4	4
SE4B1	4	4	4	4	4	4	4
SKE1B1	3/4	3/4	4	3/4	3/4	4	4
SME1B1	4	4	4	4	4	4	4

Table 7. Color change and color staining rating of color fastness to acid perspiration.

Sample	Color Change	Color Staining					
		Acetate	Cotton	Nylon	Polyester	Acrylic	Wool
SE1B1	4	4	4	4	4	4	4
SE2B1	4/5	4	4/5	4	4	4/5	4
SE3B1	4.5	4/5	4	4/5	4	4	4
SE4B1	4	4/5	4	4	4	4	4
SKE1B1	4	4	4	4	4/5	4	4
SME1B1	4	4	4	4	4/5	4	4

Table 8. Color change and color staining rating of color fastness to alkali perspiration.

Sample	Color Change	Color Staining					
		Acetate	Cotton	Nylon	Polyester	Acrylic	Wool
SE1B1	4	4	4.5	4	4	4	4
SE2B1	4	4	4	4	4	4	4
SE3B1	4/5	4	4.5	4	4/5	4	4
SE4B1	4/5	4/5	4	4	4/5	4	4
SKE1B1	4	3/4	3/4	4	3/4	4	4
SME1B1	4	4	4	4	4	4	4

Table 9. Blue Scale rating of color fastness to light.

Sample	Color Change
SE1B1	4
SE2B1	4
SE3B1	4
SE4B1	4
SKE1B1	3/4
SME1B1	4

3.4. Antimicrobial Analysis

The dye solutions showed excellent antimicrobial activity, with inhibition zones ranging from 14 to 20 mm for *E. coli* (Gram-negative bacteria) and 10 to 15 mm for *S. aureus* (Gram-positive bacteria) on the petri plates (Figure 9). The antimicrobial activity of extracted solutions showed better results against *E. coli* compared to *S. aureus* which depended on the activity of active substance (quercetin - a polyphenolic flavonoid) in dye extraction against bacteria. The results demonstrate (Figure 10) that different extracted solutions exhibited different inhibition zones against tested bacteria, depending on the optimization of liquor extractions and dry matter concentration with the extraction medium. The extracted solution, E1 (1:4) had the highest absorbance of 7.73 at 490 nm among other extracted solutions and had the highest antimicrobial activity, i.e., 20 mm inhibition zone against *E. coli* and 15 mm inhibition zone against *S. aureus* whereas the extracted solution, E2 had 18 mm inhibition zone against *E. coli* and 13 mm inhibition zone against *S. aureus*. Again E3 and E4 extracted dye solutions had 17 mm and 14 mm inhibition zone against *E. coli* and 12 mm and 10 mm inhibition zone against *S. aureus* respectively. A 16 mm inhibition zone was observed against *E. coli* and a 15 mm inhibition zone was observed against *S. aureus* in the mordant solution, E5. The test results revealed that extracts from neem leaves showed better activity against Gram-negative bacteria. It could be said that the microbial activity of extracted solution will be dependent on the presence of active bio components and the concentration of extracted solution [51,52].

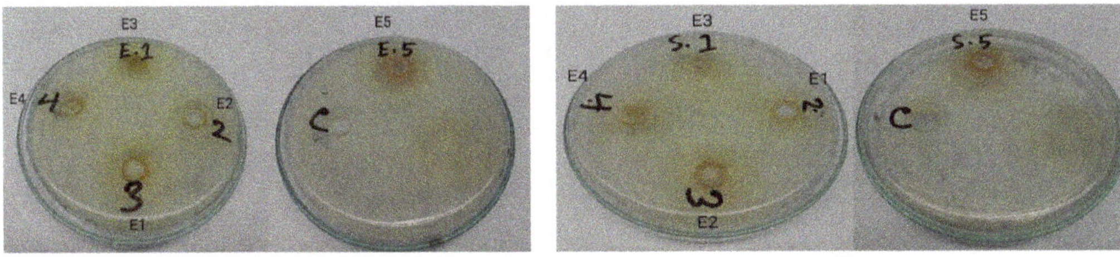

(a) (b)

Figure 9. Analysis of agar diffusion test in terms of zone of inhibition in mm of extracted dye and mango bark mordant solutions on petri plates against bacteria: (**a**) Zone of inhibition (mm) against *Escherichia coli*; (**b**) Zone of inhibition (mm) against *Staphylococcus aureus*.

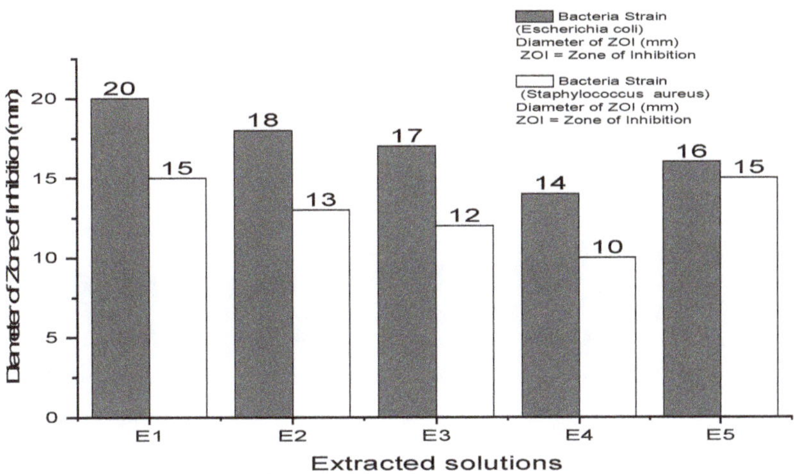

Figure 10. Bar diagram of antimicrobial test in terms of inhibition diameter of extracted solutions against bacteria of *Escherichia coli* and *Staphylococcus aureus*.

Antimicrobial analysis of dyed samples on petri plates after agar diffusion test determined that samples dyed with concentrated dye bath solution with the extracted solution had better inhibition zone against Gram-negative bacteria (Figure 11) and Gram-positive bacteria (Figure 12). Samples dyed with 25 g/L dye bath concentration using different extracted solutions exhibited the diameter of zone of inhibition (ZOI) ranges 0.1 to 0.67 mm against Gram-negative bacteria, *E. coli* and 0.1 to 0.53 mm against Gram-positive bacteria, *S. aureus*. It expressed that a layer of neem extract was formed on dyed samples during dyeing that might have present of phytoconstituents alkaloids, flavonoids, glycosides, and saponins considered the antibiotic elements of plant extract [53]. According to test results shown in Table 10, the SE1B1 sample exhibited better diameter of zone of inhibition against bacteria (0.65 mm against *E. coli* and 0.52 mm against *S. aureus*). The SE2B1 sample exhibited 0.32 mm ZOI diameter against *E. coli* and 0.2 mm ZOI diameter against *S. aureus*. Again, SE1B2 and SE2B2 showed antibacterial activity in terms of diameter of inhibition zone against bacteria that was 0.3 mm and 0.1 mm, respectively against *E. coli* and 0.15 mm and 0.05 mm against *S. aureus*. Samples SE3B2 and SE4B2 had no antimicrobial activity against *E. coli* and *Staphylococcus aureus* but samples, SE3B1, and SE4B1 had poor antimicrobial activity (ZOI: 0.25 mm and 0.2 mm, respectively, against *E. coli* and 0.15 mm and 0.1 mm, respectively, against *S. aureus*).

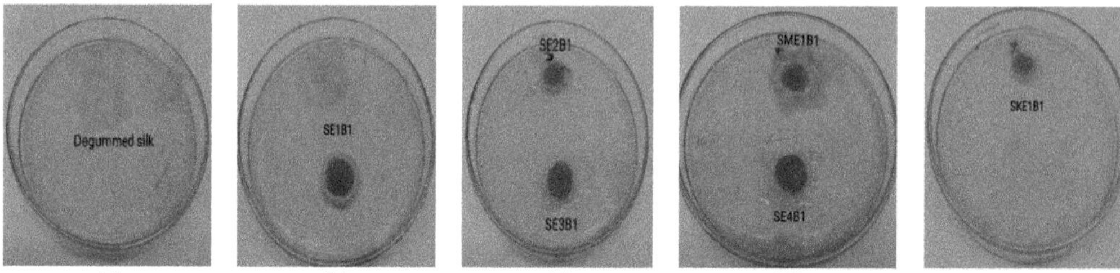

(a) (b) (c) (d) (e)

Figure 11. Analysis of agar diffusion test in terms of zone of inhibition in mm of degummed and dyed silk fabrics on petri plates against Gram-negative bacteria, *Escherichia coli*: (**a**) degummed silk fabric; (**b**) silk fabric dyed with 1:4 extracted dye solution of 25 g/L dye bath concentration (SE1B1); (**c**) silk fabric dyed with 1:6 and 1:8 extracted dye solution of 25 g/L dye bath concentration, respectively (SE2B1, SE3B1); (**d**) silk fabric dyed with 1:10 extracted dye solution of 25 g/L dye bath concentration (SE4B1) and silk fabric dyed with 1:4 dye extraction and Mango bark mordant solution of 25 g/L dye bath concentration (SME1B1) and (**e**) silk fabric dyed with 1:4 dye extraction and Potassium dichromate of 25 g/L dye bath concentration (SKE1B1).

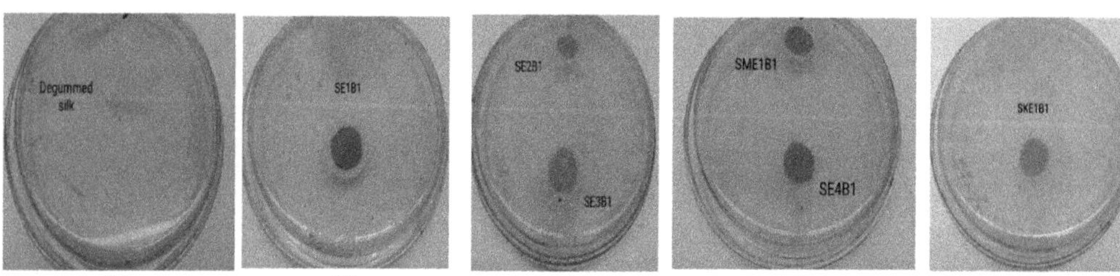

(a) (b) (c) (d) (e)

Figure 12. Analysis of agar diffusion test in terms of zone of inhibition in mm of degummed and dyed silk fabrics on petri plates against Gram-positive bacteria, *Staphylococcus aureus*: (**a**) degummed silk fabric; (**b**) silk fabric dyed with 1:4 extracted dye solution of 25 g/L dye bath concentration (SE1B1); (**c**) silk fabric dyed with 1:6 and 1:8 extracted dye solution of 25 g/L dye bath concentration respectively (SE2B1, SE3B1); (**d**) silk fabric dyed with 1:10 extracted dye solution of 25 g/L dye bath concentration (SE4B1) and silk fabric dyed with 1:4 dye extraction and Mango bark mordant solution of 25 g/L dye bath concentration (SME1B1) and (**e**) silk fabric dyed with 1:4 dye extraction and Potassium dichromate of 25 g/L dye bath concentration (SKE1B1).

Table 10. Diameter of inhibition zone against *Escherichia coli* and *Staphylococcus aureus* of dyed fabrics.

Sample	Bacteria Strain (*Escherichia coli*) Diameter of Inhibition Zone (ZOI) in mm	Bacteria Strain (*Staphylococcus aureus*) Diameter of Inhibition Zone (ZOI) in mm
Degummed silk (undyed)	nil	nil
SE1B1	0.65	0.52
SE1B2	0.3	0.15
SE2B1	0.32	0.2
SE2B2	0.1	0.05
SE3B1	0.25	0.15
SE3B2	nil	nil
SE4B1	0.2	0.1
SE4B2	nil	nil
SKE1B1	0.05	nil
SKE1B2	nil	nil
SME1B1	0.67	0.53
SME1B2	0.34	0.25

In addition, samples dyed with synthetic mordant of Potassium dichromate SKE1B1 showed poor antimicrobial activity against *E. coli* but had no inhibitory zone against *S. aureus*. The synthetic mordanted sample, SKE1B2, did not show any inhibition zone against bacteria. It might be said synthetic mordant could destroy or reduce the antimicrobial action of natural extract during dyeing. A naturally mordanted sample, SME1B1, showed highest inhibitory zone among all dyed samples. Samples, SME1B1 and SME1B2 showed 0.67 mm and 0.34 mm inhibition diameter against *E coli* and 0.53 mm and 0.25 mm inhibition diameter against *S. aureus*. A natural mordant, Mango bark extracted solution also may contain bioactive agents themselves and influence the antimicrobial action of extracted dye solutions during dyeing [54,55].

3.5. Mosquito Repellency Analysis

This research study focused on repellency against in-house flies (mosquitoes) on dyed silk samples with different concentrated neem *(Azadirachta indica)* extract along with mordant. The experimental results determined that dyed samples of different extractions have good mosquito repellent percentage ranges 40 to 66.67 % except SE4B2 sample (Table 11). Sample SE1B1 dyed with E1 (1:4) extracted solution showed an excellent mosquito repellency percentage that was 66.67%. Sample SE2B1 dyed with E2 (1:6) extracted solution showed moderate mosquito repellency percentage that was 60% whereas SE3B1 showed 50% mosquito repellency. SE2B2, SE3B2, SE4B1, and SE4B2 samples were showed an average mosquito repellent percentage that was 40%, 40%, 50%, and 25%, respectively. SE1B1 was dyed with the highest concentrated dye solution that is why it gave better repellency (66.67) compared to other samples dyed with E2, E3, and E4 extraction. Moreover, Samples (SME1B1 and SME1B2) dyed with E1 extraction along with Mango bark mordant solution showed the highest repellency (75% and 66.67%, respectively) as natural mordant mango bark contains anti-insecticide compound itself and it increased the antimicrobial and anti-insecticide propertied on dyed sample [56]. However, synthetic mordant Potassium dichromate destroys the antibacterial and anti-insecticide properties on dyed samples. This is the reason that SKE1B1 and SKE1B2 exhibited the lowest mosquito repellent properties. It could be concluded that probably more concentrated dye solution contains more active bio-compound i.e., Azadiradirachtin and provide better repellency power to mosquitoes and this could be understood through the cage test results, such that as soon as tested mosquitoes identify any discomfort and sense lethal a bioactive compound in dyed samples they tried to fly and escape from the tested area, but when they failed to escape from the cage they tried to sit on an untreated area or eventually died on treated samples due to activity of bio-compound [19,57].

Table 11. Experimental results of mosquito repellent test by cage method.

Sample	U	T	MR% = $\frac{U-T}{U} \times 100$
SE1B1	3	1	66.67
SE1B2	5	2	60
SE2B1	5	2	60
SE2B2	5	3	40
SE3B1	6	3	50
SE3B2	5	3	40
SE4B1	4	2	50
SE4B2	4	3	25
SKE1B1	3	2	33.33
SKE1B2	4	3	25
SME1B1	4	1	75
SME1B2	3	1	66.67

3.6. Allergy Test Analysis

The experimental result determined that the sample dyed with neem *(Azadirachta indica)* extract created no allergic reaction on human skin (Table 12). In the research, the sample was treated with plant extracts as a natural repellent agent that is why no allergic reaction occurred.

Table 12. Experimental allergic test results of the dyed sample, SE1B1.

Age Group	Number of Samples (Person)	Allergic Reaction (Observation)
Teenager (13–19 years)	01	No symptom
Adulthood (18–35 years)	01	No symptom
Older adulthood (56-older)	01	No symptom

3.7. *FT-IR Test Analysis*

In the characteristic pick of silk fabric by FTIR analysis, it is illuminating that untreated degummed silk was identified by four prominent characteristic peaks at 3279.74 cm^{-1} of 95.51 T% which proved the presence of a strong and broad spectrum of O-H stretching of carboxylic acid usually centered on 3000 cm^{-1} [58], 1621.89 cm^{-1} (amide I [59]) of 89.60 T% attributed strong C=C alkene stretching vibrations of the beta-sheet structure of unsaturated ketone [60], 1513.89 cm^{-1} (Amide II [59]) of 89.17 T% proved N-H bending vibrations and 1228.02 cm^{-1} of 93.17 T% attributed medium C-N stretching vibrations represent to amide III [61]. Again peaks at around 3343.12 cm^{-1} are assigned to N-H stretching vibrations and 3071.9 cm^{-1} C-H aromatic stretching vibrations, 2924.4 cm^{-1} assigned to medium C-H aliphatic asymmetry stretching alkane vibrations, 2871.85 cm^{-1} C-H aliphatic symmetry stretching vibrations, 1166.4 attributed to strong C-O stretching tertiary alcohol, and peaks at 500–650 cm^{-1} assigned to C-C stretching vibration [62]. The FTIR spectrum of the dyed sample was nearly resembling with untreated degummed silk fiber with four prominent peaks [63] demonstrated in Figure 13. It is clearly expressed that there are no chemical changes or bond formations occurring on dyed silk fabric, only dye particles deposited on the fabric layer. If neem extract creates bond or made any chemical changes it would give the characteristic peak of the bond. That is the reason that dyed silk samples were quite similar to untreated silk during FTIR analysis [64,65]. In addition, the experimental report noted that major dye absorption occurred in the amide I and amide II bands and due to the concentration a variety of dye extractions, the band shifted a little (1621.89 to 1622.69 cm^{-1} and 1512.72 to 1514.44 cm^{-1}), which proved neem dye does not change the characteristics peaks of mulberry filament silk fabric [64,66]. Therefore, bioactive compounds alkaloid isolate has the N-H, C-H, C=C, C=O, C-N, C-O-C group and flavanone and flavonol [67] focus on the heterocyclic ring carbon, and limonoids are highly oxygenated triterpenes as tetranortriterpenoids grouped into glycones and glucosides [68]. Therefore, the presence of the bioactive compounds in the layer dyed samples was proved by FTIR analysis.

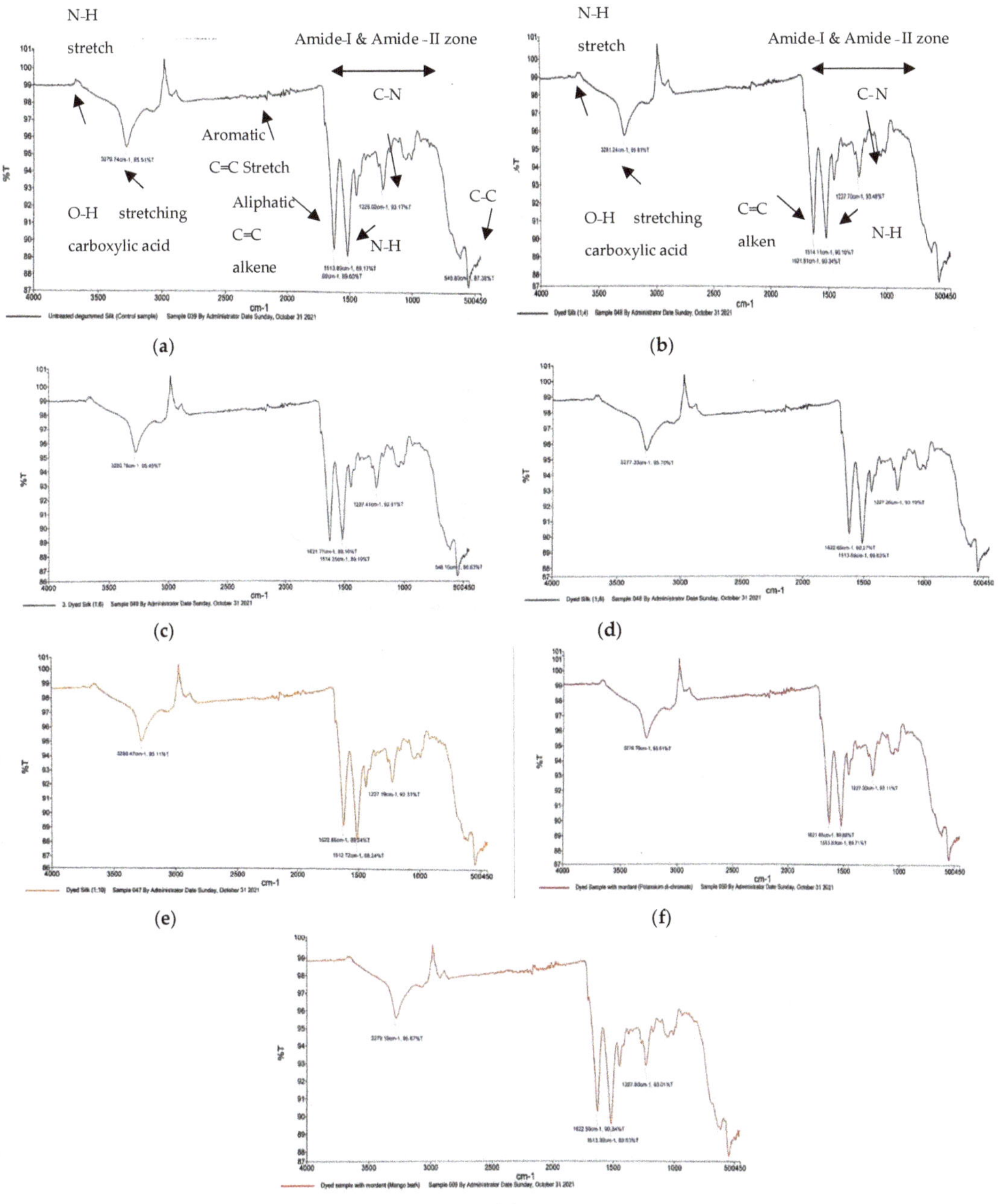

Figure 13. FT-IR spectrum of fabric: (**a**) degummed silk fabric (untreated); (**b**) silk fabric dyed with 1:4 extracted dye solution (SE1B1); (**c**) silk fabric dyed with 1:6 extracted dye solution (SE2B1); (**d**) silk fabric dyed with 1:8 extracted dye solution (SE3B1); (**e**) silk fabric dyed with 1:10 extracted dye solution of 25 g/L dye bath concentration (SE4B1); (**f**) silk fabric dyed with 1:4 extracted dye solution and Potassium di-chromate (SKE1B1). (**g**) silk fabric dyed with 1:4 extracted dye solution and Mango bark mordant solution (SME1B1).

4. Conclusions

As demonstrated in this study, neem *(Azadirachta indica)* leaves extract can be used as a potential source of natural dye for silk fabric. Antimicrobial activity and mosquito repellency of dyed mulberry filament silk swatches with neem leaf extracts were investigated. FTIR analysis showed that dyed samples with bioactive components and neem-treated silk fabric did not cause an allergic reaction on human skin, as shown by an allergy test. The experimental results revealed that extracted dye solutions exhibited significant antimicrobial and anti-mosquito activity. Additionally, the dyed samples demonstrated excellent mosquito repellency activity, but moderate activity against bacterial contamination due to the natural origin, and no binding agent was used. Sample, SE1B1 dyed in the presence of concentrated 1:4 extracted dye solution with 7.73 absorbance at 490 nm without mordant had a 0.65 mm zone of inhibition diameter against *E. coli* and 0.52 mm zone of inhibition diameter against *S. aureus* and was calculated to have 66.67% mosquito repellency. The sample dyed with natural mordant (SME1B1) exhibited the highest antimicrobial activity (0.67 mm ZOI diameter against *E. coli* and 0.53 mm ZOI diameter against *S. aureus*) and excellent mosquito repellent activity (75%) in comparison to dyed samples with extracted dye solutions and synthetic mordant. It could be said that the natural mordant extract itself contains bioactive agents that influence the antimicrobial activity and mosquito-repellent property of the extracted dye solution during dyeing. In contrast, the synthetic mordant somehow destroys or slows down the activity of the extracted dye solution. The inhibition zones indicated that the neem-dyed silk fabrics offered antibacterial activity against *E. coli* and *S. aureus*, and the cage test determined excellent mosquito-repellent properties. Further study of this research would be the efficacy of those properties after several wash cycles. Concisely, neem extracts could be an alternative to synthetic anti-mosquito and antibacterial chemicals and could be commercialized in protective textiles.

Supplementary Materials: The following supporting information can be downloaded at: https://www.mdpi.com/article/10.3390/su142215071/s1, Figure S1: Pictorial view of different dyed samples (a) Degummed silk fabric; (b) Dyed silk fabric with extracted dye solutions from neem leaves; (c) Dyed silk fabric with extracted dye solutions using Potassium dichromate mordant; (d) Dyed silk fabric with extracted dye solutions using Mango bark mordant solution; Table S1: Sample's identification with sample description; Table S2: Color co-ordinates value of dyed samples.

Author Contributions: Conceptualization, methodology, data analysis and writing, N.J. and Supervision, S.N.A. All authors have read and agreed to the published version of the manuscript.

Funding: There is no funding for this work. It is a part of the M.Sc. research work of Bangladesh University of Textiles (BUTEX).

Institutional Review Board Statement: Not applicable.

Informed Consent Statement: Not applicable.

Data Availability Statement: Not applicable.

Acknowledgments: Heartfelt thanks to ASM Hafizur Rahman, Executive Director, CHT laboratory of RH corporation of Aziz group and Johny Yasmin Kanta, General Manager, Hohenstein Laboratories Bangladesh Ltd. for providing lab facilities for the execution of the research. The authors wish to thank Tahmid Hasan, student of the Department of Environmental Science and Engineering, Bangladesh University of Textiles for drawing the schematic diagrams of this manuscript.

Conflicts of Interest: All authors have declared that there is no conflict of interest with this work.

References

1. Datta, R.K.; Nanavaty, M. *Global Silk Industry: A Complete Source Book*; Universal-Publishers: Irvine, CA, USA, 2005.
2. Peng, Z.; Yang, X.; Liu, C.; Dong, Z.; Wang, F.; Wang, X.; Hu, W.; Zhang, X.; Zhao, P.; Xia, Q. Structural and mechanical properties of silk from different instars of Bombyx mori. *Biomacromolecules* **2019**, *20*, 1203–1216. [CrossRef] [PubMed]
3. Chen, X.; Hou, D.; Tang, X.; Wang, L. Quantitative physical and handling characteristics of novel antibacterial braided silk suture materials. *J. Mech. Behav. Biomed. Mater.* **2015**, *50*, 160–170. [CrossRef] [PubMed]

4. Alexander, J.W.; Kaplan, J.Z.; Altemeier, W. Role of suture materials in the development of wound infection. *Ann. Surg.* **1967**, *165*, 192. [CrossRef] [PubMed]
5. Zuo, B.; Dai, L.; Wu, Z. Analysis of structure and properties of biodegradable regenerated silk fibroin fibers. *J. Mater. Sci.* **2006**, *41*, 3357–3361. [CrossRef]
6. Andreeßen, C.; Steinbüchel, A. Recent developments in non-biodegradable biopolymers: Precursors, production processes, and future perspectives. *Appl. Microbiol. Biotechnol.* **2019**, *103*, 143–157. [CrossRef]
7. Hashino, T.; Otsuka, K. The rise and fall of industrialization: The case of a silk weaving district in modern Japan. *Aust. Econ. Hist. Rev.* **2020**, *60*, 46–72. [CrossRef]
8. Popescu, A. Considerations upon the trends in the world silk trade. *People* **2018**, *7*, 10.
9. Girish, K.; Shankara, B.S. Neem–a green treasure. *Electron. J. Biol.* **2008**, *4*, 102–111.
10. Conrick, J. *Neem: The Ultimate Herb*; Lotus Press: Dallas, MI, USA, 2001.
11. Gowda, M.; Sheetal, A.; Kole, C. *The Neem Genome*; Springer: Berlin/Heidelberg, Germany, 2019.
12. Lilot, L.S. The Neem Tree: The Village Pharmacy. *Ethnobot. Leafl.* **2000**, *2000*, 9.
13. Biswas, K.; Chattopadhyay, I.; Banerjee, R.K.; Bandyopadhyay, U. Biological activities and medicinal properties of neem (*Azadirachta indica*). *Curr. Sci.* **2002**, *82*, 1336–1345.
14. Mahmoud, D.A.; Hassanein, N.M.; Youssef, K.A.; Abou Zeid, M.A. Antifungal activity of different neem leaf extracts and the nimonol against some important human pathogens. *Braz. J. Microbiol.* **2011**, *42*, 1007–1016. [CrossRef] [PubMed]
15. Al-Hashemi, Z.S.S.; Hossain, M.A. Biological activities of different neem leaf crude extracts used locally in Ayurvedic medicine. *Pac. Sci. Rev. A Nat. Sci. Eng.* **2016**, *18*, 128–131. [CrossRef]
16. Teklemedhin, T.; Gopalakrishnan, L. Environmental friendly dyeing of silk fabric with natural dye extracted from *Cassia singueana* plant. *J. Text. Sci. Eng.* **2018**, *3*, 2. [CrossRef]
17. Saxena, S.; Raja, A. Natural dyes: Sources, chemistry, application and sustainability issues. In *Roadmap to Sustainable Textiles and Clothing*; Springer: Berlin/Heidelberg, Germany, 2014; pp. 37–80.
18. Wangatia, L.M.; Tadesse, K.; Moyo, S.; Wangatia, L.M.; Tadesse, K.; Moyo, S. Mango bark mordant for dyeing cotton with natural dye: Fully eco-friendly natural dyeing. *Int. J. Text. Sci.* **2015**, *4*, 36–41.
19. Kumar, R.; Mehta, S.; Pathak, S.R. Bioactive constituents of neem. In *Synthesis of Medicinal Agents from Plants*; Elsevier: Amsterdam, The Netherlands, 2018; pp. 75–103.
20. Krishnan, Y.; Wong, N. Cytotoxicity and antimicrobial properties of neem (*Azadirachta indica*) leaf extracts. *Int. J. Pharm. Pharm. Sci.* **2015**, *7*, 179–182.
21. Susmitha, S.; Vidyamol, K.K.; Ranganayaki, P.; Vijayaragavan, R. Phytochemical extraction and antimicrobial properties of *Azadirachta indica* (Neem). *Glob. J. Pharmacol.* **2013**, *7*, 316–320.
22. Adeel, S.; Zia, K.M.; Abdullah, M.; Rehman, F.U.; Salman, M.; Zuber, M. Ultrasonic assisted improved extraction and dyeing of mordanted silk fabric using neem bark as source of natural colourant. *Nat. Prod. Res.* **2019**, *33*, 2060–2072. [CrossRef]
23. Zuber, M.; Adeel, S.; Rehman, F.U.; Anjum, F.; Muneer, M.; Abdullah, M.; Zia, K.M. Influence of microwave radiation on dyeing of bio-mordanted silk fabric using neem bark (*Azadirachta indica*)-based tannin natural dye. *J. Nat. Fibers* **2019**, *17*, 1410–1422. [CrossRef]
24. Sahoo, T.; Dash, S.K.; Das, P. Colouring characteristics of neem (*Azadirachta indica*) bark dye on silk. *Int. J. Agric. Environ. Biotechnol.* **2013**, *6*, 695–702. [CrossRef]
25. Vankar, P.S.; Shanker, R.; Verma, A. Enzymatic natural dyeing of cotton and silk fabrics without metal mordants. *J. Clean. Prod.* **2007**, *15*, 1441–1450. [CrossRef]
26. Abd El Aty, A.A.; El-Bassyouni, G.T.; Abdel-Zaher, N.A.; Guirguis, O.W. Experimental study on antimicrobial activity of silk fabric treated with natural dye extract from neem (*Azadirachta indica*) leaves. *Fibers Polym.* **2018**, *19*, 1880–1886. [CrossRef]
27. Tan, Q.-G.; Luo, X.-D. Meliaceous limonoids: Chemistry and biological activities. *Chem. Rev.* **2011**, *111*, 7437–7522. [CrossRef] [PubMed]
28. Jahan, T.; Begum, Z.A.; Sultana, S. Effect of neem oil on some pathogenic bacteria. *Bangladesh J. Pharmacol.* **2007**, *2*, 71–72. [CrossRef]
29. Anuar, A.A.; Yusof, N. Methods of imparting mosquito repellent agents and the assessing mosquito repellency on textile. *Fash. Text.* **2016**, *3*, 12. [CrossRef]
30. Raja, A.S.; Kawlekar, S.; Saxena, S.; Arputharaj, A.; Patil, P.G. Mosquito protective textiles—A review. *Int. J. Mosq. Res.* **2015**, *2*, 49–53.
31. Banupriya, J.; Maheshwari, V. Effects of Mosquito Repellent finishes by Herbal Method on Textiles. *Int. J. Pharm. Life Sci.* **2013**, *4*, 3133–3134.
32. Wannang, N.N.; Ajayi, V.F.; Ior, L.D.; Dapar, L.M.; Okwori, V.A.; Ohemu, T. Mosquito repellant property of *Azadirachta indica* extract (fruit bark and seed kernel). *Sci. Res. J.* **2015**, *3*, 1–4.
33. Shukla, D.; Wijayapala, S.; Vankar, P.S. Effective mosquito repellent from plant based formulation. *Int. J. Mosq. Res.* **2018**, *5*, 19–24.
34. Gupta, A.; Singh, A. Development of mosquito repellent finished cotton fabric using eco friendly mint. *Int. J. Home Sci.* **2017**, *3*, 155–157.
35. Joshi, M.; Ali, S.W.; Rajendran, S. Antibacterial finishing of polyester/cotton blend fabrics using neem (*Azadirachta indica*): A natural bioactive agent. *J. Appl. Polym. Sci.* **2007**, *106*, 793–800. [CrossRef]

36. Eid, B.M.; El-Sayed, G.M.; Ibrahim, H.M.; Habib, N.H. Durable antibacterial functionality of cotton/polyester blended fabrics using antibiotic/MONPs composite. *Fibers Polym.* **2019**, *20*, 2297–2309. [CrossRef]
37. Liang, Z.; Zhou, Z.; Li, J.; Zhang, S.; Dong, B.; Zhao, L.; Wu, C.; Yang, H.; Chen, F.; Wang, S. Multi-functional silk fibers/fabrics with a negligible impact on comfortable and wearability properties for fiber bulk. *Chem. Eng. J.* **2021**, *415*, 128980. [CrossRef]
38. Mozumder, S.; Majumder, S. Comparison between cotton and silk fabric dyed with turmeric natural colorant. *Daffodil Int. Univ. J. Sci. Technol.* **2016**, *11*, 73.
39. Sonwalkar, T.N. *Handbook of Silk Technology*; Taylor & Francis: Abingdon, UK, 1993.
40. Singhee, D.; Samanta, P. Studies on dyeing process variables for application of Tesu (*Butea monosperma*) as natural dye on silk fabric. *J. Nat. Fibers* **2019**, *16*, 1098–1112. [CrossRef]
41. Ke, G.; Zhao, Z.; Shuhui, C.; Li, J. Green natural dye from Buddleja officinalis and its ultrasonic dyeing on cotton fabric. *Pigment Resin. Technol.* **2021**, *51*, 110–117. [CrossRef]
42. Hardesty, J.H.; Attili, B. Spectrophotometry and the Beer-Lambert Law: An important Analytical Technique in Chemistry; Collin College, Department of Chemistry. 2010. Available online: http://vfsilesieux.free.fr/1Seuro/BeerLaw.pdf (accessed on 12 October 2022).
43. Yang, L.; Kruse, B. Revised Kubelka–Munk theory. *I. Theory and application. JOSA A* **2004**, *21*, 1933–1941.
44. Pinho, E.; Magalhães, L.; Henriques, M.; Oliveira, R. Antimicrobial activity assessment of textiles: Standard methods comparison. *Ann. Microbiol.* **2011**, *61*, 493–498. [CrossRef]
45. Barnard, D.R.; Bernier, U.R.; Xue, R.; Debboun, M. Standard methods for testing mosquito repellents. In *Insect Repellents Principles, Methods, Uses*; Debboun, M., Frances, S.P., Strickman, D., Eds.; CRC Press: Boca Raton, FL, USA, 2006.
46. Bano, R. Use of chitosan in mosquito repellent finishing for cotton textiles. *J. Text. Sci. Eng.* **2014**, *4*, 1.
47. Phasomkusolsil, S.; Soonwera, M. Comparative mosquito repellency of essential oils against *Aedes aegypti* (Linn.), *Anopheles dirus* (Peyton and Harrison) and *Culex quinquefasciatus* (Say). *Asian Pac. J. Trop. Biomed.* **2011**, *1*, S113–S118. [CrossRef]
48. Chang, K.S.; Tak, J.H.; Kim, S.I.; Lee, W.J.; Ahn, Y.J. Repellency of Cinnamomum cassia bark compounds and cream containing cassia oil to *Aedes aegypti* (Diptera: Culicidae) under laboratory and indoor conditions. *Pest Manag. Sci. Former. Pestic. Sci.* **2006**, *62*, 1032–1038. [CrossRef]
49. Patra, A.; Raja, A.; Shah, N. Current developments in (Malaria) mosquito protective methods: A review paper. *Int. J. Mosquito Res.* **2019**, *6*, 38–45.
50. Bukhari, H.; Heba, M.; Khadijah, Q. Eco-friendly dyeing textiles with neem herb for multifunctional fabrics. Part 1: Extraction standardization. *Int. J. Tech. Res. App.* **2014**, *2*, 51–55.
51. Chia, M.A.; Akinsanmi, J.T.; Tanimu, Y.; Ladan, Z. Algicidal effects of aqueous leaf extracts of neem (*Azadirachta indica*) on *Scenedesmus quadricauda* (Turp.) de Brébission. *Acta Bot. Bras.* **2016**, *30*, 1–8. [CrossRef]
52. Mohammed, H.A.; al Fadhil, A.O. Antibacterial activity of *Azadirachta indica* (Neem) leaf extract against bacterial pathogens in Sudan. *Afr. J. Med. Sci.* **2017**, *3*, 246–2512.
53. Win, S.; Ni, K.T. Study on Extraction and Application of Natural Dye from Neem Bark (Melia azadirachta L.). *Dagon Univ. Res. J.* **2020**, *11*, 263–269.
54. Venthodika, A.; Chhikara, N.; Mann, S.; Garg, M.K.; Sofi, S.A.; Panghal, A. Bioactive compounds of *Aegle marmelos* L.; medicinal values and its food applications: A critical review. *Phytother. Res.* **2021**, *35*, 1887–1907. [CrossRef]
55. Saha, S.; Sadhukhan, P.; Sil, P.C. Mangiferin: A xanthonoid with multipotent anti-inflammatory potential. *Biofactors* **2016**, *42*, 459–474. [CrossRef]
56. Quintana, S.E.; Salas, S.; García-Zapateiro, L.A. Bioactive compounds of mango (*Mangifera indica*): A review of extraction technologies and chemical constituents. *J. Sci. Food Agric.* **2021**, *101*, 6186–6192. [CrossRef]
57. Sarah, R.; Tabassum, B.; Idrees, N.; Hussain, M.K. Bio-active Compounds Isolated from Neem Tree and Their Applications. In *Natural Bio-Active Compounds*; Springer: Singapore, 2019; pp. 509–528.
58. Hemmalakshmi, S.; Priyanga, S.; Devaki, K. Fourier Transform Infra-Red Spectroscopy Analysis of *Erythrina variegata* L. *J. Pharm. Sci. Res.* **2017**, *9*, 2062–2067.
59. Grdadolnik, J. Saturation effects in FTIR spectroscopy: Intensity of amide I and amide II bands in protein spectra. *Acta Chim. Slov.* **2003**, *50*, 777–788.
60. Ţucureanu, V.; Matei, A.; Avram, A.M. FTIR spectroscopy for carbon family study. *Crit. Rev. Anal. Chem.* **2016**, *46*, 502–520. [CrossRef] [PubMed]
61. Ragavendran, P.; Sophia, D.; Arul Raj, C.; Gopalakrishnan, V.K. Functional group analysis of various extracts of *Aerva lanata* (L.,) by FTIR spectrum. *Pharmacologyonline* **2011**, *1*, 358–364.
62. Fan, L.; Ziegler, T. Application of density functional theory to infrared absorption intensity calculations on main group molecules. *J. Chem. Phys.* **1992**, *96*, 9005–9012. [CrossRef]
63. Ling, S.; Qi, Z.; Knight, D.P.; Shao, Z.; Chen, X. Synchrotron FTIR microspectroscopy of single natural silk fibers. *Biomacromolecules* **2011**, *12*, 3344–3349. [CrossRef]
64. Boulet-Audet, M.; Lefèvre, T.; Buffeteau, T.; Pézolet, M. Attenuated total reflection infrared spectroscopy: An efficient technique to quantitatively determine the orientation and conformation of proteins in single silk fibers. *Appl. Spectrosc.* **2008**, *62*, 956–962. [CrossRef]

5. Devi, D.; Sarma, N.S.; Talukdar, B.; Chetri, P.; Baruah, K.C.; Dass, N.N. Study of the structure of degummed *Antheraea assamensis* (muga) silk fibre. *J. Text. Inst.* **2011**, *102*, 527–533. [CrossRef]
6. Kong, X.D.; Cui, F.Z.; Wang, X.M.; Zhang, M.; Zhang, W. Silk fibroin regulated mineralization of hydroxyapatite nanocrystals. *J. Cryst. Growth* **2004**, *270*, 197–202. [CrossRef]
7. Kumar, S.; Pandey, A.K. Chemistry and biological activities of flavonoids: An overview. *Sci. World J.* **2013**, *2013*, 162750. [CrossRef]
8. Tundis, R.; Loizzo, M.R.; Menichini, F. An overview on chemical aspects and potential health benefits of limonoids and their derivatives. *Crit. Rev. Food Sci. Nutr.* **2014**, *54*, 225–250. [CrossRef]

sustainability

MDPI

Article

Sustainable Extraction of Colourant from Harmal Seeds (*Peganum harmala*) for Dyeing of Bio-Mordanted Wool Fabric

Shahid Adeel [1,*], Fozia Anjum [1], Muhammad Zuber [2], Muhammad Hussaan [3], Nimra Amin [4] and Meral Ozomay [5]

1 Department of Chemistry, Govt. College University, Faisalabad 38000, Pakistan
2 Department of Chemistry, Riphah International University Faisalabad Campus, Faisalabad 38000, Pakistan
3 Department of Botany, Govt. College University, Faisalabad 38000, Pakistan
4 Department of Applied Chemistry, Govt. College University, Faisalabad 38000, Pakistan
5 Department of Textile Engineering, Marmara University, Istanbul 34722, Turkey
* Correspondence: shahidadeel@gcuf.edu.pk

Abstract: The recent pandemic scenario has caused demand for green products that have medicinal aspects, as well as greener approaches for global health. Natural dye from plants, particularly from harmal seeds, is an excellent alternative to carcinogenic yellow synthetic dyes. The current study has been conducted to isolate natural colorants from harmal seeds in methanolic medium through Gamma-Assisted Extraction (GAE). The dyeing variables that are necessary for shade development before and after mordanting were selected. It has been found that 6 kGy is the optimal absorbed dose for extraction of colorant from 6 g of powder to isolate the colorant in the methanolic medium through the Gamma-irradiated extraction mode (GAE). To get excellent results, 30 mL of methanolic extract containing 6 g/100 mL of Glauber salt was sued for dyeing of irradiated wool at 45 °C for 65 min. For improving the color strength and acceptable rating of fastness, 9% of henna, 3% of acacia, 10% of turmeric, and 7% of pomegranate extracts as pre-bio-mordants as well as 7% of acacia, 3% of pomegranate, 9% of henna, and 10% of turmeric extracts as post-mordants have given high results compared to when chemical mordants have been used. It was concluded that Gamma-ray treatment has excellent color strength in the dyeing of bio-mordanted wool using harmal seed extracts under mild conditions, and has good fastness ratings after using chemical and bio-mordanting methods as well.

Keywords: bio-mordants; chemical mordants; harmal seeds; sustainability; wool fabric

check for updates

Citation: Adeel, S.; Anjum, F.; Zuber, M.; Hussaan, M.; Amin, N.; Ozomay, M. Sustainable Extraction of Colourant from Harmal Seeds (*Peganum harmala*) for Dyeing of Bio-Mordanted Wool Fabric. *Sustainability* **2022**, *14*, 12226. https://doi.org/10.3390/su141912226

Academic Editor: Abu Naser Md Ahsanul Haque

Received: 5 August 2022
Accepted: 14 September 2022
Published: 27 September 2022

Publisher's Note: MDPI stays neutral with regard to jurisdictional claims in published maps and institutional affiliations.

1. Introduction

The inclusion of advanced technologies in various fields, such as textiles, food, flavors, pharmaceutical, wood, plastic, glass, etc., has urged researchers and traders to develop methods that are not only sustainable in nature but also add value to the existence of green products [1,2]. These advanced technologies, such as ultraviolet [3], ultrasonic [4], microwave [5], Gamma, and plasma radiation [6], have provided a new horizon for traders, researchers, and consumers on account of their cost, time, and effective nature.

Of these tools, Gamma irradiation is the most powerful tool, which is gaining widespread utilization in textiles, isolation of natural products, medical textiles, anti-bacterial textiles, pharmaceuticals, and food processing [7,8]. It also has great use in the functionalization of fabrics and wastewater treatment, curing of fabrics bioremediation, and isolation of functional biological components (colorant) from plants without harming their physiological characteristics [9,10]. It also helps in tunning the surface of fabrics, which, in turn, helps to improve the uptake ability and diffusion power so that the effluent load is minimized [11–13]. Hence, due to the energy and labor-effective nature, the use of Gamma-ray treatment isolation of natural dyes and the improvement of their dyeing behavior onto natural fabric is on the rise [14]. Currently, the revival of natural dyes in different fields has been warmly welcomed due to their sustainable and soothing nature [15]. These colorants are not only easily biodegradable but also have no effluent problem. On account of their shades and structural compatibility, these dyes are now being considered as an alternative to synthetic dyes in various applied fields, particularly in textiles, food, flavors, solar cells, pharmaceuticals, cosmetics, etc. [16,17]. Due to their beneficial role in the environment, the global community has been given awareness of their herbal and ayurvedic nature and their possible use in every field [18]. However, plant-derived natural colorants, particularly those with an ayurvedic nature, have been explored for coloring natural fabric since the last decade [19]. Thus overall, the revival of such colorants has not only helped us to save cultural heritage but also to make the globe fresher, cleaner, and healthier.

Of these plants, harmal seeds (*Peganum harmala* L.), as a source of natural colorants, are of great value. Its seeds contain alkaloids, such as harmine, harmaline, harmalol, and harmol [20–22]. Traditionally, its seeds have been used as a cure for evil eyes, skin inflammation activities, and parasitic worms [23]. The seed extracts also exhibit antimicrobial, antiprotozoal, anti-insect, anticancer, and anti-lice action [24–26]. Its extract is also considered an alternative to Turkey red dye (obtained from madders) and is used to dye carpet and wool yarns, make tattoo stains and inks, and help with the bioremediation of synthetic dyes [27].

To the author's knowledge, there are no such extensive studies that have been done to explore the dyeing behavior of harmal seeds, whereas our group is exploring the dyeing ability of harmal seeds for the first time, starting from extraction under the Gamma-ray-assisted extraction mechanism (GAE), dyeing, and mordanting conditions. The utilization of plant-based anchors (bio-mordants), such as henna, turmeric, acacia, and pomegranate [22], has made the coloring of wool using harmal seed extracts more sustainable and therapeutic [28,29]. Hence, given the excellent nature of harmal seeds and the promising role of Gamma radiation in textiles, this research work has been designed not only to extract the colorant from harmal seeds under the influence of Gamma irradiation but also to enhance the color strength and fastness properties using chemical and bio-mordants.

2. Materials and Methods

2.1. Material Collection

Harmal seeds (*Penganum harmala* L.) have been purchased from the local herbal market in Faisalabad. Seeds have been washed with water to remove the dirt and dried under shade. The dried seeds were ground finely and sieved to get powder of homogeneous particle size. Pakistan-made commercial chemicals such as methanol (CH_3OH), hydrochloric acid (HCL), sodium hydroxide (NaOH), Glauber's salt (Na_2SO_4), and chemical mordants (salts of Al, Fe etc.) were purchased. For bio-mordanting, powders of turmeric rhizomes, henna leaves, acacia bark, and pomegranate rinds were purchased from the local superstore. The crude source of bio-mordants was also sieved to get powder of equal particle size. Pre-treated wool (GSM = 100 g/m^2) was purchased from the Textile Market, Faisalabad, Pakistan.

2.2. Irradiation and Extraction Process

The absorbed dose of 2–10 kGy was given to wool fabric, and, for surface tunning, Gamma-ray-assisted extraction (GAE) was performed using a Cs-137 Gamma irradiator at N.I.A.B. (Nuclear Institute of Agriculture and Biology), Faisalabad, Pakistan. The unirradiated powder (NRP) was used for the isolation of colorant by employing 4 g with 100 mL of methanolic medium (60%) at reflux for 60 min—keeping a solid-to-liquid ratio of 1:25. After refluxing, methanolic crude mixture was filtered and the filtrate was used to dye untreated wool (NRW) at 75 °C for 65 min—keeping the fabric-to-dye-bath volume ratio of 1:25 (Khan et al., 2014; Adeel et al., 2017). This process was named control in Figure 1. In another experiment, Gamma-ray-assisted extraction (GAE) was done by proving a Gamma ray absorbed a dose of up to 10 kGy with an interval of 2 kGy, and, after extraction, the crude mixtures were filtered and used to dye irradiated (RW) and unirradiated wool fabric (NRW) at 750 °C for 65 min.

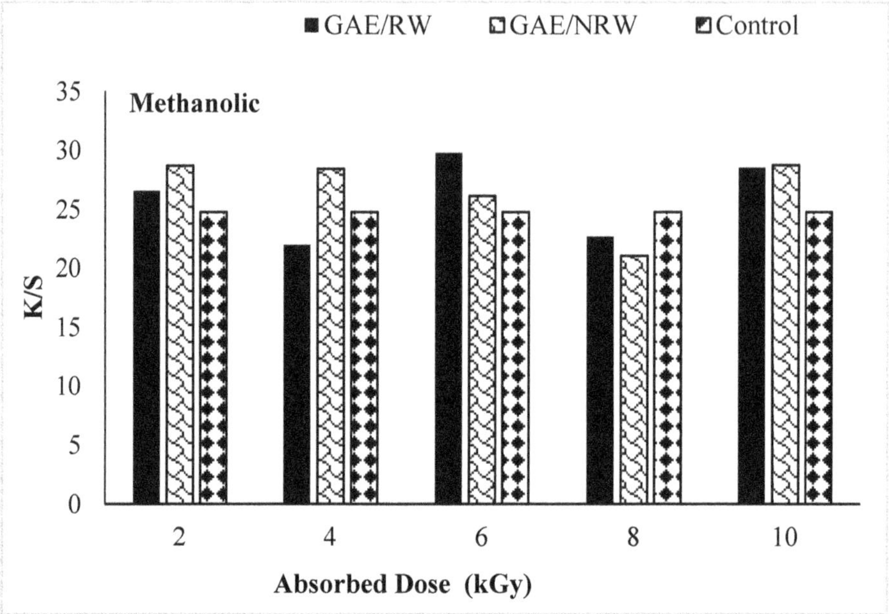

Figure 1. Gamma-ray-assisted methanolic extraction of colorant from harmal seeds and its application for wool dyeing.

2.3. Optimization of Mordanting Condition

After the optimization of the extraction condition at a particular absorbed dose (6 KGy) it was necessary to find a particular volume obtained from a particular amount of powder and then vary the amount of powder for isolation. For this purpose, 2, 4, 6, 8, and 10 g of powder were used to isolate the colorant using 100 mL of methanolic medium upon refluxing for 60 min. In another set of experiments that utilized an optimal amount of powder (6 g), 10–70 mL of methanolic extract for 1 g of fabric was employed to keep the pH of the dye bath at 1–7. To select the optimum contact levels for the colorfast dyeing process, the contact levels (time) and heating levels (temperature) as dyeing variables from 25–85 min/°C were employed. In another set of experiments, to achieve maximum exhaustion, 1–10 g/100 mL of Glauber's salt (GS) were employed in optimal conditions.

2.4. Shade Development Methods

For improvement in shade, 1–10% of the chemical mordants, such as the salt of Al Fe, Cu, Sn and Co, and tannic acid (TA), have been utilized before (pre-) and after (post-) the dyeing process at 65 °C for 65 min, thereby keeping the mordant-to-fabric ratio of 1:25 For making the dyeing process more soothing and sustainable, 1–10% of the extracts of turmeric containing curcumin, henna containing Lawson, acacia bark containing quercetin and pomegranate rind containing tannin have been employed before and after dyeing at 65 °C for 65 min, thereby keeping the bio-mordant –to- fabric ratio of 1:25 [22]. For the isolation of the colorants from plant-based mordants, curcumin from turmeric, Lawson from henna, quercetin from acacia, and tannin from pomegranate have been extracted using already documented methods from Adeel et al. (2021).

2.5. Evaluation of Characteristics of Dyed and Undyed Fabrics

For the investigation of the changes in color strength (K/S), all the fabrics dyed before or after irradiation have been evaluated through the Kubelka–Munk equation, which was computed in a Data color SF 600 (USA) equipped with a D 6510° observer. To evaluate the fastness, all the selected dyed, mordanted fabrics were exposed to ISO standards, such as ISO 105 CO3 for laundering (washing), ISO 105 BO5 for light, 1SO 105 X12 for crocking, and ISO 105 D01 for dry cleaning (DCF) at the Q.A. & Q.C. Lab. of Noor Fatima Fabrics, Faisalabad, Pakistan.

3. Result and Discussion

Gamma-ray treatment has given profound results in the extraction of colorant from harmal seeds in methanolic medium. The results shown in Figure 1 indicate that Gamma-ray treatment has good color depth (K/S) at low-absorbed doses (6 kGy). Depending upon the nature of the colorant (alkaloid), methanol medium has been found to be the best solvent to isolate the colorant, followed by its application onto wool fabrics at 6 kGy. Below the optimal absorbed dose, the extraction process is slow, whereas, above the optimal absorbed dose, the other phytochemicals are also isolated, which, upon coloration of the fabrics, influence the shade. Previously, it has been observed that methanol is the best solvent to extract the colorant because it has more dissipation power than aqueous, alkaline, or acidic medium [30]. The mass transfer process has occurred (solid–liquid interaction) through Gamma irradiation via rupture of the cell wall where the colorant evolved into the solvent [31]. Moreover, Gamma-ray treatment via producing free radicals modifies the isolation of functional bioactive components through the low consumption of the solvent at a reduced time. The other factor is the irradiation of wool fabrics, which shows that there is some etching occurring onto fibers—which might enhance its substantivity [32]. Hence, overall, the Gamma-ray-assisted extraction of colorant (GAE) in methanolic medium has given good color strength on irradiated wool (RW) when an absorbed dose of 6 kGy is employed.

After getting the extraction medium, the amount of powder needed is necessary to find out because of the above selected amount (>6 g), whereas the other phytochemicals are also more significantly involved in such a way that actual colorant has a low chance of absorbing onto the fabric. The results given in Figure 2a show that methanolic extract obtained from 6 g of harmal powder upon irradiation up to 6 kGy has an excellent yield on irradiated wool fabric (RW). After getting the optimal powder amount (6 g), it was also necessary to find out the volume of extract (30 mL) that was suitable for the coloration of wool fabrics. It has been observed that 30 mL of irradiated methanolic extract (RE) obtained from 6 g of powder (Figure 2b) has good color strength on irradiated wool (RW). A low volume does not impart good color depth, whereas, above the optimal volume, the clusters of molecules may get diffused into the modified wool fabric (RW) that remains on the surface. During this washing, unfixed dye is stripped off, which in turn results in low color strength. Hence, after reducing the amount of powder, the lowering of the extract volume reveals that Gamma-ray treatment has proven its cost-effective nature. As the wool contains a large number of polar groups, it has a greater ability to form a stable interaction with colorant in an acidic medium. If the pH of the extract is lowered, the anion groups of the fabric get protonated, wherein the fabrics have a significant chance of interacting with the −OH of the colorant via ionic bonding, which will give a good color yield [33]; this is because the wool keratin has amide linkages that are sensitive to the medium of the dye bath. Moreover, towards alkalinity, the -COOH of wool keratin may face a weak interaction with colorant -OH, thereby resulting in a low color strength. However, in an acidic medium, the amino group of keratins gets protonated and amido linkages are easily available for binding with the functional site [34,35]. Hence, the results displayed in Figure 2c reveal that the dyeing of irradiated wool (RW) should be done using an irradiated methanolic extract with a pH of 3 to get excellent results.

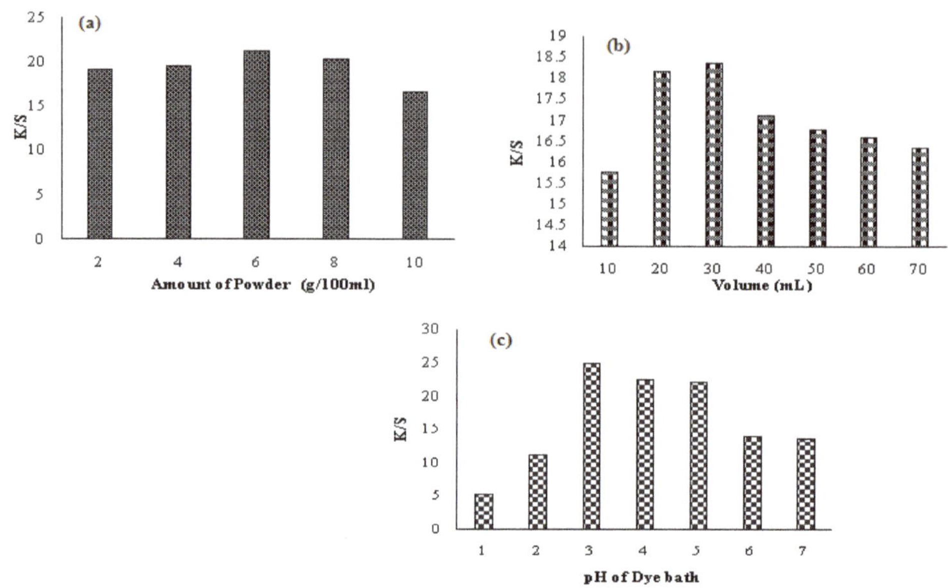

Figure 2. Impact of powder amount (**a**), volume (**b**), and pH of the extract (**c**) on the isolation o colorant from harmal seeds in a methanolic medium.

The other factors are contact variables, which play a great role in the dyeing of woo with plant-based colorants. Gamma-ray treatment has reduced not only the heating leve (dyeing temperature) but also the contact time. This is because Gamma radiation is a powerful tool that has reduced the particle size of the colorant, which in turn has less time to absorb onto fabrics. Moreover, the kinetic energy of the colorant is increased when the dyeing variable is slowly increased up to 45 °C for 65 min. (Figure 3a). After these levels the equilibrium of the dye bath is distributed, and the colorant starts stripping from the fabric to rush toward the dye bath. Hence, in order to get good coloring properties on irradiated wool (RW), irradiated methanolic extract (RE) should be used at 45 °C for 65 mir

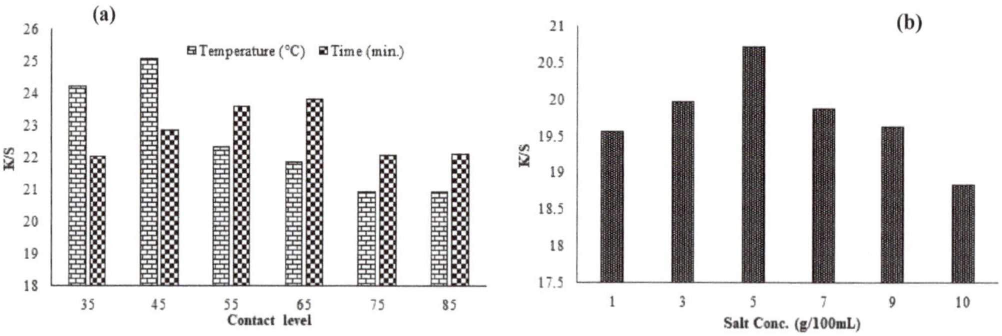

Figure 3. Effect of heating levels (**a**) and salt concentration (**b**) on dyeing of irradiated wool using irradiated methanolic extract of harmal seeds.

The amount of salt to achieve maximum exhaustion needs to be optimized. This is because the natural dyeing process is dependent on the amount of salt for the interaction of the colorant with the wool via short-range attractive forces [36]. It was found that 5 g of Glauber's salt (GS) has an excellent exhaustion of the colorant to rush towards wool fabrics (Figure 3b). Above that amount (>5 g), over exhaustion has led to over aggregation, which in turn, upon washing, unfixed the colorant and removed it, and a low color strength was observed. The reduction in the amount of salt for exhaustion during the dyeing of wool with harmal seed extract shows that Gamma-ray treatment is a cost-effective tool. Overall, it can be seen that Gamma-ray extraction has taken less extract volume, energy, time, and salt to produce better results.

Bio-mordanting is a newly-introduced process, so called bio-dyeing, because through a special interaction, new colorfastness tints are developed; furthermore, due to their therapeutic nature, these molecules impart biological characteristics towards fabrics before and after dyeing. These bio-molecules have not only given new shades but also enhanced the color characteristics [36–38]. These functional potent isolates (as shown in Figure 4), via forming intermolecular H-bonding with colorant and functional sites of fabric, have given a high color yield with newly-developed shades [39,40]. It has been found from Table 1 that during pre-bio-mordants, 9% of Lawson from henna (structure given below), 10% of curcumin from turmeric (structure given below), 3% of quercetin from acacia (structure given below), and 7% of tannin from pomegranate (structure given below) has given a high color strength (K/S) on irradiated wool (RW) when employed before dyeing using methanolic extracts of harmal seed. A low amount of bio-mordants, when used, have given a lower yield (K/S) whereas, above the optimum amount, the aggregates formed may have difficulty diffusing into the fabric and remain unfixed on the fabric. Upon washing, the overcrowded aggregates are stripped, and a lower yield (K/S) is observed. Similarly, during post-bio-mordanting, 9% of Lawson from henna, 10% of curcumin from turmeric, 7% of quercetin from acacia, and 3% of tannin from pomegranate has given a high color yield through their application onto dyed fabric (Table 1). The tonal variation given in Table 1 shows that most fabrics that are dyed before and after bio-mordanting are darker in tone and reddish-yellow in hue, because there is a joint effect between two biomolecules, i.e., alkaloids from harmal in combination with curcumin from turmeric, quercetin from acacia, tannin from pomegranate, and Lawson from henna, which can be seen forming through additional H-bonding.

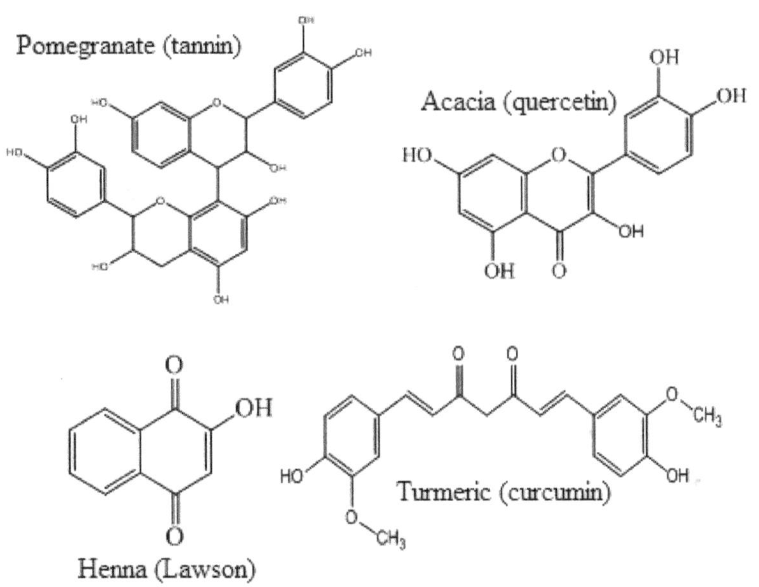

Figure 4. Functional Potent isolates of bio-mordants.

Table 1. Color characteristics of chemical and bio-mordanted wool fabrics dyed before, after, and during dyeing with Gamma-treated harmal seed extracts.

Mordant Concentration	K/S	L^*	a^*	b^*	Mordant Concentration	K/S	L^*	a^*	b^*
Al 10% (Pre)	16.341	70.80	2.83	30.04	T. A 10% (Pre)	21.281	58.45	7.66	30.69
Al 7% (Post)	7.1222	71.08	3.58	24.77	T. A 7% (Post)	8.666	54.57	8.77	22.60
Fe 10% (Pre)	19.299	59.43	6.48	30.87	Acacia 3% (Pre)	12.612	60.33	9.12	30.21
Fe 7% (Post)	7.6815	60.00	10.40	28.73	Acacia 7% (Post)	6.6678	58.62	12.19	27.35
Co 7% (Pre)	19.77	61.55	10.51	34.49	Pomegranate 7%(Pre)	19.527	61.52	7.12	30.76
Co 7% (Post)	7.2014	60.69	7.52	20.69	Pomegranate 3%(Post)	6.4293	68.68	4.78	32.29
Sn 10% (Pre)	17.439	69.86	6.50	33.30	Henna 9% (Pre)	21.578	49.70	9.46	28.65
Sn 7% (Post)	8.0339	74.57	4.29	30.94	Henna 9% (Post)	12.547	51.74	12.37	30.13
Cu 10% (Pre)	16.924	54.69	2.23	31.69	Turmeric 10% (Pre)	23.645	51.35	13.23	43.86
Cu 7% (Post)	7.6831	63.06	7.05	23.25	Turmeric 10% (Post)	23.576	59.21	16.14	64.98

The role of chemical mordants is reversed as compared to bio-mordants because during chemical treatment, the involvement of d-orbitals of the metal dye complex reduction power of metals works via coordinating the covalent bonding through metal dye complex formation [41,42]. It has been observed that 10% of tannic acid (TA) has given a high color yield, as compared to the other pre-mordants used. Similarly, 10% of Al, Fe, Cu, and Sn, and 7% of Co have a good color yield (K/S) upon dyeing of irradiated wool (RW) with irradiated methanolic extract (Table 1). This is because the metal ions have been absorbed evenly onto the wool fabric, which, upon dyeing with irradiated extract, has formed a stable metal dye complex through a coordinated covalent bond [43,44]. Similarly, during post-mordanting, 7% of TA has given a good color yield (K/S) as compared to the other metal mordants used (Table 1). Hence, 7% of Al, Cu, Co, Fe, and Sn, as post-mordants have excellent yields. This good color strength is attributed to the even dyeing of wool after Gamma-ray treatment. The tonal variation given in Table 1 shows that most fabrics dyed before and after chemical mordanting are brighter in shade and are reddish-yellow in hue. Thus, the reduction of the chemical mordant amount used after Gamma-ray treatment reveals that this powerful tool is sustainable and cost-effective.

The rating results of colorfastness properties given in Table 2 for pre- and post-chemical mordanting, as well as bio-mordants, show that most fabrics dyed after mordanting have excellent ratings for light, washing, rubbing, and dry cleaning. This is because the conjugation system present in bio-mordants and colorants plays a greater role when the fabric is in contact via a covalent bond. The extra bonding given by bio-mordants may create more stable interactions with fabrics and colorants [45,46], which, upon exposure to agents such as heat, detergents, light, crocking, and dry-cleaning agents, resist detaching. Similarly, the rating results given in Table 2 for chemical pre- and post-mordanting reveal that the stable metal dye complex formation on fabric resists fading of the colorant [47–49]. The greater the potential efficiency of metal to form a covalent bond with an amide linkage of wool and –OH of colorant isolated from harmal seeds, the more stable the complex will be that is formed [50,51]. Hence, the overall rating results show that bio-mordants have a promising resistance to heat, light, detergent, crocking, and dye cleaning agents. The shades that were made are represented in Table 3. It can be concluded that the utilization of bio-mordants has not only made the dyeing of irradiated wool greener and more sustainable but also given improved fastness characteristics.

Table 2. Colourfastness rating of mordanted and dyed Gamma-treated wool fabric using harmal seed extracts.

Mordant Concentration	LF	WF		RF		DCF	PF	
		c. c	c.s	DRF	WRF		Acidic	Alkaline
Control	3/4	3	3	$\frac{3}{4}$	$\frac{3}{4}$	3/4	3	3
Al 10% (Pre)	4	4/5	4/5	4/5	4	4	4/5	4/5
Al 7% (Post)	4/5	5	5	5	5	4/5	5	5
Fe 10% (Pre)	4/5	4/5	4/5	4/5	4	4	5	5
Fe 7% (Post)	4/5	5	4/5	5	4/5	4	4/5	4/5
Co 7% (Pre)	4/5	5	4/5	4	$\frac{3}{4}$	4	4/5	4/5
Co 7% (Post)	4/5	5	5	5	4	4/5	5	5
Sn 10% (Pre)	4	4/5	4/5	4/5	4	4	4/5	4/5
Sn 7% (Post)	4/5	5	5	5	5	4	4/5	4/5
Cu 10% (Pre)	4/5	5	4/5	4/5	4	4/5	4/5	4/5
Cu 7% (Post)	4/5	5	5	4/5	4	5	5	5
T. A 10% (Pre)	4/5	4/5	4/5	4/5	4	4	4/5	4/5
T. A 7% (Post)	4/5	5	5	4/5	4	5	4/5	4/5
Acacia 3% (Pre)	4	4/5	4/5	4/5	4	4	4/5	4/5
Acacia 7% (Post)	4/5	4/5	4/5	4	4	4/5	4/5	4/5
Pomegranate 7% (Pre)	4/5	4/5	4/5	4/5	4	4/5	4/5	4/5
Pomegranate3%(Post)	4/5	4/5	4/5	4/5	4	4/5	4/5	4/5
Henna 9% (Pre)	4/5	4/5	5	4/5	4/5	4	4/5	4/5
Henna 9% (Post)	4/5	4/5	5	4/5	4/5	4	4/5	4/5
Turmeric 10% (Pre)	4	4/5	4/5	4	$\frac{3}{4}$	4	4/5	4/5
Turmeric 10% (Post)	4	4/5	4/5	4	$\frac{3}{4}$	4	4/5	4/5

LF = Light fastness, WF = wash fastness, c.s = color stain, c.c= color change, RF = Rub fastness, DRF = dry rub fastness, WRF = wet rub fastness, DCF = dry clean fastness, PF= perspiration fastness.

Table 3. Tonal variation of Gamma-treated dyed and mordanted wool fabrics using harmal seed extracts.

WITH OUT MORDANT

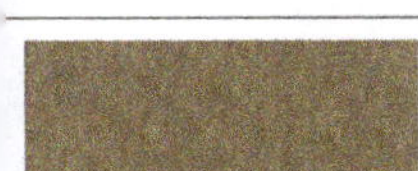	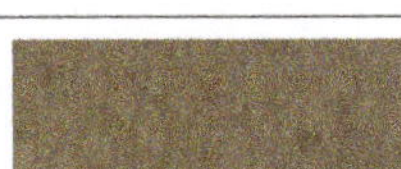		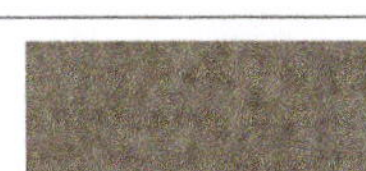	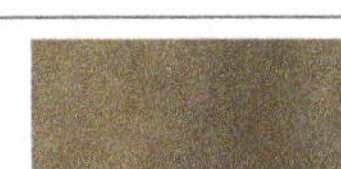
Al 10% Pre	Al 7% Post	Fe 10% Pre	Fe 7% Post	Co 7% Pre

Table 3. *Cont.*

Co 7% Post	Sn 10% Pre	Sn 7% Post	Cu 10% Pre	Cu 7% Post

T.A 10% Pre	T.A 7% Post	Acacia 3% Pre	Acacia 7% Post	Pomegranate 7%Pre
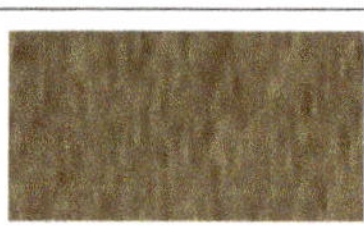				
Pomegranate 3% Post	Henna 9% Pre	Henna 9% Post	Turmeric 10% Pre	Turmeric 10% Post

4. Conclusions

Sustainability is one of the greater demands of the global community because of global warming and global heat, and unexpected seasonal changes have been found due to carcinogenic, effluent loads from industries. Among sustainable products, natural plant pigments and colorants have taken a special place and particular attention of the people due to their excellent biological and herbal characteristics. This study is one a series of explorations of plants that have ayurvedic and biological natures and have potential to impart color onto fabric. Additionally, among modern techniques for their extraction Gamma-ray treatment in textiles has shown promising effects not only in the exploration of new dye-yielding plants, such as harmal seeds, but also in enhancing extraction yield and color characteristics. An absorbed dose of 6 kGy was optimized to get extraction of the colorant in methanolic medium. The dyeing wool fabric for 65 min at 45 °C using 30 mL of dye bath of pH 3 using 5 g/100 mL of the Glaubar's salt as an exhausting agent was done to get colorfast shades before and after the selected amount of chemical and bio-mordants. It is concluded that this novel modern tool can be applied to explore new dye-yielding plants as a source of natural dyes for the coloration of natural fabrics, and the addition of new bio-mordants can be done to assess new shades with excellent colorant characteristics.

Author Contributions: Formal analysis, F.A.; Investigation, M.H.; Methodology, S.A. and M.O.; Supervision, M.Z.; Writing – review & editing, N.A. All authors have read and agreed to the published version of the manuscript.

Funding: This research received no external funding.

Institutional Review Board Statement: Not applicable.

Informed Consent Statement: Not applicable.

Data Availability Statement: Not applicable.

Acknowledgments: The work is a part of PhD Studies.

Conflicts of Interest: The authors declare no conflict of interest.

References

1. Barani, H.; Haji, A.; Maleki, H. Analysis of lecithin treatment effects on the structural transformation of wool fiber using vi-brational spectroscopy. *Int. J. Biol. Macromol.* **2018**, *108*, 585–590. [CrossRef] [PubMed]
2. Shokry, H.; Elkady, M.; Hamad, H. Nano activated carbon from industrial mine coal as adsorbents for removal of dye from simulated textile wastewater: Operational parameters and mechanism study. *J. Mater. Res. Technol.* **2019**, *8*, 4477–4488. [CrossRef]
3. Singh, M.; Vajpayee, M.; Ledwani, L. Eco-friendly surface modification of natural fibres to improve dye uptake using natural dyes and application of natural dyes in fabric finishing: A review. *Mater. Today: Proc.* **2021**, *43*, 2868–2871. [CrossRef]
4. Kandasamy, N.; Kaliappan, K.; Palanisamy, T. Upcycling sawdust into colorant: Ecofriendly natural dyeing of fabrics with ultrasound assisted dye extract of Pterocarpus indicus Willd. *Ind. Crop. Prod.* **2021**, *171*, 113969. [CrossRef]
5. Sharma, A.; Mazumdar, B.; Keshav, A. Valorization of unsalable Amaranthus tricolour leaves by microwave-assisted extrac-tion of betacyanin and betaxanthin. *Biomass Convers. Biorefinery* **2021**, *163*, 1–17.
6. Naebe, M.; Haque, A.N.M.A.; Haji, A. Plasma-Assisted Antimicrobial Finishing of Textiles: A Review. *Engineering* **2021**, *12*, 145–163. [CrossRef]
7. Silva, D.; Rocha, R.; Silva, C.J.; Barroso, H.; Botelho, J.; Machado, V.; Mendes, J.J.; Oliveira, J.; Loureiro, M.V.; Marques, A.C.; et al. Gamma radiation for sterilization of textile based materials for personal protective equipment. *Polym. Degrad. Stab.* **2021**, *194*, 109750. [CrossRef]
8. Islam, M.T.; Repon, M.R.; Liman, M.L.R.; Hossain, M.M.; Al Mamun, M.A. Functional modification of cellulose by chitosan and gamma radiation for higher grafting of UV protective natural chromophores. *Radiat. Phys. Chem.* **2021**, *183*, 109426. [CrossRef]
9. Khan, A.A.; Iqbal, N.; Adeel, S.; Azeem, M.; Batool, F.; Bhatti, I.A. Extraction of natural dye from red calico leaves: Gamma ray assisted improvements in colour strength and fastness properties. *Dye. Pigment.* **2014**, *103*, 50–54. [CrossRef]
10. Rehman, F.U.; Adeel, S.; Haddar, W.; Bibi, R.; Azeem, M.; Mia, R.; Ahmed, B. Microwave-Assisted Exploration of Yellow Natural Dyes for Nylon Fabric. *Sustainability* **2022**, *14*, 5599. [CrossRef]
11. Chirila, L.; Popescu, A.; Stanculescu, I.R.; Cutrubinis, M.H.S.; Cerempei, A.N.A.; Sandu, I. Gamma irradiation effects on natural dyeing performances of wool fabrics. *Rev. Chim.* **2016**, *67*, 2628–2633.
12. Vujcic, I.; Masic, S.; Medic, M.; Milicevic, B.; Dramicanin, M. The influence of gamma irradiation on the color change of wool, linen, silk, and cotton fabrics used in cultural heritage artifacts. *Radiat. Phys. Chem.* **2018**, *156*, 307–313. [CrossRef]
13. Thinkohkaew, K.; Piroonpan, T.; Jiraborvornpongsa, N.; Potiyaraj, P. Radiation induced graft polymerization of fluorinated methacrylate onto polypropylene spunbond nonwoven fabric. *Surfaces Interfaces* **2021**, *24*, 101125. [CrossRef]
14. Chirila, L.; Popescu, A.; Cutrubinis, M.; Stanculescu, I.; Moise, V.I. The influence of gamma irradiation on natural dyeing properties of cotton and flax fabrics. *Radiat. Phys. Chem.* **2018**, *145*, 97–103. [CrossRef]
15. Ansari, T.N.; Iqbal, S. Antibacterial efficiency of naturally occurring dyes and mordants. *Proc. Indian Natl. Sci. Acad.* **2021**, *87*, 408–419. [CrossRef]
16. Bujak, T.; Zagórska-Dziok, M.; Ziemlewska, A.; Nizioł-Łukaszewska, Z.; Wasilewski, T.; Hordyjewicz-Baran, Z. Antioxidant and Cytoprotective Properties of Plant Extract from Dry Flowers as Functional Dyes for Cosmetic Products. *Molecules* **2021**, *26*, 2809. [CrossRef] [PubMed]
17. Yang, R.; Zhang, Y.; Ranjitkar, S.; Li, M.; Guo, Y.; Yan, X.; Wang, C.; Stepp, J.R.; Yang, L. Reusing wasteroot of Rubia wallichiana dyeing from Monpa of Tibet in China. *Sci. Rep.* **2021**, *11*, 1–15. [CrossRef]
18. Arifeen, W.-U.; Rehman, F.-U.; Adeel, S.; Zuber, M.; Ahmad, M.N.; Ahmad, T. Environmental friendly extraction of walnut bark-based juglone natural colorant for dyeing studies of wool fabric. *Environ. Sci. Pollut. Res.* **2021**, *28*, 49958–49966. [CrossRef]
19. Thakker, A.M. Sustainable processing of cotton fabrics with plant-based biomaterials *Sapindus mukorossi* and *Acacia concinna* for health-care applications. *J. Text. Inst.* **2020**, *112*, 718–726. [CrossRef]
20. Pozzi, F.; Shibayama, N.; Leona, M.; Lombardi, J.R. TLC-SERS study of Syrian rue (*Peganum harmala*) and its main alkaloid constituents. *J. Raman Spectrosc.* **2012**, *44*, 102–107. [CrossRef]
21. Fatma, B.; Fatiha, M.; Elattafia, B.; Noureddine, D. Phytochemical and antimicrobial study of the seeds and leaves of Peganum harmala L. against urinary tract infection pathogens. *Asian Pac. J. Trop. Dis.* **2016**, *6*, 822–826. [CrossRef]
22. Adeel, S.; Zuber, M.; Rehman, F.U.; Zia, K.M. Microwave-assisted extraction and dyeing of chemical and bio-mordanted cotton fabric using harmal seeds as a source of natural dye. *Environ. Sci. Pollut. Res.* **2018**, *25*, 11100–11110. [CrossRef] [PubMed]
23. Li, S.-P.; Wang, Y.-W.; Qi, S.-L.; Zhang, Y.-P.; Deng, G.; Ding, W.-Z.; Ma, C.; Lin, Q.-Y.; Guan, H.-D.; Liu, W.; et al. Analogous β-Carboline Alkaloids Harmaline and Harmine Ameliorate Scopolamine-Induced Cognition Dysfunction by Attenuating Acetylcholinesterase Activity, Oxidative Stress, and Inflammation in Mice. *Front. Pharmacol.* **2018**, *9*, 346. [CrossRef]
24. Ren, Y.; Gong, J.; Fu, R.; Li, Z.; Li, Q.; Zhang, J.; Yu, Z.; Cheng, X. Dyeing and antibacterial properties of cotton dyed with prodigiosins nanomicelles produced by microbial fermentation. *Dye. Pigment.* **2017**, *138*, 147–153. [CrossRef]
25. Soosaraei, M.; Fakhar, M.; Teshnizi, S.H.; Hezarjaribi, H.Z.; Banimostafavi, E.S. Medicinal plants with promising antileishmanial activity in Iran: A systematic review and meta-analysis. *Ann. Med. Surg.* **2017**, *21*, 63–80. [CrossRef]
26. Ayoob, I.; Hazari, Y.M.; Lone, S.H.; Rehman, S.U.; Khuroo, M.A.; Fazili, K.M.; Bhat, K.A. Phytochemical and Cytotoxic Evaluation of Peganum Harmala: Structure Activity Relationship Studies of Harmine. *ChemistrySelect* **2017**, *2*, 2965–2968. [CrossRef]
27. Miraj, S. A review study of therapeutic effects of Peganum harmala. *Pharm. Lett.* **2016**, *8*, 161–166.

28. Adeel, S.; Kiran, S.; Shahid, M.; Habib, S.R.; Habib, N.; Hussaan, M. Ecofriendly application of coconut coir (*Cocos nucifera*) extract for silk dyeing. *Environ. Sci. Pollut. Res.* **2021**, *29*, 564–572. [CrossRef]
29. Hasan, M.U.; Adeel, S.; Batool, F.; Ahmad, T.; Tang, R.-C.; Amin, N.; Khan, S.R. Sustainable application of Cassia obovata–based chrysophanic acid as potential source of yellow natural colorant for textile dyeing. *Environ. Sci. Pollut. Res.* **2021**, *29*, 10740–10753. [CrossRef]
30. Hussaan, M.; Iqbal, N.; Adeel, S.; Azeem, M.; Javed, M.T.; Raza, A. Microwave-assisted enhancement of milkweed (*Calotropis procera* L.) leaves as an eco-friendly source of natural colorants for textile. *Environ. Sci. Pollut. Res.* **2016**, *24*, 5089–5094. [CrossRef]
31. Abolhasani, A.; Barzegar, M.; Sahari, M.A. Effect of gamma irradiation on the extraction yield, antioxidant, and antityrosinase activities of pistachio green hull extract. *Radiat. Phys. Chem.* **2018**, *144*, 373–378. [CrossRef]
32. Munna, M.K.H.; Chinyerenwa, A.C.; Kamruzzaman, M.; Hossain, M.A.; Ahamed, M.K.; Wahab, M.A. Effect of gamma radiation on cotton fabric with chitosan to improve the mechanical properties. *Int. J. Text. Sci.* **2017**, *6*, 1–6.
33. Yusuf, M.; Khan, S.A.; Shabbir, M.; Mohammad, F. Developing a Shade Range on Wool by Madder (*Rubia cordifolia*) Root Extract with Gallnut (*Quercus infectoria*) as Biomordant. *J. Nat. Fibers* **2016**, *14*, 597–607. [CrossRef]
34. Haji, A.; Qavamnia, S.S.; Nasiriboroumand, M. The use of D-optimal design in optimization of wool dyeing with Juglans regia bark. *Ind. Text.* **2018**, *69*, 104–110.
35. Lachguer, K.; El Ouali, M.; Essaket, I.; El Merzougui, S.; Cherkaoui, O.; Serghini, M.A. Eco-Friendly Dyeing of Wool with Natural Dye Extracted from Moroccan Crocus sativus L. Flower Waste. *Fibers Polym.* **2021**, *22*, 3368–3377. [CrossRef]
36. Islam, S.U.; Rather, L.J.; Shabbir, M.; Sheikh, J.; Bukhari, M.N.; Khan, M.A.; Mohammad, F. Exploiting the potential of polyphenolic biomordants in environmentally friendly coloration of wool with natural dye from *Butea monosperma* flower extract. *J. Nat. Fibers* **2018**, *16*, 512–523. [CrossRef]
37. Ghaheh, F.S.; Moghaddam, M.K.; Tehrani, M. Comparison of the effect of metal mordants and bio-mordants on the colorimetric and antibacterial properties of natural dyes on cotton fabric. *Color. Technol.* **2021**, *137*, 689–698. [CrossRef]
38. Rather, L.J.; Zhou, Q.; Ali, A.; Haque, Q.M.R.; Li, Q. Valorization of Agro-industrial Waste from Peanuts for Sustainable Natural Dye Production: Focus on Adsorption Mechanisms, Ultraviolet Protection, and Antimicrobial Properties of Dyed Wool Fabric. *ACS Food Sci. Technol.* **2021**, *1*, 427–442. [CrossRef]
39. Rather, L.J.; Zhou, Q.; Li, Q. Re-use of Cinnamomum camphora natural dye generated wastewater for sustainable UV protective and antioxidant finishing of wool fabric: Effect of Fe(II) sulfate. *Sustain. Chem. Pharm.* **2021**, *21*, 100422. [CrossRef]
40. Zhang, W.; Liu, Y.; Song, D.; Guo, H.; Hu, J.; Wang, Y.; Xu, W. Dyeing behavior and mechanism of Crocein Orange G on carboxymethyl cotton fabric. *Cellulose* **2021**, *28*, 5911–5922. [CrossRef]
41. Doty, K.; Haar, S.; Kim, J. Black walnut, Osage orange and eastern redcedar sawmill waste as natural dyes: Effect of aluminum mordant on color parameters. *Fash. Text.* **2016**, *3*, 1–16. [CrossRef]
42. Shabbir, M.; Rather, L.J.; Islam, S.U.; Bukhari, M.N.; Shahid, M.; Khan, M.A.; Mohammad, F. An eco-friendly dyeing of woolen yarn by Terminalia chebula extract with evaluations of kinetic and adsorption characteristics. *J. Adv. Res.* **2016**, *7*, 473–482. [CrossRef] [PubMed]
43. Zhang, Y.; Zhou, Q.; Rather, L.J.; Li, Q. Agricultural waste of Eriobotrya japonica L. (Loquat) seeds and flora leaves as source of natural dye and bio-mordant for coloration and bio-functional finishing of wool textile. *Ind. Crop. Prod.* **2021**, *169*, 113633. [CrossRef]
44. Yan, X.; Hong, L.; Pei, S.; Hamilton, A.; Sun, H.; Yang, R.; Liu, A.; Yang, L. A natural yellow colorant from Buddleja officinalis for dyeing hemp fabric. *Ind. Crop. Prod.* **2021**, *171*, 113968. [CrossRef]
45. Lykidou, S.; Pashou, M.; Vouvoudi, E.; Nikolaidis, N. Study on the Dyeing Properties of Curcumin on Natural and Synthetic Fibers and Antioxidant and Antibacterial Activities. *Fibers Polym.* **2021**, *22*, 3336–3342. [CrossRef]
46. Hosseinnezhad, M.; Gharanjig, K.; Jafari, R.; Imani, H.; Razani, N. Cleaner colorant extraction and environmentally wool dyeing using oak as eco-friendly mordant. *Environ. Sci. Pollut. Res.* **2020**, *28*, 7249–7260. [CrossRef] [PubMed]
47. Datta, M.K.; Talukder, E.; Faisal, A.; Sarker, A.; Jiang, H. The Sustainable Coloration of Wool Fabric Using Naturally Extracted Dyes from Sappan Heartwood. *J. Nat. Fibers* **2021**, 1–15. [CrossRef]
48. Önal, A.; Özbek, O.; Tombul, K.C.; Nached, S. Investigation of the dyeing properties of cotton fabrics and wool yarns using Prunus persica leaf extract. *J. Indian Chem. Soc.* **2021**, *98*, 100092. [CrossRef]
49. Islam, M.T.; Liman, L.R.; Roy, M.N.; Hossain, M.; Repon, R.; Al Mamun, A. Cotton dyeing performance enhancing mechanism of mangiferin enriched bio-waste by transition metals chelation. *J. Text. Inst.* **2021**, *113*, 567–579. [CrossRef]
50. Sharma, A.; Kadam, S.; Mathur, P.; Islam, S.U.; Sheikh, J. Re-using henna natural dyeing wastewater for coloration and multifunctional finishing of linen fabric. *Sustain. Chem. Pharm.* **2018**, *11*, 17–22. [CrossRef]
51. Glogar, M.; Tancik, J.; Brlek, I.; Sutlovic, A.; Tkalec, M. Optimisation of process parameters of Alpaca wool printing with Juglans regia natural dye. *Color Technol.* **2020**, *136*, 188–201.

sustainability

Article

Utilization of Colored Extracts for the Formulation of Ecological Friendly Plant-Based Green Products

Shahid Adeel [1], Maryam Habiba [2], Shumaila Kiran [3], Sarosh Iqbal [3], Shazia Abrar [3,*] and Ch Moazzam Hassan [3]

[1] Department of Chemistry, Government College University, Sir Syed Block, New Campus Jhang Road, Faisalabad 38000, Pakistan
[2] Minhaj University, Main Campus Hamdard Chowk Township, Lahore 54770, Pakistan
[3] Department of Applied Chemistry, Government College University, Sir Syed Block, New Campus Jhang Road, Faisalabad 38000, Pakistan
* Correspondence: shaziaabrar@gcuf.edu.pk

Abstract: Green or sustainable cosmetics are products that contain natural ingredients obtained from renewable raw materials. Fruit peels represent a sustainable source of bioactive compounds. Polyphenols, e.g., flavonoids, have the ability to scavenge free radicals; thus they exhibit antioxidant activity. Recently, natural antioxidants have been in the limelight as being safe, effective, and versatile. In this study, antioxidant effects and the sun protection ability of apple (*Malus domestica*), banana (*Musa sapientum*), and orange (*Citrus reticulata*) peel extracts were evaluated in skincare formulations. The extraction of phenolic compounds was performed in three different solvents, i.e., ethanol, methanol, and acetone. Total phenolic contents, antioxidant activity, and sun protection factor were determined for the fruit peel extracts. The acetone extract of apple and ethanol extract of banana peels contained polyphenols, i.e., 24.3 ± 1.5 and 26.7 ± 0.6 mg GAE per gram of the extracts, respectively. These extracts showed DPPH radical scavenging activity and were incorporated into oil-in-water (O/W) cosmetic emulsions. All the formulated samples were found to be stable when subjected to centrifuging and thermal stress. Antioxidant activities of cream samples were above 80%, and the sun protection factor was above 15. The results have confirmed the applications of fruit peel waste in the formulation of photostable, antioxidant, and sun screen formulations. These creams would help to maintain skin health, protect it from UV radiation, and reduce the aging effect. Thus, fruit peel waste could present an ecofriendly and sustainable source of natural antioxidants for the personal care industry.

Keywords: natural antioxidants; antiaging; sun screening; phenolic compounds; cosmetic formulation

Citation: Adeel, S.; Habiba, M.; Kiran, S.; Iqbal, S.; Abrar, S.; Hassan, C.M. Utilization of Colored Extracts for the Formulation of Ecological Friendly Plant-Based Green Products. *Sustainability* **2022**, *14*, 11758. https://doi.org/10.3390/su141811758

Academic Editor: Abu Naser Md Ahsanul Haque

Received: 10 August 2022
Accepted: 14 September 2022
Published: 19 September 2022

1. Introduction

Due to the green revolution, the utilization of natural ingredients in personal care products has extensively increased in recent decades [1]. Plant-based biochemicals show various unique characteristics, i.e., they are biodegradable, non-toxic, eco-friendly, and sustainable. Natural ingredients such as colorants have been studied for their application in the textile and food industries [2–4]. The cosmetic industry is one of the important fields where phytochemicals can be applied to obtain various health benefits [5]. Cosmetic products, such as creams, lotions, lipsticks, etc., are applied directly to external body parts for the purpose of skin cleansing and beautification. The use of synthetic chemicals in cosmetics may expose our body system to various harmful chemicals and toxic heavy metals. Therefore, the replacement of synthetic chemicals with plant-based natural ingredients helps to obtain safe, ecofriendly, and sustainable cosmetic products.

Fruit peels are the waste by-product of the fruit processing industry and have been studied as a sustainable potential source of various bioactive compounds. Polyphenols such as phenolic acids, flavonoids, anthocyanins, tannins, and stilbenes are organic compounds responsible for the plant's defensive system for fighting various types of stresses owing to their strong therapeutic properties [6,7]. Fruit peels have been reported to contain a much higher concentration of plant acids compared to edible fruit flesh, thus exhibiting higher antioxidant potential [8,9]. Phenolic compounds can prevent oxidation reactions by scavenging the free radicals to form more stable phenoxy radicals, which results in a decrease in oxidative damage at the cellular level [10,11]. The important phenolic compounds present in some common fruit peels are caffeic acid, ferulic acid, sinapinic acid, coumaric acid, quercetin, epicatechin, etc. [12,13]. These compounds show several therapeutic properties such as anti-inflammatory, antibacterial, antityrosinase, antiaging, antiallergic, anticarcinogenic, etc. [14–17]. Therefore, fruit peel extracts present an excellent sustainable source of natural compounds required for antiacne and antiaging, dermo protective skincare formulation, thus boosting overall skin quality. Caffeic acid and ferulic acid may help to alleviate the photosensitization and thus protect the skin from UV-induced photo-damaging reactions [18,19]. Thus, phenolics reduce the photoaging, pigmentation, and carcinogenic reactions in the skin [20,21]. The therapeutic properties of phenolic compounds in fruit peels of banana, apple, and orange can be utilized for the formulation of ecofriendly and multipurpose skincare cosmetics. The best way to achieve the benefits of these natural ingredients is to incorporate them into a stable system such as emulsions. Emulsions are active vehicles with a chemical composition suitable for skin. Emulsions provide stability to the natural extracts with enhanced permeability as compared to solutions and gels. The main components of creams are fatty alcohols/fatty acids, emollients, and moisturizers that can increase the absorption of the active ingredient into the lipid layer of the skin. The characterization of these formulations, especially the physical and chemical stability of the extracts in the formulation, is required to accomplish the desired efficacy of the product.

The main objective of the current research work was to determine polyphenol contents, antioxidant activity, and sun protection factor (SPF) of cosmetic formulations containing peel extracts of apple (*Malus Domestica*), banana (*Musa sapientum*), and orange (*Citrus reticulata*). The extraction of phenolic compounds was accomplished using solvents such as ethanol, methanol, and acetone. The total phenolic contents and antioxidant activities of all the extracts were measured using a DPPH assay and compared with a controlled set of experiments. The main objective was to incorporate the antioxidants from fruit peels into stable oil in water (O/W) type skin care products. The SPF factor was measured for the estimation of their UV-protecting ability. The physical and photostability of the formulations were determined. To the best of our knowledge, the utilization of fruit peel antioxidants for the formulation of multipurpose, stable skincare formulations has not been studied so far.

2. Materials and Methods

2.1. Materials and Reagent

All the solvents and reagents were of analytical grade. Deionized water, ethanol, methanol, and acetone were purchased from Pakistan Chemical Store, Jinnah colony, Faisalabad, Pakistan. All the reagents and chemicals, i.e., DPPH (2,2-diphenyl-1-picrylhydrazyl), Folin-Ciocalteu reagent, sodium carbonate, gallic acid, ascorbic acid, polawax, shea butter, EDTA, parabens, and stearic acid, were purchased from Sigma Aldrich.

The apple (*Malus domestica)*, banana (*Musa)*, and orange (*Citrus X sinensis)* were obtained from the local market. The fruits were peeled, and the peels were air dried, ground to a fine powder, and stored in an air-tight container for further use.

2.2. Preparation of Fruit Peels Extracts

The fruit peel extracts were prepared by following the method reported in the literature with slight modifications [22]. The fruit peel powder (25 g) of each sample was soaked in 500 mL of solvent (80% and 100% acetone, ethanol, methanol) and incubated at room temperature overnight. It was stirred on a hot plate using a magnetic stirrer at 200–250 rpm at 35 °C for 3 h, filtered using Whatman No. 1 filter paper and then centrifuged at 6000 rpm for 10 min at −4 °C. The supernatant was filtered, separated, and then concentrated using a rotary evaporator. The extracts were dried up to 40 °C and stored in the refrigerator, thus preserving the phenolic compounds present in the extracts.

2.3. Determination of Total Phenolic Content (TPC)

Total phenolic content (TPC) was determined by following the spectrophotometric method [23]. A double-beam UV-VIS spectrophotometer LX212DS was used. It has a wavelength range of 190–1100 nm, with a tungsten lamp as a source of visible light. The absorbance of the resulting mixture was measured at 750 nm using distilled water as the blank. The calibration curve was prepared using gallic acid as standard with a concentration range of (25–500 mg/mL). Total phenolic content was measured as mg gallic acid equivalent per gram of the extract (mg of GA/g of extract).

2.4. Determination of DPPH Radical Scavenging Activity

Antioxidant activity was determined in terms of DPPH radical scavenging activity [24]. The extract was prepared (1 mg/mL), and 3 mL was added to 3 mL of 0.1 mM DPPH solution and incubated in the dark for 20 min. The absorbance was measured at 518 nm using a double-beam UV/VIS spectrophotometer. Ethanol was used as the blank, and ethanolic DPPH as the control. Ascorbic acid was used as standard. The results were expressed as % age DPPH radical scavenging activity using the formula:

$$\%\ \text{age scavenging} = \frac{\text{Ac} - \text{Ae}}{\text{Ac}} \times 100$$

where Ac is the absorbance of the control without the sample, and Ae is the absorbance of the solution with the sample.

2.5. Development of O/W Emulsion with Fruit Peel Extracts

Oil-in-water (O/W) emulsions with dried extracts of fruit peel were prepared. A control base without extract was also formulated with oil and water phases, as mentioned in Table 1. The oil phase consists of petrolatum, shea butter, stearic acid, and almond oil, and was heated up to 70 °C and then added to the preheated (75 °C) water phase, along with preservatives and emulsifier (polawax). The mixture was continuously stirred vigorously for 15 min and then cooled down at room temperature. In the next step, fruit peel extracts were added (4.0% to the cooled emulsion). A control cream base was prepared without the addition of fruit peel extracts.

Table 1. Composition of formulations.

Ingredients	Apple	Banana	Orange	Control Base
Water	Q.S.	Q.S.	Q.S.	Q.S.
EDTA	0.05%	0.05%	0.05%	0.05%
Glycerine	2.5%	2.5%	2.5%	2.5%
Polawax	5%	5%	5%	5%
Shea butter	1%	1%	1%	1%
Stearic acid	0.3%	0.3%	0.3%	0.3%
Petrolatum	1%	1%	1%	1%
Extract	4%	4%	4%	0%
Almond Oil	1%	1%	1%	1%
Perfume	1%	1%	1%	1%
Parabens	0.2%	0.2%	0.2%	0.2%

2.6. Basic Characterization of Formulations

The prepared samples of emulsions were evaluated for basic parameters such as physical appearance, homogeneity, stability, and pH.

2.6.1. pH and Stability Testing

The pH of all the cream samples was measured immediately after formulation and after an interval of 30 days using HANNA HI-2211 benchtop pH meter.

The stability testing was performed in triplicate by measuring the accelerated stability centrifuge test, freezing, and thawing cycles. The accelerated stability was measured at three different temperatures (8 °C, 25 °C, and 40 °C) with a 30-day interval for 90 days. All the samples were subjected to the freezing and thawing cycle by placing them in the freezer at a temperature of −5 °C for one day and then in an oven at 40 °C for the next day, completing the six consecutive cycles for 12 days. All the samples were centrifuged at 3000 rpm for 10 min.

2.6.2. Spreadability

Spreadability was measured by the method reported in the literature. Each cream sample (1 g) was placed evenly on a circular glass disc, and the disc was covered with glass and a weight (100, 200, 300, 400 in grams) was placed on it for 5 min. The distance covered by the sample was measured, and the spreadability was calculated. It was expressed in millimeters of increased diameter. The spreadability factor (Sf) was calculated by the following equation:

$$Sf = A/W$$

2.6.3. Sun Protection Factor

The sun protection factor was determined using a double-beam UV-VIS spectrophotometer. The sample (1 g) was diluted with 100 mL of ethanol and sonicated for 10 min. It was filtered, and the absorbance of the filtrate was measured from 290 to 320 nm at an interval of 5 nm. The results were calculated with the help of the normalized Mansur equation [25].

$$SPF = \mathrm{CF} \times \sum_{290}^{320} EE\ (\lambda) \times I\ (\lambda) \times Abs\ (\lambda)$$

In this equation: CF—correction factor (CF = 10); *EE (l)*—erythemal effect spectrum; *I (l)*—solar intensity spectrum; *Abs (l)*—absorbance of sunscreen product.

2.6.4. Photostability

The photostability was measured by the reported method [26]. The sample (1 g) was applied evenly on a glass disc and placed under a UV light source from a Philip lamp at 254 nm (30 W, 100% emission) for 2 h. The SPF of the sample was measured before and after the exposure. It was expressed as the percentage and calculated by using the following equation.

$$\text{Photostability} = \frac{\text{SPF of cream}}{\text{SPF of UV irradiated cream}} \times 100$$

2.7. Antioxidant Activity of Cream

The cream samples (1 g) were diluted with 100 mL of ethanol and sonicated for 10 min. The mixture (10 mL) was further diluted with 100 mL of ethanol. The liquid extract, i.e., 3 mL (1 mg/mL), was added to 3 mL of 0.1 mM DPPH solution and incubated in the dark for 20 min. The absorbance was measured at 518 nm using a UV-VIS Spectrophotometer. Ethanol was used as blank and ethanolic DPPH as control. The result was expressed as %-age DPPH radical scavenging activity using the formula:

$$\% \text{ age scavenging } = \frac{\text{Ac} - \text{Ae}}{\text{Ac}} \times 100$$

where Ac is the absorbance of the control without the sample, and Ae is the absorbance of the solution with the sample.

2.8. Statistical Analysis

All the measurements were performed in triplicates for statistical analysis, i.e., means, standard deviation (S.D.), and standard error (S.E.), was performed using IBM SPSS Statistics (version 26.0) software. It was installed from Department of Information Technology, Government College University Faisalabad, Pakistan.

3. Results and Discussion

3.1. Extraction and Determination of Total Phenolic Contents (TPC)

The extracts of fruit peels were prepared in different solvents, i.e., ethanol, methanol, and acetone. The concentration of plant phenolics varies according to the nature of the solvent and the method used for extraction. The Folin–Ciocalteu (F-C) reagent is generally used for the determination of phenolic contents in the extracts. This F-C can also react with other non-phenolic components, i.e., folic acid, ascorbic acid, thiamine, nucleic acids, ascorbic acid, and some metal ions [27]. Since the most abundant antioxidants present in the extracts are phenolics, the values obtained by the F-C reagent are due to the phenolic contents. Gallic acid was used as standard, and a calibration curve was prepared using different concentrations (15, 30, 50, 100, 200 mg/L) of gallic acid. The linear relationship between the concentrations of gallic acid and absorbance is shown as a standard curve in Figure 1.

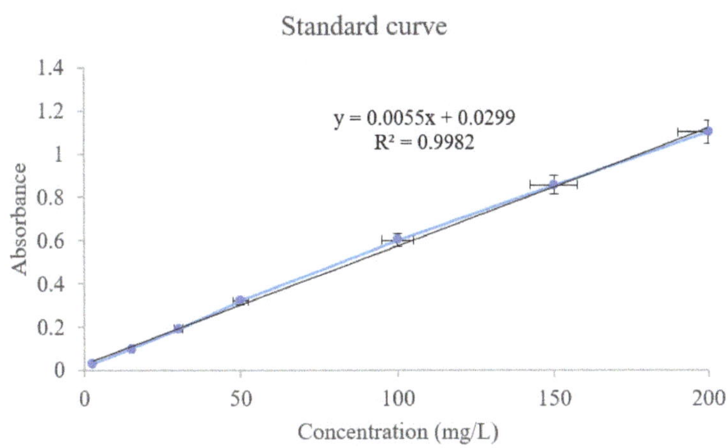

Figure 1. Linear curve of standard gallic acid (mg/L).

The concentration of total phenolic contents (TPC) of apple, orange, and banana peel extracts was expressed in mg of gallic acid equivalent per gram of the extract (mg GAE/g). The concentration of phenolic compounds in apple peels ranges from 5.3 ± 0.5 to 24.3 ± 1.5 mg GAE/g, as shown in Figure 2. Quantitative analysis of phenolic contents from different apple varieties has confirmed the highest level of phenolics, i.e., catechins, flavonol glycosides, and rutin, are in apple peels compared to pulp [28]. Phenolic compounds are mainly responsible for the antioxidant activity of apple peels. The results showed maximum TPC in pure acetone while the minimum was found in pure ethanol. The acetone can better penetrate the plant cells and extract the phenolics efficiently [29]. The values of TPC were comparable to the values reported [16].

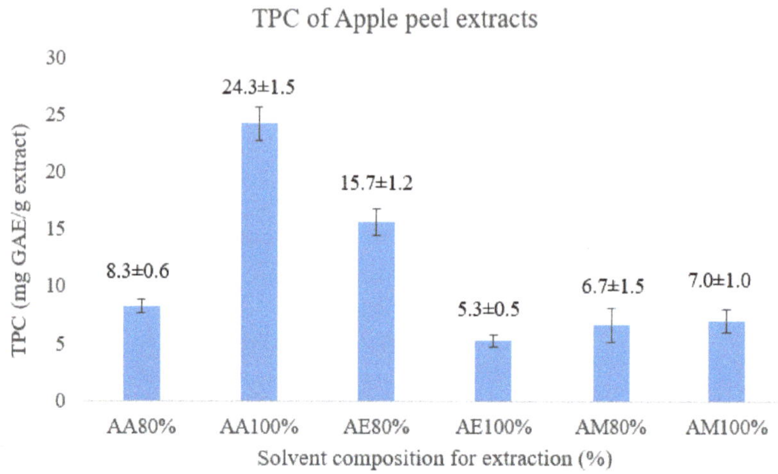

Figure 2. Total phenolic content (mg GAE/g of extract) in apple peel. (AA = Apple Acetone, AE = Apple Ethanol, AM = Apple Methanol).

Banana peels make a significant contribution to agro-industrial waste, which cannot be utilized efficiently. The best way to consume these is to extract their phenolic contents for use in the food or cosmetic industries. In banana peel extract, TPC ranged from 21.3 ± 1.5 to 26.7 ± 0.6 mg GAE/g, as shown in Figure 3. Maximum phenolics were found in pure ethanol followed by pure acetone, while minimum phenolics were in 80% methanol. The literature has confirmed maximum phenolic contents in the banana peels as compared to the pulp and gallocatechin was most abundant among other polyphenols [30]. The concentration of phenolics in banana peels has been reported in the literature [7]. Orange peels showed TPC values from 11.7 ± 1.5 to 29.0 ± 1.0 mg GAE/g, as shown in Figure 4. A high level of TPC was found in acetone (100%), followed by 80% methanol, while minimum phenolics were depicted in pure ethanol and pure methanol.

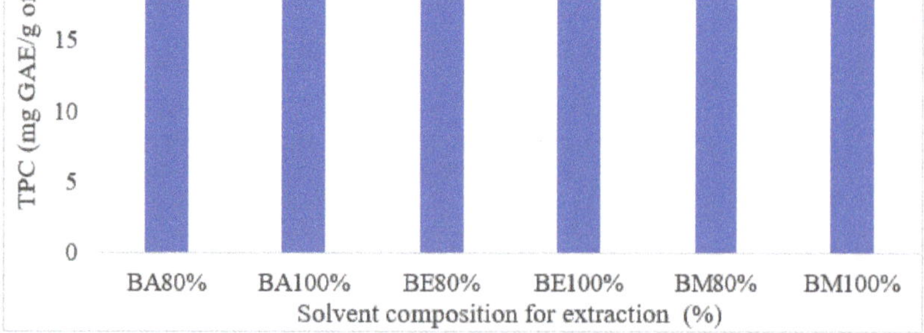

Figure 3. Total phenolic content (mg GAE/g of extract) in banana peel. (BA = Banana Acetone, BE = Banana Ethanol, BM = Banana Methanol).

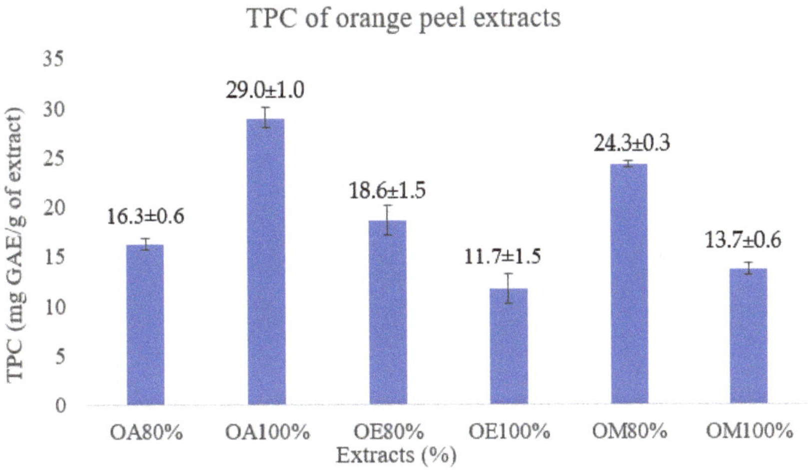

Figure 4. Total phenolic content (mg GAE/g of extract) in orange peel. (OA = Orange Acetone, OE = Orange Ethanol, OM = Orange Methanol).

The concentration of TPC in different extracts was influenced by the nature of the solvent. Varying the solvent polarity changed the total phenolic content expressing different solubility of phenols in the different solvent environments. The addition of water decreased the total phenolic content, which suggests that water may solubilize other compounds such as carbohydrates leading to lower TPC values [31]. One-hundred percent acetone was found more appropriate under our experimental conditions to prepare apple and orange peel extracts, while 100% ethanol yielded higher TPC in banana peel extract. Consequently, acetone extracts of fruit peels were preferred for further study.

3.2. Antioxidant Activity

Antioxidant activity is related to the concentration of phenolics in fruit peel extracts [32]. It was measured in terms of DPPH radical scavenging activity. All the extracts showed different radical scavenging patterns, as shown in Figure 5. Antioxidant activity of peel extracts ranged from 60.2% to 75.6% free radical scavenging, with apple peel extracts exhibiting the best antioxidant activity in terms of DPPH radical scavenging. It has been investigated that apple peels showed higher antioxidant potential than apple flesh [33]. These peel extracts also exhibited antimicrobial and anticancer activities. The phenolic extracts of peels can represent a valuable source of antioxidants for reducing photooxidation. The antioxidant activity of standard ascorbic acid as a function of increasing concentration is shown in Figure 6. One of the important extrinsic factors that induce skin aging is the oxidative stress caused by UV radiation [34]. In prolonged exposure to sunlight, free radicals are produced in the skin. These reactive oxygen species (ROS) not only cause the oxidation of biomolecules but also damage the dermal connective tissues, ultimately affecting the skin elasticity that leads to aging effects [35]. Antioxidants scavenge free radicals and hence minimize oxidative stress [36]. The free radical scavenging mechanism can be utilized to formulate the antiaging formulation by incorporating the natural extracts of phenolic compounds. Hence the free radical scavenging activities of fruit peel extracts make them a potent ingredient for skincare cosmetic formulations.

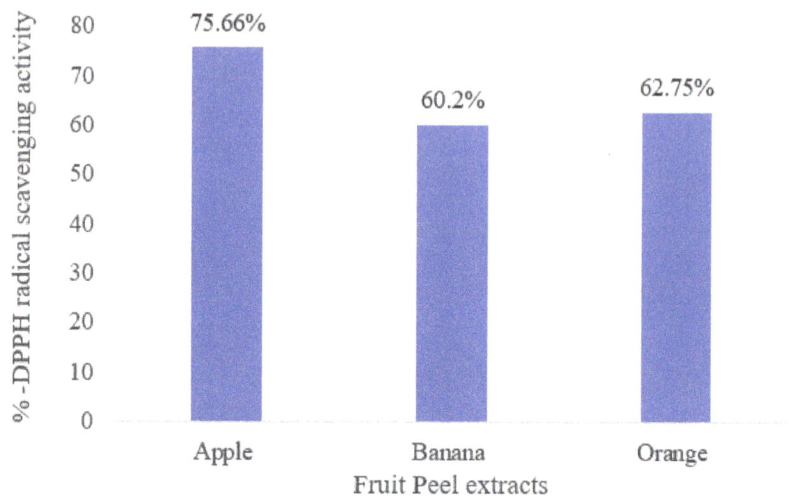

Figure 5. DPPH radical scavenging activity of fruit peel extracts.

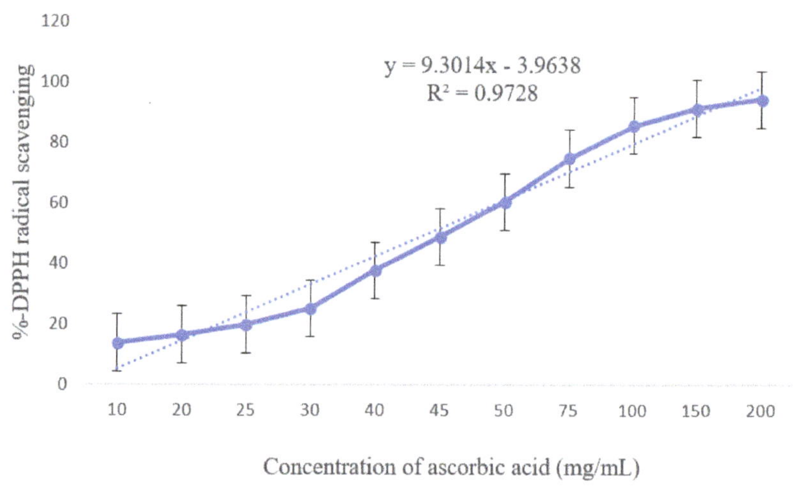

Figure 6. DPPH radical scavenging activity of standard ascorbic acid.

The antioxidant activity of all the formulations has been presented in Table 2 This determines the stability and efficacy of extracts in terms of their antioxidant activity after being incorporated into emulsions. The formulations showed significant DPPH radical scavenging activities, which depict that these fruit peel extracts are a stable source of antioxidants, and they maintained their effect even in cosmetic emulsions. Furthermore, the antioxidant activity of the cream formulations is directly related to the antioxidant potential of the extract. Hence, these fruit peel extracts are consistent enough to act as a source of antioxidants in cosmetic formulations. To the best of our knowledge, no previous studies have reported the utilization of fruit peel extracts for the formulation of stable skincare formulations.

Table 2. SPF, photostability, and antioxidant activity of cream base and formulations with fruit peel extracts (ACR—cream with apple peel extract; BCR—cream with banana peel extract; OCR—cream with orange peel extract).

Sample	Extract	Color	SPF	Photostability	Antioxidant Activity
CRB	None	White	4.0 ± 0.2	-	None
ACR	Apple	Cream	20.0 ± 0.4	95%	88.0%
BCR	Banana	Gold brown	19.0 ± 0.2	98%	87.2%
OCR	Orange	Orange	18.3 ± 0.1	97%	80.0%

3.3. Incorporation of Extracts into Emulsion

Emulsions are vehicles that enhance the absorption of both hydrophilic and lipophilic active ingredients into the cutaneous layer of skin. Creams are semisolid emulsions consisting of water and oil phases homogenized by using emulsifiers. The delivery is influenced by the interaction between the active ingredients, the vehicle, and the skin. Either the active ingredients are incorporated in the dispersed phase or continuous phase of the emulsion, its absorption is enhanced by increasing interaction with the skin epidermis [37]. In the current study, oil-in-water (O/W) emulsions were prepared as the vehicle for delivering antioxidants. Moreover, in oil-in-water type emulsion (O/W), water as the external phase increases the hydration level of the stratum corneum, thus increasing the absorption of hydrophilic compounds. Extracts rich in antioxidants were incorporated into the emulsions for their delivery to the skin epidermis. In order to reduce the reactive oxygen species, natural antioxidants must be able to permeate and penetrate through the strum corneum and reach the deeper layers of the skin. The emulsion (O/W) was prepared using the ingredients listed in Table 1. The peel extracts were added to the cream base at a concentration of 4%. All the extracts were colored, which imparted a beautiful appearance to the emulsions, which is an important aspect of cosmetic products. The visually attractive products are more liked by consumers. ACR (cream with apple peel extract) was light yellow, BCR (cream with banana peel extract) appeared golden brown, and OCR (orange peel extract) was light orange. Fruit peel extracts were completely homogenous in formulations without any aggregation or lumps. The basic parameters of the formulations are expressed in the above Table 2.

3.4. Emulsion Stability Studies

To ensure the quality, efficacy, and safety of the products, it is essential to perform stability studies of cosmetic formulations. Such studies involve the monitoring of the physicochemical characterization of the formulations throughout the study period. pH is an important characteristic of the formulation that determines the effectiveness of the ingredients and stability of the formulation. The normal skin pH range is 4.5–6.5, and cosmetic formulations that have pH in this range are considered safe for the skin [38]. A change in pH represents the change in the efficacy of the product. The pH of all the formulations and cream bases was determined immediately after their manufacturing and then after 30 days. There was no significant change in pH observed throughout the study period. After the centrifuge testing, the pH of all the samples remained unchanged, as shown in Table 3, which confirmed the stability of the product.

The stability of the emulsions containing extracts was assessed by subjecting them to thermal stress, photo radiation, and centrifugation. It was found that all the samples of creams were stable with no phase separation and liquefaction. There was no change in the physical state, color, and pH after the stability testing. During the centrifuge test, the samples were subjected to a change in gravity conditions through the differential density of the oil and water phases. However, no physical instability was observed in terms of phase change after the centrifugation.

Table 3. pH of the cream formulation.

Sample	pH (0 Time)	pH (30 Days)	pH (After Centrifuge)
CRB	5.07 ± 0.1	5.03 ± 0.2	5.06 ± 0.1
ACR	5.45 ± 0.3	5.41 ± 0.1	5.45 ± 0.1
BCR	5.53 ± 0.1	5.52 ± 0.2	5.53 ± 0.3
OCR	4.89 ± 0.2	4.85 ± 0.4	4.87 ± 0.1

(CRB: Cream base; ACR: Apple Cream; BCR: Banana Cream; OCR: Orange cream).

3.5. Characterization of the Emulsions

3.5.1. Spreadability

The term spreadability of the emulsion is used to express the extent of ease by which an emulsion is applied to the skin. It is an important characteristic as it determines the even distribution of the active ingredients in the emulsion. Different factors can affect the spreadability of the emulsion, i.e., the temperature, composition of the emulsion, storage conditions, and viscosity of the emulsion. In order to see the difference, the spreadability of the emulsion prepared with extracts and that of the control base were compared. Under the similar condition of temperature, there was no change in the spreadability of different emulsion samples except one, which contained orange peel extracts. A good spreadability value defines the distribution and enhances the absorption of active ingredients in the skin. The values of the spreadability of the formulations samples (ACR, BCR, OCR) are presented in Figure 7. There was a linear increase in the spreadability of the sample with the increase in applied weight. All the samples showed good spreadability values. In the case of cream formulated with orange peel extract, the spreadability increased significantly as compared to other cream samples. Orange peel extracts contain anthocyanins, in addition to other flavonoids [39]. Due to their increased interaction of anthocyanins with the water phase of the O/W type emulsion, this cream sample has low viscosity and, thus, higher spreadability than other samples.

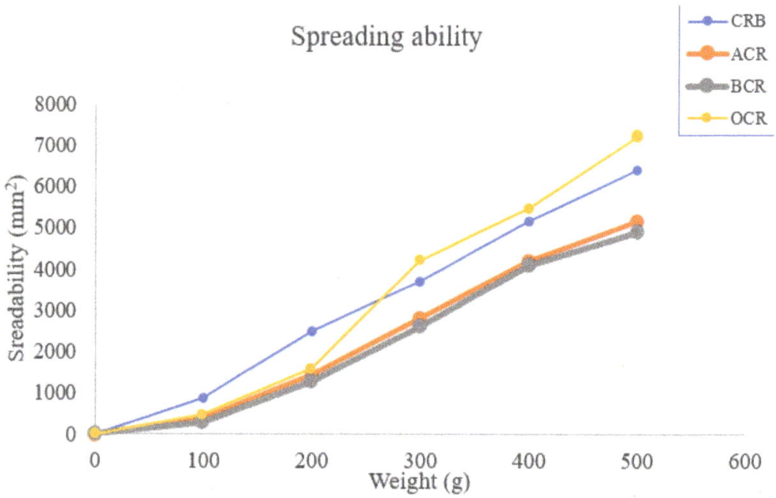

Figure 7. Spreadability of cream base (CRB) and formulations containing fruit peel extracts (ACR—cream with apple peel extract; BCR—cream with banana peel extract; OCR—cream with orange peel extract).

3.5.2. Sun Protection Factor (SPF)

Sun protection factor (SPF) is a measure of the efficiency of the sun screening formulation to absorb ultraviolet radiation from the sun, thus preventing the skin from UV-induced damage and sunburn. SPF is determined quantitatively by measuring the absorbance of UV radiation by the cosmetic emulsion, between 290 and 400 nm. It gives an estimate of the effectiveness of the sunscreen formulation. The higher the value of SPF, the higher will be the effect of the formulation against UV radiation. The results of the sun protection factor (SPF) of all the formulated samples and cream bases are given in Table 2. SPF is an important parameter for predicting the sunscreen ability of a cosmetic product. SPF above 15 is considered sufficient to prevent the incidence of skin cancer caused by UV radiation [40]. There are different factors that can affect the SPF value, such as the solvent, pH, the type of emulsion and composition of emulsion, and the interaction of the active ingredient with the skin. Other components of the emulsion can also interfere by absorbing UV radiation. For maximum efficacy of the extract, it must be uniformly spread on the skin as a thin film and remain close to the surface of the skin. Plant polyphenols exhibit antioxidant activity along with the absorption of UV radiation. These can be used as active ingredients in sun screening formulation. All the prepared samples showed an SPF above 15 with 4% extract. Hence these fruit peel extracts are a natural approach to required sun protection. It has been recommended by the Food and Drug Administration (FDA) that for getting appropriate UV-protection or SPF value, it is important to precisely apply the amount of sun protection vehicle (2.0 mg/cm^2) on the skin [41].

3.5.3. Photostability

In order to produce the desired effects, the antioxidants must remain on the skin without any degradation or change in chemical structure. Ultraviolet radiation of high intensity has the ability to degrade the active agents in cosmetic formulations, which could result in a decrease in their activity. For their optimum efficacy, the sun screening agents must remain stable during the period of exposure to sunlight. Therefore, a study of the photostability of the cosmetic product is very important. Many sunscreens, upon interaction with UV radiation, degrade and fail to protect the skin [42]. Further, the determination of photostability is important as it defines the UV-protection capacity of the sun screening formulations. It was found that all the formulations (ACR, BCR, and OCR) of fruit peel extracts were more than 95% photostable when irradiated with UV light, as shown in Table 2. Photostability makes the fruit peel extracts a reliable natural choice for sun screening formulations. These formulations being highly photostable, safe, and effective, are far better than various commercial sunscreen agents.

4. Conclusions

Fruit peels represent a valuable source of important phenolic compounds that are wasted away as useless material of the food industry. The current work found that acetone extract of apple and orange peels and ethanol extract of banana peels contain phenolic contents, i.e., 24.3 $\pm$ 1.5, 29.0 $\pm$ 1.0, and 26.7 $\pm$ 0.6 mg GAE, respectively. These fruit peel extracts also showed significant antioxidant activity. The apple peels showed higher antioxidant potential, i.e., 75.6%, when compared with standard ascorbic acid. These extracts were successfully incorporated into (O/W) emulsions at a concentration of 4% and were stable under thermal stress, UV radiation, and centrifugation. The pH of the cream samples ranges from 4.89 $\pm$ 0.2 to 5.53 $\pm$ 0.1. All the samples showed no change in their texture and pH throughout the study period. The formulated creams exhibited high sun protection factor (SPF), i.e., above 15. Since the creams samples were photostable, these can be applied to the skin when exposed to sunlight and to reduce skin aging effects. Fruit peels are abundantly available as a byproduct of the food industry; their utilization is useful for waste management. Their consumption will provide a cost-effective, ecofriendly, and sustainable alternative to synthetic antioxidants for the skincare and cosmetic industry.

Author Contributions: Data curation, C.M.H.; Formal analysis, M.H.; Investigation, S.K.; Methodology, S.I.; Supervision, S.A. (Shazia Abrar); Writing—review & editing, S.A. (Shahid Adeel). All authors have read and agreed to the published version of the manuscript.

Funding: This research received no external funding.

Institutional Review Board Statement: Not applicable.

Informed Consent Statement: Not applicable.

Data Availability Statement: Not applicable.

Conflicts of Interest: The authors declare no conflict of interests.

References

1. Goyal, N.; Jerold, F. Biocosmetics: Technological advances and future outlook. *Environ. Sci. Pollut. Res.* **2021**. [CrossRef] [PubMed]
2. Habib, N.; Akram, W.; Adeel, S.; Amin, N.; Hosseinnezhad, M.; Haq, E.U. Environmental-friendly extraction of Peepal (Ficus Religiosa) bark-based reddish brown tannin natural dye for silk coloration. *Environ. Sci. Pollut. Res.* **2022**. [CrossRef] [PubMed]
3. Hussaan, M.; Iqbal, N.; Adeel, S.; Azeem, M.; Tariq Javed, M.; Raza, A. Microwave-assisted enhancement of milkweed (*Calotropis procera* L.) leaves as an eco-friendly source of natural colorants for textile. *Environ. Sci. Pollut. Res.* **2017**, *24*, 5089–5094. [CrossRef] [PubMed]
4. Kaderides, K.; Kyriakoudi, A.; Mourtzinos, I.; Goula, A.M. Potential of pomegranate peel extract as a natural additive in foods. *Trends Food Sci. Technol.* **2021**, *115*, 380–390. [CrossRef]
5. Karray, A.; Krayem, N.; Saad, H.B.; Sayari, A. Spirulina platensis, Punica granatum peel, and moringa leaves extracts in cosmetic formulations: An integrated approach of in vitro biological activities and acceptability studies. *Environ. Sci. Pollut. Res.* **2021**, *28* 8802–8811. [CrossRef]
6. Choi, D.-Y.; Lee, Y.-J.; Hong, J.T.; Lee, H.-J. Antioxidant properties of natural polyphenols and their therapeutic potentials for Alzheimer's disease. *Brain Res. Bull.* **2012**, *87*, 144–153. [CrossRef]
7. Mileo, A.M.; Nisticò, P.; Miccadei, S. Polyphenols: Immunomodulatory and Therapeutic Implication in Colorectal Cancer. *Front. Immunol.* **2019**, *10*, 729. [CrossRef]
8. Ajila, C.M.; Naidu, K.A.; Bhat, S.G.; Rao, U.J.S.P. Bioactive compounds and antioxidant potential of mango peel extract. *Food Chem.* **2007**, *105*, 982–988. [CrossRef]
9. Singh, B.; Singh, J.P.; Kaur, A.; Singh, N. Phenolic composition, antioxidant potential and health benefits of citrus peel. *Food Res Int.* **2020**, *132*, 109114. [CrossRef]
10. Aseervatham, G.S.B.; Sivasudha, T.; Jeyadevi, R.; Arul Ananth, D. Environmental factors and unhealthy lifestyle influence oxidative stress in humans—An overview. *Environ. Sci. Pollut. Res.* **2013**, *20*, 4356–4369. [CrossRef]
11. Mathew, S.; Abraham, T.E.; Zakaria, Z.A. Reactivity of phenolic compounds towards free radicals under in vitro conditions. *J. Food Sci. Technol.* **2015**, *52*, 5790–5798. [CrossRef] [PubMed]
12. Vu, H.T.; Scarlett, C.J.; Vuong, Q.V. Phenolic compounds within banana peel and their potential uses: A review. *J. Funct. Foods* **2018**, *40*, 238–248. [CrossRef]
13. Lee, K.H.; Cho, J.-Y.; Lee, H.J.; Park, K.Y.; Ma, Y.-K.; Lee, S.-H.; Cho, J.A.; Kim, W.-S.; Park, K.-H.; Moon, J.-H. Isolation and identification of phenolic compounds from an Asian pear (Pyrus pyrifolia Nakai) fruit peel. *Food Sci. Biotechnol.* **2011**, *20* 1539–1545. [CrossRef]
14. Bouarab-Chibane, L.; Forquet, V.; Lantéri, P.; Clément, Y.; Léonard-Akkari, L.; Oulahal, N.; Degraeve, P.; Bordes, C. Antibacterial Properties of Polyphenols: Characterization and QSAR (Quantitative Structure–Activity Relationship) Models. *Front. Microbiol* **2019**, *10*, 829. [CrossRef] [PubMed]
15. Kazłowska, K.; Hsu, T.; Hou, C.-C.; Yang, W.-C.; Tsai, G.-J. Anti-inflammatory properties of phenolic compounds and crude extract from Porphyra dentata. *J. Ethnopharmacol.* **2010**, *128*, 123–130. [CrossRef] [PubMed]
16. Nagendra Prasad, K.; Yang, B.; Yang, S.; Chen, Y.; Zhao, M.; Ashraf, M.; Jiang, Y. Identification of phenolic compounds and appraisal of antioxidant and antityrosinase activities from litchi (*Litchi sinensis* Sonn.) seeds. *Food Chem.* **2009**, *116*, 1–7. [CrossRef]
17. Simic, M.G.; Jovanovic, S.V. Inactivation of Oxygen Radicals by Dietary Phenolic Compounds in Anticarcinogenesis. In *Food Phytochemicals for Cancer Prevention II*; ACS Symposium Series; American Chemical Society: Washington, DC, USA, 1994, Volume 547, pp. 20–32.
18. González-Montelongo, R.; Lobo, G.; Gonzalez, M. Antioxidant activity in banana peel extracts: Testing extraction conditions and related bioactive compounds. *Food Chem.* **2010**, *119*, 1030–1039. [CrossRef]
19. Correia, P.; Araújo, P.; Ribeiro, C.; Oliveira, H.; Pereira, A.R.; Mateus, N.; de Freitas, V.; Brás, N.F.; Gameiro, P.; Coelho, P.; et al. Anthocyanin-Related Pigments: Natural Allies for Skin Health Maintenance and Protection. *Antioxidants* **2021**, *10*, 1038. [CrossRef]
20. Hubner, A.; Sobreira, F.; Vetore Neto, A.; Pinto, C.A.S.d.O.; Dario, M.F.; Díaz, I.E.C.; Lourenço, F.R.; Rosado, C.; Baby, A.R.; Bacchi E.M. The Synergistic Behavior of Antioxidant Phenolic Compounds Obtained from Winemaking Waste's Valorization, Increased the Efficacy of a Sunscreen System. *Antioxidants* **2019**, *8*, 530. [CrossRef]

1. Hano, C.; Tungmunnithum, D. Plant Polyphenols, More than Just Simple Natural Antioxidants: Oxidative Stress, Aging and Age-Related Diseases. *Medicines* **2020**, *7*, 26. [CrossRef]
2. Pothitirat, W.; Chomnawang, M.T.; Supabphol, R.; Gritsanapan, W. Free radical scavenging and anti-acne activities of mangosteen fruit rind extracts prepared by different extraction methods. *Pharm. Biol.* **2010**, *48*, 182–186. [CrossRef] [PubMed]
3. Derakhshan, Z.; Ferrante, M.; Tadi, M.; Ansari, F.; Heydari, A.; Hosseini, M.S.; Conti, G.O.; Sadrabad, E.K. Antioxidant activity and total phenolic content of ethanolic extract of pomegranate peels, juice and seeds. *Food Chem. Toxicol.* **2018**, *114*, 108–111. [CrossRef] [PubMed]
4. Bandonienė, D.; Murkovic, M.; Pfannhauser, W.; Venskutonis, P.; Gruzdienė, D. Detection and activity evaluation of radical scavenging compounds by using DPPH free radical and on-line HPLC-DPPH methods. *Eur. Food Res. Technol.* **2002**, *214*, 143–147. [CrossRef]
5. Sharma, T.; Tyagi, V.; Bansal, M. Determination of sun protection factor of vegetable and fruit extracts using UV–Visible spectroscopy: A green approach. *Sustain. Chem. Pharm.* **2020**, *18*, 100347. [CrossRef]
6. Jarzycka, A.; Lewińska, A.; Gancarz, R.; Wilk, K.A. Assessment of extracts of Helichrysum arenarium, Crataegus monogyna, Sambucus nigra in photoprotective UVA and UVB; photostability in cosmetic emulsions. *J. Photochem. Photobiol. B* **2013**, *128*, 50–57. [CrossRef] [PubMed]
7. Everette, J.D.; Bryant, Q.M.; Green, A.M.; Abbey, Y.A.; Wangila, G.W.; Walker, R.B. Thorough study of reactivity of various compound classes toward the Folin−Ciocalteu reagent. *J. Agric. Food Chem.* **2010**, *58*, 8139–8144. [CrossRef]
8. Escarpa, A.; González, M.C. High-performance liquid chromatography with diode-array detection for the determination of phenolic compounds in peel and pulp from different apple varieties. *J. Chromatogra A* **1998**, *823*, 331–337. [CrossRef]
9. Alberti, A.; Zielinski, A.A.F.; Zardo, D.M.; Demiate, I.M.; Nogueira, A.; Mafra, L.I. Optimisation of the extraction of phenolic compounds from apples using response surface methodology. *Food Chem.* **2014**, *149*, 151–158. [CrossRef]
10. Tongkaew, P.; Tohraman, A.; Bungaramphai, R.; Mitrpant, C.; Aydin, E. Kluai Hin (*Musa sapientum* Linn.) peel as a source of functional polyphenols identified by HPLC-ESI-QTOF-MS and its potential antidiabetic function. *Sci. Rep.* **2022**, *12*, 4145. [CrossRef]
11. Abozed, S.S.; El-kalyoubi, M.; Abdelrashid, A.; Salama, M.F. Total phenolic contents and antioxidant activities of various solvent extracts from whole wheat and bran. *Ann. Agric. Sci.* **2014**, *59*, 63–67. [CrossRef]
12. Agatonovic-Kustrin, S.; Kustrin, E.; Morton, D.W. Phenolic acids contribution to antioxidant activities and comparative assessment of phenolic content in mango pulp and peel. *S. Afr. J. Bot.* **2018**, *116*, 158–163. [CrossRef]
13. Wolfe, K.; Wu, X.; Liu, R.H. Antioxidant Activity of Apple Peels. *J. Agric. Food Chem.* **2003**, *51*, 609–614. [CrossRef]
14. de Jager, T.L.; Cockrell, A.E.; Du Plessis, S.S. Ultraviolet Light Induced Generation of Reactive Oxygen Species. *Adv. Exp. Med. Biol.* **2017**, *996*, 15–23. [CrossRef] [PubMed]
15. Rinnerthaler, M.; Bischof, J.; Streubel, M.K.; Trost, A.; Richter, K. Oxidative stress in aging human skin. *Biomolecules* **2015**, *5*, 545–589. [CrossRef]
16. Birben, E.; Sahiner, U.M.; Sackesen, C.; Erzurum, S.; Kalayci, O. Oxidative stress and antioxidant defense. *World Allergy Organ. J.* **2012**, *5*, 9–19. [CrossRef] [PubMed]
17. Abla, M.J.; Banga, A.K. Quantification of skin penetration of antioxidants of varying lipophilicity. *Int. J. Cosmet. Sci.* **2013**, *35*, 19–26. [CrossRef] [PubMed]
18. Blaak, J.; Staib, P. The Relation of pH and Skin Cleansing. *Curr. Probl. Dermatol.* **2018**, *54*, 132–142. [CrossRef] [PubMed]
19. Afsharnezhad, M.; Shahangian, S.; Panahi, E.; Sariri, R. Evaluation of the antioxidant activity of extracts from some fruit peels. *Casp. J. Environ. Sci.* **2017**, *15*, 213–222.
20. Nesseem, D. Formulation of sunscreens with enhancement sun protection factor response based on solid lipid nanoparticles. *Int. J. Cosmet. Sci.* **2011**, *33*, 70–79. [CrossRef] [PubMed]
21. Kim, S.M.; Oh, B.H.; Lee, Y.W.; Choe, Y.B.; Ahn, K.J. The relation between the amount of sunscreen applied and the sun protection factor in Asian skin. *J. Am. Acad. Dermatol.* **2010**, *62*, 218–222. [CrossRef]
22. Berkey, C.; Oguchi, N.; Miyazawa, K.; Dauskardt, R. Role of sunscreen formulation and photostability to protect the biomechanical barrier function of skin. *Biochem. Biophys. Rep.* **2019**, *19*, 100657. [CrossRef] [PubMed]

MDPI
St. Alban-Anlage 66
4052 Basel
Switzerland
www.mdpi.com

Sustainability Editorial Office
E-mail: sustainability@mdpi.com
www.mdpi.com/journal/sustainability

www.ingramcontent.com/pod-product-compliance
Lightning Source LLC
Chambersburg PA
CBHW040809250926
51850CB00007BA/1122